Everything You Always Wanted to Know About...

Chemistry

3rd edition

STERLING
Education

Customer Satisfaction Guarantee

Your feedback is important because we strive to provide the highest quality educational materials. Email us comments or suggestions.

info@sterling-prep.com

We reply to emails – check your spam folder

3   2   1

ISBN-13: 979-8-8855703-1-2

Sterling Education materials are available at quantity discounts.

Contact info@sterling–prep.com

Sterling Education
6 Liberty Square #11
Boston, MA 02109

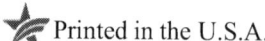

STERLING
Education

From the foundations of chemical reactions to complex mechanisms of atomic particles, this comprehensive guide is an excellent resource for learning about chemistry. It provides an essential overview of fundamental mechanisms of chemical and physical processes.

Created by highly qualified chemistry instructors and education specialists, this book takes readers on a journey through the fascinating and complex world of chemistry to develop their understanding of the subject.

It explains chemical properties of matter and uses clear and comprehensible language to describe essential reaction mechanisms. Readers will learn about electronic structure, chemical bonding, phases, reaction rates and dynamics, solutions, acids and bases, empirical and molecular formulae, thermodynamics, and electrochemistry.

The editors sincerely hope this guide is a valuable resource for you learning.

250424akp

Featured on

Everything You Always Wanted to Know About…

Chemistry

Physics

Cell & Molecular Biology

Organismal Biology

Human Anatomy & Physiology

American History

American Law

American Government & Politics

Comparative Government & Politics

World History

European History

Psychology

Sociology

Environmental Science

Human Geography

Visit our Amazon store

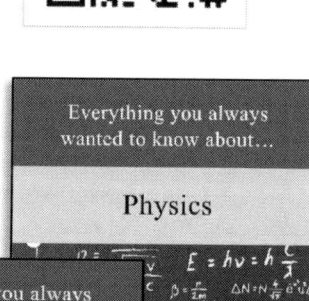

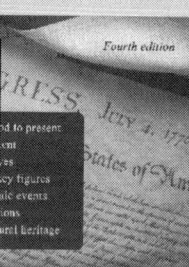

Table of Contents

Table of Contents (*continued*)

CHAPTER 1: Electronic and Atomic Structure; Periodic Table (*continued*)

Table of Contents (*continued*)

Table of Contents (*continued*)

CHAPTER 2: Chemical Bonding (*continued*)

Table of Contents (*continued*)

Table of Contents (*continued*)

Table of Contents (*continued*)

Table of Contents (*continued*)

Table of Contents (*continued*)

Table of Contents (*continued*)

Table of Contents (*continued*)

Table of Contents (*continued*)

Table of Contents (*continued*)

17

Table of Contents (*continued*)

Table of Contents (*continued*)

Table of Contents (*continued*)

CHAPTER 8: Thermochemistry (*continued*)

Table of Contents (*continued*)

CHAPTER 8: Thermochemistry (*continued*)

Table of Contents (*continued*)

Table of Contents (*continued*)

Table of Contents (*continued*)

Table of Contents (*continued*)

Table of Contents (*continued*)

CHAPTER 1

Electronic and Atomic Structure
Periodic Table

- Periodic Table of the Elements

- Electronic Structure

- Electron Configuration

- Quantum Numbers

- Orbital Shapes

- Electron Phenomena

- Chemical Properties within Groups and Rows

- Ionization Energy

- Electrostatic Energy

- Atomic Size

- Classification of Elements by Valence Structure

- Classification of Elements by Electronic Structure

- Practice Questions and Detailed Explanations

Electronic Structure

Atomic structure

The *atom* is the smallest unit of an element that retains that element's characteristics.

An atom consists of *subatomic particles*, including protons, neutrons, and electrons.

The *nucleus* is the densely packed region at the center of an atom consisting of protons and neutrons. The nucleus's diameter is about 10,000 times smaller than the atom's overall diameter.

The *proton* is the positively charged particle located in the atom's nucleus.

Each proton has a charge of +1 and a mass approximately equal to a neutron.

The *neutron* is an uncharged particle in the nucleus of the atom.

The *electron* is a small, negatively charged particle located in the electron cloud. Each electron has a charge of –1, and its mass is about 2,000 times smaller than a proton or neutron.

Most of an atom's volume comes from its *electron cloud*, the outer region surrounding the nucleus.

The *electrostatic attraction* (i.e., opposite electrostatic charges) between the positive core protons and orbiting negative electrons holds electrons around the nucleus.

The *repulsion* between neighboring electrons spreads them over the electron cloud's entire volume.

Atomic number and mass number

The *atomic number* (Z) equals the number of protons in an atom. If the atom has no charge, then Z equals the number of electrons. On the periodic table, elements are arranged by atomic numbers.

The *mass number* (A) is an atom's total number of protons and neutrons (i.e., nucleons).

Cations and anions

An *ion* forms when an atom loses or gains electrons, causing the atom to have either a net negative or a net positive charge.

The loss of electrons yields positively charged *cations*, and gaining electrons produces negatively charged *anions*.

Cations and anions are represented by a superscript of a positive or negative sign after a chemical symbol.

Isotopes

Isotopes are atoms of the element with the same number of protons (i.e., same atomic number Z) but a different number of neutrons.

Therefore, isotopes have different mass numbers A.

The isotope of an atom has virtually identical chemical properties because they have the same number of protons and therefore are the same element.

Most elements naturally occur as a mixture of two or more stable isotopes.

For example, the element carbon ($Z=6$) includes the isotopes ^{12}carbon, ^{13}carbon, and ^{14}carbon. The 12, 13, and 14 are the mass numbers (A) of the isotopes.

^{12}Carbon has six neutrons, ^{13}carbon has seven neutrons, and ^{14}carbon has eight neutrons.

Relative atomic mass

The *relative atomic mass* of an element (also known as *atomic weight*) is the weighted average of the masses of its stable isotopes.

Mass spectrometry is an experimental method used to determine the atomic masses of isotopes. In mass spectrometry, a sample is ionized by bombarding it with electrons, breaking it into charged fragments.

Ions separate when subjected to an electric or magnetic field. The amount of deflection the ions experience is proportional to the ions' mass.

The unit of measurement used for atomic weight is the *atomic mass unit* (amu) or *dalton* (Da), which is approximately equivalent to the mass of one nucleon (a single proton or neutron).

The dalton is based on the atomic mass of the carbon-12 isotope, meaning 1 amu equals 1/12 the mass of a ^{12}C atom or 1.66×10^{-27} g.

^{12}Carbon is the only atomic species with an atomic mass strictly a whole number.

The atomic masses of elements are remarkably close to the atomic mass units' whole numbers.

Atomic mass is expressed by the equation:

$$avg.\ atomic\ mass = (mass_1) \cdot (abundance_1) + (mass_2) \cdot (abundance_2) + \ldots$$

Hydrogen has three isotopes; masses and abundances are shown below:

Isotope	Mass	Abundance
^{1}H (protium)	1.0078 amu	99.985 %
^{2}H (deuterium)	2.0140 amu	0.0156 %
^{3}H (tritium)	3.01605 amu	*trace amounts*

The relative atomic mass of hydrogen (i.e., atomic weight) is:

$$(0.99985 \times 1.0078 \text{ amu}) + (0.000156 \times 2.0140 \text{ amu})$$

$$1.0076 + 0.000314 = 1.0079 \text{ amu for H}$$

Orbital structure of hydrogen atoms

An atom's *electron configuration* is the arrangement of electrons around the nucleus. Electron configurations describe electrons as moving independently in a predefined orbital around the nucleus.

Nobel laureate Niels Bohr (1885-1962), a Danish physicist of the 20th century, was the first to apply quantum physics, using wave functions, to define the electrons' energy discrete values.

A single-electron atom is an elementary particle consisting of one electron around its core nucleus of a proton and a neutron.

The *Bohr model*, discussed later, focuses on the hydrogen atom and the single electron that orbits its nucleus.

In quantum mechanics, the hydrogen electron exists in a spherical probability density cloud around its nucleus.

Electron shells

The *principal quantum number* (n = 1, 2, 3. . .) defines which shell the electron occupies and describes the orbital's *size* and distance from the nucleus.

Electron shells (or principal energy levels) are labeled by principal quantum numbers (n = 1, 2, 3. . .) but can be labeled alphabetically as n = K, L, M...

Higher n shells indicate larger orbitals further from the nucleus and higher energy levels.

The maximum number of electrons per shell is $2n^2$. For example, the second shell can hold up to $2(2)^2 = 8$ electrons. Electrons usually occupy outer shells after inner shells have filled.

However, some outer shell electrons are promoted to a higher shell if it imparts stability to the atom.

Notes for active learning

Electronic Configuration

Aufbau principle

The *Aufbau principle* states that electrons fill orbitals in the order of lowest to highest energy.

Orbitals fill according to the diagonal lines shown.

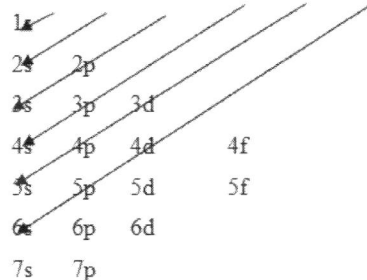

$$1s^2\ 2s^2\ 2p^6\ 3s^2\ 3p^6\ 4s^2\ 3d^{10}\ 4p^6\ 5s^2\ 4d^{10}\ 5p^6\ 6s^2\ 4f^{14}\ 5d^{10}\ 6p^6\ 7s^2\ 5f^{14}\ 6d^{10}\ 7p^6$$

In multi-electron atoms, the energy level for an electron is affected by the n and l quantum numbers.

The l quantum number (discussed below) is represented by the letters s, p, d, and f and describes a subshell. Subshells are comprised of atomic orbitals, and each orbital can hold two electrons.

The $4s$ orbital is lower in energy than the $3d$ orbitals. From the *Aufbau principle,* electrons populate the $4s$ orbital before the $3d$ orbital, even though the $4s$ orbital has a higher principal (n) value than the $3d$ orbital.

Hund's rule

Hund's rule is another essential guideline for writing electron configurations.

Hund's rule states that an electron must occupy every orbital in a sublevel (e.g., s, p, d, f) before a second electron can occupy an orbital in that sublevel.

Hund's rule maximizes the number of electrons with the same electron magnetic spin (discussed below) to increase the atom's stability. Double-occupied orbitals are higher in energy and less stable than orbitals with a single electron.

Octet rule

The *octet rule* refers to atoms' tendency to gain, lose or share electrons to achieve a full orbital of eight electrons in its *valence* (i.e., outermost) orbital. The octet rule applies to the *s* and *p* orbital electrons. Therefore, the octet rule is particularly useful for *representative elements.*

Representative elements are not in the transition block (*d* orbitals) or the inner-transition metal block (*f* orbitals, known as *lanthanides* and *actinides*).

A complete octet is typically represented as an electron configuration ending in s^2p^6.

When atoms have either an excess or deficiency in the number of valences (outermost) electrons, they react to satisfy their octets and form stable compounds.

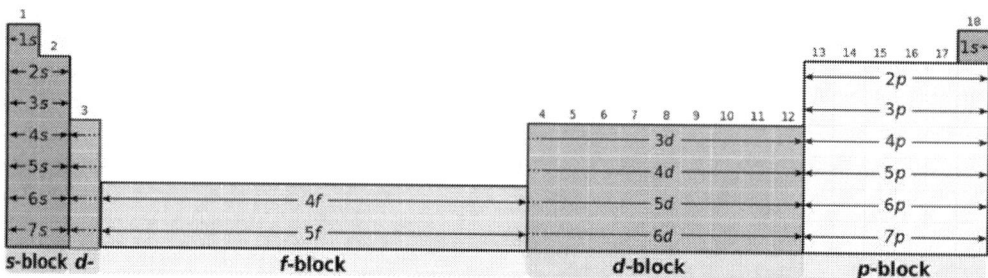

The vertical columns are groups (families). The representative elements are the left two groups with s shells (metals) and six right p orbitals (nonmetals)

Expressing electronic configuration

Using the Aufbau Principle and Hund's Rule, the electron configuration for zinc (Zn) atom with 30 electrons is:

$$1s^2\, 2s^2\, 2p^6\, 3s^2\, 3p^6\, 4s^2\, 3d^{10}$$

- The first number indicates the principal energy level (n).

- The letter (s, p, d, f) indicates the subshell (ranging from 0 to $n-1$).

- The superscript indicates the number of electrons in the subshell.

As the number of electrons increases, the electron configurations become long.

There is a short-hand approach to writing electron configurations referencing the noble gases (inert gases).

The noble gases have complete octets in the periodic table's far-right column (group or family) (Group 18).

The electronic configurations of Na (atomic number or $Z = 11$):

1. Identify the noble gas in the previous row (i.e., *period*) before the element.

    The noble gas before Na ($Z = 11$) is Ne ($Z = 10$)

    The noble gas before Cl ($Z = 17$) is Ne ($Z = 10$), not Ar ($Z = 18$)

2. Write the noble gas in square brackets:

    [Ne]

    Ne has the electron configuration of $1s^2 2s^2 2p^6$

3. Write the electron configuration for Na:

    Na: $1s^2 2s^2 2p^6 3s^1$

4. Abbreviate the electron configuration by referencing Ne:

    [Ne] $3s^1$

Abbreviated electronic configuration

For example, the electronic configurations of Zn (atomic number or $Z = 30$):

Zinc: $1s^2 2s^2 2p^6 3s^2 3p^6 4s^2 3d^{10}$

The noble gas argon has a Z value = 18

The short-hand electronic configuration for Zn is:

Zn: [Ar] $4s^2 3d^{10}$

When writing electron configurations for ions of an element, add or subtract the number of electrons gained or lost from the atom when filling the subshells ($s. p, d, f$).

For sodium (Na), the electron configuration is:

Na: $1s^2 2s^2 2p^6 3s^1$

When sodium reacts to form the Na^+ cation, it loses one electron, and its electron configuration becomes:

Na^+: $1s^2 2s^2 2p^6$

The electron configuration for the sodium cation (Na^+) is identical to Ne with 10 electrons.

Note a few exceptions to the guidelines above because of the stability rule when writing electron configurations.

Stability rule

The stability rule states that *a sublevel is more stable when it has a half-filled configuration* (one electron in every orbital) or a *full configuration* (two electrons in every orbital).

If an atom is one electron short of achieving a half-filled or full configuration, it takes one from a neighboring *s* orbital to increase its stability.

This electron-promoting from the *s* orbital is most common for transition metals.

Chromium (Cr, 24 electrons) has the electron configuration:

Cr: [Ar] $4s^2 3d^4$

According to the stability rule, the Cr electron configuration is more stable with a half-filled configuration:

Cr: [Ar] $4s^1 3d^5$

The $3d$ subshell took an electron from the neighboring $4s$ subshell, and the $4s$ subshell has a partially filled orbital.

Copper (Cu) has 29 electrons with the expected electron configuration:

Cu: [Ar] $4s^2 3d^9$

From the stability rule, the electron configuration changes to the more stable:

[Ar] $4s^1 3d^{10}$

The $3d$ subshell of Cu has the electron configuration of $4s^1$, which is now partially filled.

Orbital diagrams

An *orbital diagram* is another method to express an atom's electron configuration.

Orbital diagrams are a visual representation using fishhook (single-headed) arrows to indicate electrons in orbitals, left side below.

$\underset{1s}{\uparrow\downarrow}\ \ \underset{2s}{\uparrow\downarrow}\ \ \underset{2px}{\uparrow}\ \ \underset{2py}{\uparrow}\ \ \underset{2pz}{}$	$1s^2 2s^2 2p^2$
Single-headed fishhook arrow points up or down to represent an electron. Each line horizontal represents an orbital; the subshell and energy level are written below the line. Spaces separate the subshells *three horizontal lines for the 3 p orbitals (x, y, z)	Coefficient = energy level (1, 2) Letter = subshell (s, p) Exponent = # of electrons in subshell This method does not give as much information as orbital diagrams.

Dalton's atomic theory

It is vital to understand the scientific discoveries that contributed to the atom's current model.

At the beginning of the 19th century, chemists were able to show that pure compounds contained fixed and unvarying amounts of constituent components.

In 1806, John Dalton (1766-1844) provided a significant step in explaining this model of an atom with his particle theory, known as *Dalton's Atomic Theory*.

Dalton's Atomic Theory consisted of these essential principles:

1. All matter is composed of microscopic particles (i.e., atoms).

2. Atoms of one element have the same shape, size, mass, and properties but differ from other elements.

3. Atoms can neither be subdivided nor changed into another atom.

4. Atoms cannot be created nor destroyed.

5. The atom is the smallest unit of matter that undergoes a chemical reaction.

In addition to these principles, Dalton proposed the "*rule of greatest simplicity*," which suggested that atoms only combine in binary ratios (i.e., 1:1). This rule was problematic because Dalton assumed that the formula for water was HO instead of H_2O.

Dalton eventually suggested that the rule of greatest simplicity was not correct, and in 1810, he suggested that a water molecule has three atoms.

Law of multiple proportions

Following the *law of multiple proportions*, atoms combine in whole-number ratios (1:1, 1:2, 1:3, ...).

There are problems with Dalton's Atomic Theory as he claimed that atoms of an element have the same mass.

The discovery of isotopes, which are atoms of the same element that vary in mass and density (different number of neutrons), proves otherwise.

Additionally, it is now known that atoms subdivide into constituent particles in nuclear processes; Dalton's postulate that atoms cannot be subdivided remains correct within the scope of chemical reactions.

Bohr atom

In the early 1900s, German Nobel laureate physicists Max Planck (1858-1947) discovered that radiation is emitted in quantized amounts of energy instead of a single continuous ray.

Based on Planck's discovery, Nobel laureate Niels Bohr (1885-1962) conducted experiments on hydrogen atoms and revised the atom model.

Bohr suggested that electrons orbit the nucleus in fixed orbits with defined energies and sizes. Bohr's model is like how planets orbit the sun (except the attraction within the atom is by electrostatic forces rather than gravity).

When electrons transition between orbits, they emit or absorb energy equivalent to the difference in energy levels.

Bohr identified each energy level using an integer *n*, the *principal quantum number,* and those additional electrons occupy the lowest available energy level.

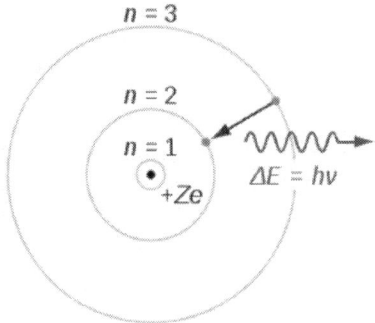

Bohr's model of electrons in discrete orbitals according to specific energy levels

Quantum Numbers

Principal quantum number *n*

The *principal quantum number* (*n*) defines which shell the electron occupies and describes the orbital's *size* and distance from the nucleus.

The principal quantum number has positive integer values of $n = 1, 2, 3...$

Electron shells (or principal energy levels) are labeled by principal quantum numbers ($n = 1, 2, 3...$) but can be labeled alphabetically as $n = $ K, L, M...

Higher *n* shells indicate larger orbitals further from the nucleus and higher energy levels.

Azimuthal quantum number *l*

The *azimuthal quantum number l* (or *angular momentum quantum number*) is the second quantum number that defines the *shape* of an orbital.

Except for the first level where $n = 1$, the principal energy levels have two or more possible subshells (*s, p, d, f*) with different energy levels.

For any *n*, the values for the azimuthal (i.e., angular momentum) quantum number *l* may range from 0 to ($n - 1$), represented by the letters *s, p, d,* and *f* (shown below).

Orbital	Angular Momentum Quantum Number
s	$l = 0$
p	$l = 1$
d	$l = 2$
f	$l = 3$

Magnetic quantum number *m_l*

The *magnetic quantum number* (*m_l*) is the third quantum number that describes the *orientation* of an orbital and defines the *number of orbitals* within a subshell.

The possible values for *m_l* range from −*l* to +*l*, including 0.

- An *s* subshell has *l* = 0; the *m_l* has a single value (0) and one orbital.

  A *s* subshell can hold 2 electrons.

- A *p* subshell has *l* = 1; the *m_l* has values of −1, 0, +1.

  A *p* subshell has three orbitals holding up to 6 electrons.

- A *d* subshell has *l* = 2; *m_l* = −2, −1, 0, +1, +2.

  A *d* subshell has five orbitals with up to 10 electrons.

- An *f* subshell has *l* = 3; *m_l* = −3, −2, −1, 0, +1, +2, +3.

  An *f* subshell has seven orbitals with up to 14 electrons.

Each succeeding subshell holds four more electrons than its predecessor.

Spin quantum number *m_s*

The *spin quantum number* (*m_s*) is the fourth quantum number and describes the spin of an individual electron occupying an orbital.

The spin quantum number distinguishes specific electrons in orbitals.

There are two opposing values for electron-spin: +½ and −½.

By convention, the first electron to occupy an orbital has a spin number of +½ while the second electron has a spin number of −½.

In orbital diagrams, the +½ electrons are represented by upward arrows, and the −½ electrons are indicated by downward arrows.

Energy Level	Number of subshells	Subshells	# of orbitals in that subshell	Total number of electrons in a subshell
1	1	*s*	1	2
2	2	*s*	1	2
		p	3	6
3	3	*s*	1	2
		p	3	6
		d	5	10
4	4	*s*	1	2
		p	3	6
		d	5	10
		f	7	14

Quantum numbers represent the locations of the electrons within the atom

Notes for active learning

Orbital Shapes

Pauli Exclusion Principle

In 1925, Nobel laureate Wolfgang Pauli (1900-1958) proposed the *Pauli Exclusion Principle*, which states that no two electrons in the same atom may have the same four quantum numbers (n, l, m_l, and m_s).

The maximum number of electrons in an orbital is two, and each must possess opposite spins (i.e., $+\frac{1}{2}$ or $-\frac{1}{2}$) because electrons of the same magnetic spin cannot be in the same subshell of an orbital.

Geometric shapes for orbitals *s, p, d, f*

There are four orbitals whose shapes were predicted using wave mechanics (described below).

The four types of orbitals are *s, p, d,* and *f* from the initial letters of **S**harp, **P**rincipal, **D**iffuse, and **F**undamental.

Each energy level has between 1 and 4 subshells.

Each subshell has an electron probability density of shape.

Each subshell has a specific orbital number as given by the m_l value.

- *s orbitals* are spheres:

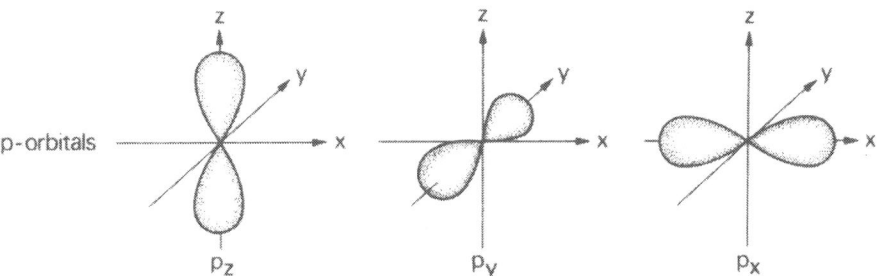

- *p orbitals* have two lobes and are dumbbell shaped.

There are three different *p* orbitals (three orientations: p_x, p_y, p_z):

- *d orbitals* have four lobes and are clover shaped.

  There are five *d* orbitals (orientations: d_{yz}, d_{z^2}, d_{xy}, d_{xz}, $d_{x^2-y^2}$):

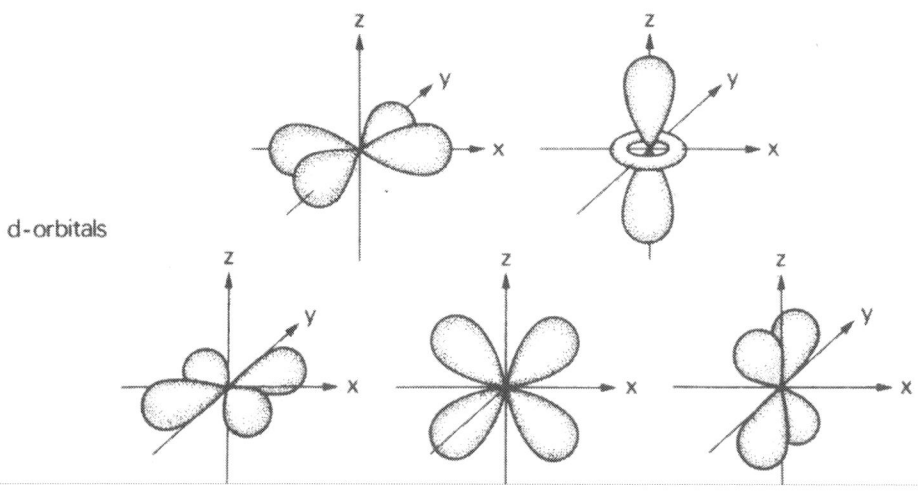

- *f orbitals* are complicated, and the shape combines the other orbitals' shapes.

  There are seven *f* orbitals:

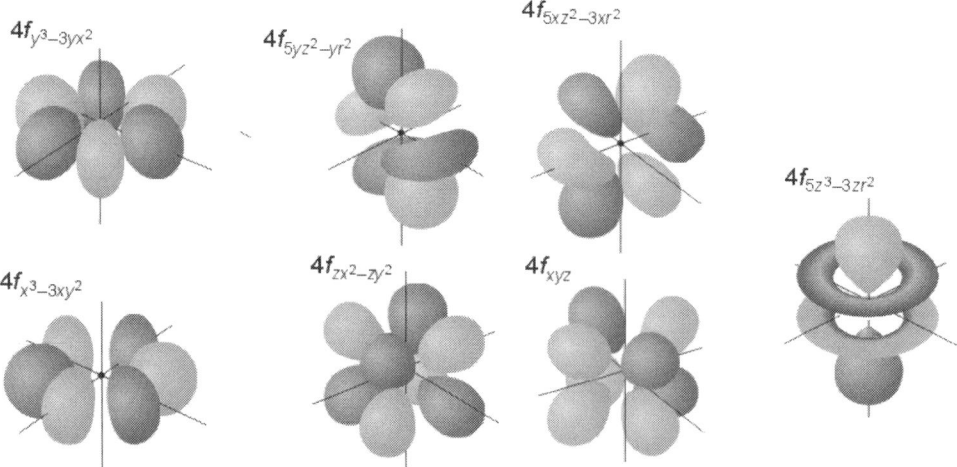

De Broglie wave-particle duality

A series of remarkable discoveries led to quantum mechanics and enabled an evolved model to understand electron behavior.

In 1924, Nobel laureate Louis de Broglie (1892-1987) published his work on *wave-particle duality* and proposed that electrons simultaneously be classified as particles and waves. He proposed that electrons are particles (with mass and velocity) and possess *wave properties* (with wavelength and frequency).

The de Broglie equation demonstrates that all objects exhibit wave behavior.

$$\lambda = h \,/\, mv$$

where λ is the wavelength, m is mass, h is Planck's constant (6.626×10^{-34} J·s), and v is velocity

However, the wave behavior of everyday objects becomes negligible as the mass increases. Subatomic particles have tiny masses:

Schrödinger's wave function

In 1926, Nobel laureate Erwin Schrödinger (1887-1961) derived a wave equation modeling electrons' movement based on wave properties.

In quantum mechanics, Schrödinger's wave equation's solution yields a *wave function*, represented by the uppercase Greek letter psi (Ψ), which, by itself, is not instructive.

The square of the absolute value of psi, $|\Psi|^2$ is the *probability density* of an electron's position.

The *wave function* $|\Psi(x)|^2$ gives the probability of locating an electron at position x.

Orbitals are electron "*clouds*" representing the probability of an electron's specific location in the atom.

Schrödinger's equation showed that electrons behave like waves. It implies that electrons do not travel in defined paths, and it is impossible to determine an electron's exact position and speed simultaneously.

Heisenberg uncertainty principle

German physicist Nobel laureate Werner Heisenberg (1901-1976) developed a theory based on a theoretical limit to how small the measurements' uncertainty can be.

The *Heisenberg Uncertainty Principle* states that an electron's position and momentum cannot be measured simultaneously, and the measurement of position or momentum distorts the value for the other.

Heisenberg Uncertainty Principle is expressed as:

$$\Delta x \times \Delta p \geq \frac{h}{4\pi}$$

where Δx is the *uncertainty in position*, Δp is the *uncertainty in momentum*, and h is Planck's constant of 6.626×10^{-34} J·s

Electron Phenomena

Paramagnetism and diamagnetism

Electrons are constantly spinning in a fixed direction, which generates *magnetic fields*.

According to the *Pauli Exclusion Principle*, if one electron is spinning clockwise, the other must spin counterclockwise. The opposing spins result in the orbital with no net spin.

Electrons occupying the same orbital have opposite values for spin quantum numbers (m_s); they have different quantum numbers.

On electron-induced magnetic behavior, atoms are *ferromagnetic, diamagnetic,* or *paramagnetic.*

1. *Diamagnetic* materials generate an induced magnetic field in a direction opposite to an externally applied magnetic field, and the applied magnetic field repels these materials.

   All electrons in diamagnetic atoms are paired.

2. *Paramagnetic* materials, when in the presence of a magnetic field, generate internally-induced magnetic fields in the same direction as the external field.

   A paramagnetic electron is an unpaired electron.

   An atom is paramagnetic if it has at least one unpaired paramagnetic electron in any orbital, regardless of the number of paired electrons.

   Paramagnetic atoms are slightly attracted to a magnetic field and cannot retain magnetization without an external magnetic field.

3. *Ferromagnetic* materials generate permanent magnetic moments without needing an applied external magnetic field.

   Certain transition metals, such as iron and nickel, have ferromagnetic properties.

   Electrons of ferromagnetic substances can align spontaneously in the same direction, reinforcing each electron's magnetic properties.

Einstein's photoelectric effect

In 1905, Albert Einstein (1879-1955) proposed the *photoelectric effect* that explained how incoming visible light interacts with electrons when light is reflected off metals' surfaces.

Discrete packets of light energy (or *photons*) are transferred to electrons on a metal surface as kinetic energy.

When electrons possess sufficient energy to escape, electrons' emission from the metal occurs.

The emission of electrons occurs if the incident light possesses energy greater than or equal to the *work function* ϕ_w of the electron.

Energy (E) of light waves is calculated by:

$$E = h\nu$$

where h is Planck's constant and ν is the frequency of the light wave (Hz)

Einstein's *photoelectric effect* model accurately predicts that a photon's energy is proportional to its frequency and not its intensity.

Ground and excited states of electrons

Every electron in an atom occupies a fixed orbital. When an electron occupies its default orbital, the electron is in its *ground state*.

A ground-state electron is at its lowest energy level and is the most stable.

If an electron absorbs energy, it has greater potential energy and occupies a higher energy level.

An electron in a higher energy level is an *excited state* electron.

An electron can remain in its excited state for a limited time before it drops back to its ground state and emits excess energy as visible electromagnetic radiation.

Atomic emission spectra

Electrons move between fixed orbitals, so the wavelength of light released or absorbed when they transition between energy levels is limited to a specific set of visible wavelengths known as the *line spectrum*.

A line spectrum is produced by gas under low pressure.

Solids, liquids, and compressed gases produce *continuous spectra* with all colors of visible light.

Each element has different numbers of electrons, which means that each element has a unique set of *atomic emission spectra* represented by discrete lines along a frequency scale.

Due to this chemical property's uniqueness, *the atomic emission spectra* identify elements in a mixture.

Bright-Line Spectra

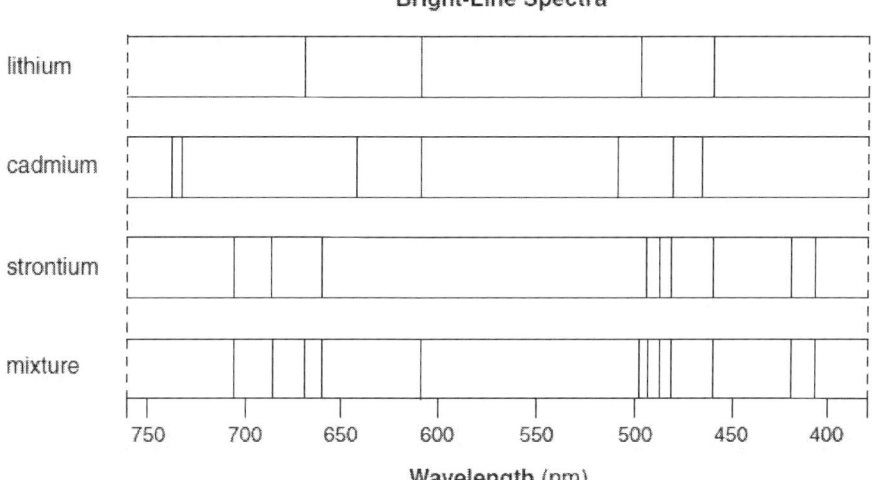

Each element produces a unique atomic emission spectrum
that permits the classification of elements within a mixture

Atomic absorption spectra

The *atomic absorption spectra* indicate the energy absorbed when electrons are promoted to orbitals with higher energy levels.

An element's absorption and emission spectrum correlate because the net difference between energy levels is the same, regardless of the direction of the electron's movement.

The atomic emission spectrum of hydrogen has been widely observed.

Hydrogen's spectrum consists of five lines, and each series is named after its discoverer.

Lyman ($n = 1$)	Balmer ($n = 2$)
Paschen ($n = 3$)	Brackett ($n = 4$)
Pfund ($n = 5$)	Humphreys ($n = 6$)

The emission series *Lyman, Balmer,* and *Paschen* correspond to *ultraviolet, visible,* and *infrared* radiation.

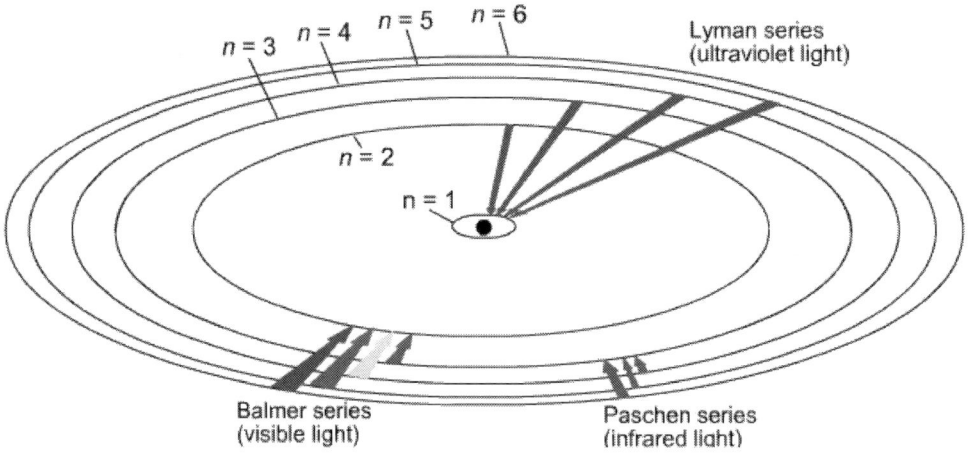

Atomic emission spectrum for a hydrogen atom

Rydberg formula

Swedish physicist Johannes Rydberg (1854-1919) proposed an equation for the relationship between spectral lines of different elements.

The *Rydberg formula* for hydrogen calculates the energy emission of hydrogen electrons expressed as:

$$E = hc / \lambda$$

$$E = -R_h \left[\frac{1}{(n_i)^2} - \frac{1}{(n_f)^2} \right]$$

where R_h is Rydberg's constant (2.179×10^{-18} J), n_i is the principal quantum number of the initial orbital, n_f the principal quantum number of the final orbital, c is the speed of light (3.0×10^8 m/s), λ is the wavelength and h is Planck's constant (6.626×10^{-34} J·s)

Chemical Properties within Groups and Rows

Periodic table

After discovering several elements in the 18th and 19th centuries, scientists proposed that similar properties could categorize elements. In 1869, the first significant development occurred when Russian chemist Dimitri Mendeleev (1834-1907) and German chemist Julius Lothar Meyer (1830-1895) independently published variations for the periodic table.

Mendeleev arranged the elements by atomic masses, resulting in the similarity of physical and chemical properties between elements in the same column or group. Mendeleev's table was revolutionary as it accurately predicted the position of undiscovered elements.

Meyer arranged his table based on increasing atomic volume.

Mendeleev's table exhibited patterns between elements in the same column or group, but there were exceptions.

For example, argon exists in elemental form as an unreactive gas (noble gas or inert gas), but it was categorized with highly reactive sodium and lithium metals.

In 1913, British physicist Henry Moseley (1887-1915) improved Mendeleev's periodic table by proposing that the chemical properties of elements are related to atomic number (Z, number of protons) instead of atomic weight (A, number of protons, and neutrons).

Moseley's discovery became the foundation of the modern periodic table.

The modern periodic table arranges elements in rows and columns.

> The horizontal rows are *periods* (identified by numbers 1-7).

> The vertical columns are *groups* or *families* (1-18).

Group nomenclature

There are different standards for group nomenclature:

- In the older system, each group was assigned a Roman numeral (I-VIII) based on the number of valence electrons.

- The groups were assigned a letter (A or B) based on location on the table. However, there were two different standards on the A/B designation.

- The American A/B standard assigned A to the main (representative) group of elements on the sides and B to the transition group in the center.

- The European A/B standard assigned A to the left and B to the table's right side.

- For consistency, a new universal naming standard was devised from left to right; the groups are consecutive integers (1-18).

Elements are ordered by increasing atomic number, so there are recurring properties within the rows (periods) and columns (groups or families).

Several physical and chemical properties of elements are predicted from their position in the periodic table.

These trends include the number of valences (outermost) electrons, ionization energy, electron affinity, atomic or ionic radii, and electronegativity.

Valence electrons

Valence electrons are the outermost shell electrons responsible for forming chemical bonds.

The group numbers represent the number of valence electrons for elements in the column on the periodic table.

However, this rule applies to the main group elements and does not apply to the transition metals.

Coulomb's Law expresses electrostatic attraction

Two factors contribute to valence electrons experiencing weaker electrostatic attraction by protons in the nucleus.

The *first factor* is the *distance of the electron n shell from the nucleus*.

The increased distance of higher principal quantum numbers for the electron's orbital increases the distance from the nucleus and decreases attraction.

Coulomb's Law expresses electrostatic force:

$$F = kq_1q_2 \, / \, r^2$$

where F is the electrostatic force, k is Coulomb's constant (9×10^9 N·m^2/C^2), q_1 and q_2 are the respective charges of the two particles, and r is the distance between the particles

The *second factor* is *screening electrons*, which are inner electrons that shield the attractive force experienced by the valence electrons.

Periodic Table of the Elements

Atomic Number	Valence
Symbol	
Name	
Atomic Mass	

1 IA 1A																		18 VIIIA 8A
H [-1, +1] 1 Hydrogen 1.008	2 IIA 2A											13 IIIA 3A	14 IVA 4A	15 VA 5A	16 VIA 6A	17 VIIA 7A	**He** 0 2 Helium 4.003
Li +1 3 Lithium 6.941	**Be** +2 4 Beryllium 9.012												**B** +3 5 Boron 10.811	**C** +4,+3,+2,1 6 Carbon 12.011	**N** +5,+3,3 7 Nitrogen 14.007	**O** -2 8 Oxygen 15.999	**F** -1 9 Fluorine 18.998	**Ne** 0 10 Neon 20.180
Na +1 11 Sodium 22.990	**Mg** +2 12 Magnesium 24.305	3 IIIB 3B	4 IVB 4B	5 VB 5B	6 VIB 6B	7 VIIB 7B	8	9 VIII 8	10	11 IB 1B	12 IIB 2B	**Al** +3 13 Aluminum 26.982	**Si** +4,-4 14 Silicon 28.086	**P** +5,+3,3 15 Phosphorus 30.974	**S** +6,+4,+2,2 16 Sulfur 32.066	**Cl** +7,+5,+3,-1 17 Chlorine 35.453	**Ar** 0 18 Argon 39.948
K +1 19 Potassium 39.098	**Ca** +2 20 Calcium 40.078	**Sc** +3 21 Scandium 44.956	**Ti** +4 22 Titanium 47.88	**V** +5 23 Vanadium 50.942	**Cr** +6,+3 24 Chromium 51.996	**Mn** +7,+4,+2 25 Manganese 54.938	**Fe** +6,+3,+2 26 Iron 55.845	**Co** +3,+2 27 Cobalt 58.933	**Ni** +4,+2 28 Nickel 58.693	**Cu** +2 29 Copper 63.546	**Zn** +2 30 Zinc 65.38	**Ga** +3 31 Gallium 69.723	**Ge** +4,+2,-4 32 Germanium 72.631	**As** +5,+3,3 33 Arsenic 74.922	**Se** +6,+4,+2,2 34 Selenium 78.971	**Br** +5,+3,-1 35 Bromine 79.904	**Kr** +6,+4,+2,0 36 Krypton 84.798
Rb +1 37 Rubidium 85.468	**Sr** +2 38 Strontium 87.62	**Y** +3 39 Yttrium 88.906	**Zr** +4 40 Zirconium 91.224	**Nb** +5 41 Niobium 92.906	**Mo** +6,+4 42 Molybdenum 95.95	**Tc** +7,+4 43 Technetium 98.907	**Ru** +4,+3 44 Ruthenium 101.07	**Rh** +3 45 Rhodium 102.906	**Pd** +4,+2 46 Palladium 106.42	**Ag** +1 47 Silver 107.868	**Cd** +2,+1 48 Cadmium 112.414	**In** +3 49 Indium 114.818	**Sn** +4,+2,-4 50 Tin 118.711	**Sb** +5,+3,3 51 Antimony 121.760	**Te** +6,+4,+2,2 52 Tellurium 127.6	**I** +7,+5,+3,-1 53 Iodine 126.904	**Xe** +6,+4,+2,0 54 Xenon 131.294
Cs +1 55 Cesium 132.905	**Ba** +2 56 Barium 137.328	57-71	**Hf** +4 72 Hafnium 178.49	**Ta** +5 73 Tantalum 180.948	**W** +6 74 Tungsten 183.85	**Re** +7,+4 75 Rhenium 186.207	**Os** +4 76 Osmium 190.23	**Ir** +4,+3 77 Iridium 192.22	**Pt** +4,+2 78 Platinum 195.08	**Au** +3,+1 79 Gold 196.967	**Hg** +2,+1 80 Mercury 200.59	**Tl** +3,+1 81 Thallium 204.383	**Pb** +4,+2 82 Lead 207.2	**Bi** +3 83 Bismuth 208.980	**Po** +4,+2,2 84 Polonium [208.982]	**At** +1,-1 85 Astatine 209.987	**Rn** +2,0 86 Radon 222.018
Fr +1 87 Francium 223.020	**Ra** +2 88 Radium 226.025	89-103	**Rf** unknown 104 Rutherfordium [261]	**Db** unknown 105 Dubnium [262]	**Sg** unknown 106 Seaborgium [266]	**Bh** unknown 107 Bohrium [264]	**Hs** unknown 108 Hassium [269]	**Mt** unknown 109 Meitnerium [278]	**Ds** unknown 110 Darmstadtium [281]	**Rg** unknown 111 Roentgenium [280]	**Cn** unknown 112 Copernicium [285]	**Nh** unknown 113 Nihonium [286]	**Fl** unknown 114 Flerovium [289]	**Mc** unknown 115 Moscovium [289]	**Lv** unknown 116 Livermorium [293]	**Ts** unknown 117 Tennessine [294]	**Og** unknown 118 Oganesson [294]

Lanthanide Series

57 **La** +3 Lanthanum 138.905	58 **Ce** +4,+3 Cerium 140.116	59 **Pr** +3 Praseodymium 140.908	60 **Nd** +3 Neodymium 144.243	61 **Pm** +3 Promethium 144.913	62 **Sm** +3 Samarium 150.36	63 **Eu** +3,+2 Europium 151.964	64 **Gd** +3 Gadolinium 157.25	65 **Tb** +3 Terbium 158.925	66 **Dy** +3 Dysprosium 162.500	67 **Ho** +3 Holmium 164.930	68 **Er** +3 Erbium 167.259	69 **Tm** +3 Thulium 168.934	70 **Yb** +3,+2 Ytterbium 173.055	71 **Lu** +3 Lutetium 174.967

Actinide Series

89 **Ac** +3 Actinium 227.028	90 **Th** +4 Thorium 232.038	91 **Pa** +5 Protactinium 231.036	92 **U** +6 Uranium 238.029	93 **Np** +5 Neptunium 237.048	94 **Pu** +4 Plutonium 244.064	95 **Am** +3 Americium 243.061	96 **Cm** +3 Curium 247.070	97 **Bk** +3 Berkelium 247.070	98 **Cf** +3 Californium 251.080	99 **Es** +3 Einsteinium [254]	100 **Fm** +3 Fermium 257.095	101 **Md** +2 Mendelevium 258.1	102 **No** +2 Nobelium 259.101	103 **Lr** +3 Lawrencium [262]

Effective nuclear charge Z_{eff}

The *nuclear charge* is the charge of the positive protons in the nucleus and describes how electrons are pulled toward the nucleus.

The *effective nuclear charge* (Z_{eff}) is the net nuclear force experienced by valence electrons after accounting for electron shielding by inner orbital electrons.

The *effective nuclear charge Z_{eff}* is calculated by:

$$Z_{eff} = Z - S$$

where Z is the number of protons in the nucleus (i.e., the atomic number), and S is the average number of screening electrons between the nucleus and the valence electron

The effective nuclear charge experienced by an electron is proportional to its stability: the more stable the electron, the higher the effective nuclear charge, and the more energy is required to remove the electron; ionization energy.

Effective nuclear charge increases across a period due to increasing nuclear charge (increasing number of protons) but no accompanying increase in the shielding effect.

There is no general trend down a group because the nuclear charge (Z) and shielding effect increase as electron shells are added.

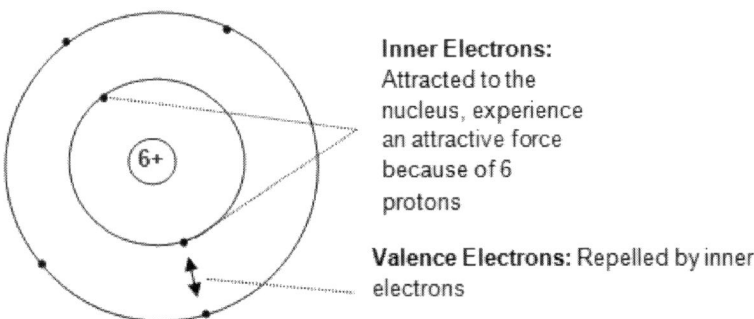

Inner Electrons: Attracted to the nucleus, experience an attractive force because of 6 protons

Valence Electrons: Repelled by inner electrons

Effective nuclear charge: valence electrons "feel" an attractive force of 4 protons.

The two inner electrons cancel the +2 of the nuclear charge due to shielding

Ionization Energy

First and second ionization energy

The *ionization energy* (SI Units kJ/mol) is required to remove an electron from a neutral atom of an element or an ion in the gaseous state.

The ionization of an electron requires energy and is an endothermic process (i.e., it absorbs heat).

While electrons are removed from the atom, the number of protons in the nucleus is constant.

Therefore, the remaining electrons experience a stronger electrostatic force, as the constant force from the nucleus is shared among the fewer remaining electrons.

Ionization energies increase for each successive removal of an electron due to the increased electrostatic force.

Thus, the first electron is the easiest to remove.

> The *first ionization energy* is the energy needed to remove an outermost electron.

> The *second ionization energy* is the energy needed to remove a second electron.

Low ionization energy indicates that an electron is easily lost. This property is most commonly found in metals, with the Group I elements (alkali metal) having the lowest IE values.

Conversely, high IE indicates that an electron is not easily removed.

High ionization energy is mainly associated with nonmetals because of their natural tendency to gain electrons to form anions instead of losing electrons.

The highest IE values are found for the inert gases (Group 8 or VIII) because octets make them highly stable.

A substantial positive nuclear charge results in stronger electrostatic attraction and higher ionization energy for valence electrons.

In contrast, a larger atomic radius means that the outermost electrons are further from the nucleus and have a weaker electrostatic attraction, increased shielding, and lower ionization energy.

If an electron is in a lower shell (closer to the nucleus), it has lower energy and is difficult to dislodge. Thus, it has higher ionization energy and requires more energy to abstract the electron from the atom.

Ionization trends

Half-filled shells are exceptions to the trend for ionization energy because these configurations are more stable than filled shell configurations and occur when an atom's valence *p* or *d* orbital is half-filled (one electron in every subshell).

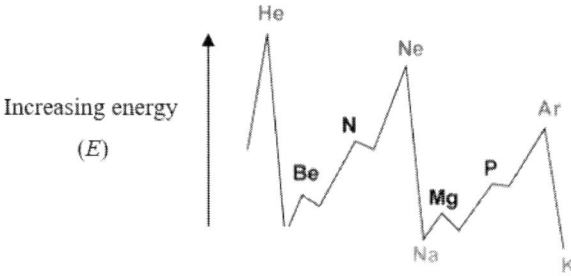

A complete octet (filled s and p subshells) increases the stability of the electrons

Ionization energies generally increase left to right across a period due to the atom's increased effective nuclear charge. However, the shielding effect from the inner electron shells remains constant.

Ionization energies generally decrease from top to bottom down a group due to the increasing distance from the nucleus by the increased principal quantum number *n* and the shielding effect of the electrons in the inner shells.

The ionization energy decreases down a group because of the increasing distance for the valence electrons to the nucleus and the increase in shielding by more shells of inner electrons.

The graph's highest peaks are noble gases (stable due to filled octets), while the lowest valleys are alkali metals with single valence electrons in the valence shell.

Local maxima occur for filled subshells and half-filled *p* subshells.

Electrostatic Energy

Electron affinity

Electron affinity is the energy released when an electron adds to a neutral atom.

Elements with high electron affinity are strongly attracted to electrons and are associated with exothermic reactions (i.e., reactions that release heat).

Electron affinity for nonmetals is indicative of their tendency to gain electrons and form negatively charged anions (e.g., Cl^- or Br^-).

The first electron affinity is the energy released when one electron adds; it is *exothermic* and releases heat.

The second electron affinity is *endothermic*.

Adding an electron to an orbital with excess electrons results in repulsion; thus, energy is needed to overcome electrons' electrostatic repulsions and stabilize the added electron.

The elements preferring electrons are the nonmetals, which have the highest electron affinities.

The elements not preferring extra electrons are the metals, which have low electron affinities.

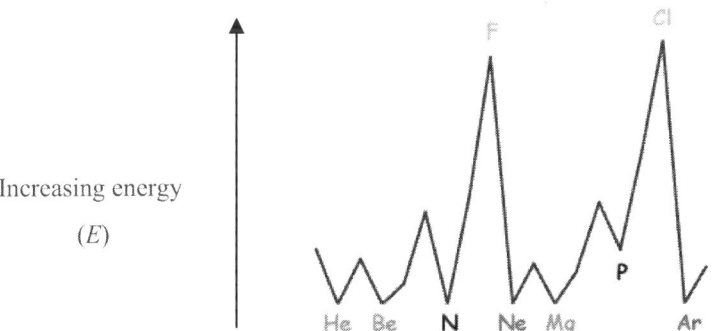

Electron affinity is highest for F, and Cl which results in complete p subshells

On the electron affinity graph, the highest peaks are for the halogens (F. Cl. Br, I, At), and the lowest peaks are for noble gases (e.g., Ne. Ar, Kr).

Local minima in energy occur for filled subshells and half-filled *p* subshells; observed for the noble gases of neon (Ne) and argon (Ar).

Like ionization energy, electron affinity is a periodic trend that generally increases from left to right and decreases from top to bottom (except the noble gases).

Electronegativity

In 1932, American chemist Nobel laureate Linus Pauling (1901-1994) derived a set of fundamental values based on *ionization energies* (i.e., an atom's tendency to lose electrons) and *electron affinities* (i.e., an atom's tendency to gain electrons).

Electronegativity is the element's tendency to gain bonding electrons.

Electronegativity measures an atom's attraction for the electrons in a bond with another atom.

On the Pauling scale, electronegativity values range from a low of 0.7 (Fr) to a high of 4.0 (F).

Electronegativity values generally increase from left to right and decrease from top to bottom on the periodic table.

The *most electronegative* elements are at the top right corner of the periodic table.

The *least electronegative* (i.e., most electropositive) elements are in the bottom left corner.

Therefore, nonmetals are typically more electronegative than metals.

There are no electronegativity values reported for the noble gases because they usually do not bond.

Electronegativity values of the elements (Pauling scale)

H 2.1																		He
Li 1.0	Be 1.5												B 2.0	C 2.5	N 3.0	O 3.5	F 4.0	Ne
Na 0.9	Mg 1.2												Al 1.5	Si 1.8	P 2.1	S 2.5	Cl 3.0	Ar
K 0.8	Ca 1.0	Sc 1.3	Ti 1.5	V 1.6	Cr 1.6	Mn 1.5	Fe 1.8	Co 1.8	Ni 1.8	Cu 1.9	Zn 1.6	Ga 1.6	Ge 1.8	As 2.0	Se 2.4	Br 2.8	Kr 3.0
Rb 0.8	Sr 1.0	Y 1.2	Zr 1.4	Nb 1.6	Mo 1.8	Tc 1.9	Ru 2.2	Rh 2.2	Pd 2.2	Ag 1.9	Cd 1.7	In 1.7	Sn 1.8	Sb 1.9	Te 2.1	I 2.5	Xe 2.6
Cs 0.7	Ba 0.9	La 1.1	Hf 1.3	Ta 1.5	W 1.7	Re 1.9	Os 2.2	Ir 2.2	Pt 2.2	Au 2.4	Hg 1.9	Tl 1.8	Pb 1.8	Bi 1.9	Po 2.0	At 2.2	Rn 2.4
Fr 0.7	Ra 0.7	Ac 1.1															

Ce 1.1	Pr 1.1	Nd 1.1	Pm 1.1	Sm 1.1	Eu 1.1	Gd 1.1	Tb 1.1	Dy 1.1	Ho 1.1	Er 1.1	Tm 1.1	Yb 1.1	Lu 1.2
Th 1.3	Pa 1.5	U 1.7	Np 1.3	Pu 1.3	Am 1.3	Cm 1.3	Bk 1.3	Cf 1.3	Es 1.3	Fm 1.3	Md 1.3	No 1.3	Lr

Pauling's electronegativity values assigned to elements on the periodic table

Bond polarity

A *covalent bond* is the relatively equal sharing of electrons between two elements.

If the two atoms' electronegativity is the same (or similar), they share the electrons equally (or almost equally).

In general, these electronegativity trends are like ionization energy and electron affinity.

As the atomic number increases, the positive nuclear charge increases, and the electrons within the same energy level (same shielding effect) are more strongly attracted to the positive nucleus.

A *polar covalent bond* results if there is a sufficient difference in electronegativity between the atoms (see the table below for value ranges).

In a polar covalent bond, the electronegative element has a greater tendency to attract electrons. This atom gets a larger share of the electron density, giving it a partial negative (δ^-) charge.

The less electronegative element in a polar covalent bond has a weaker tendency to attract electrons. The atom gets a smaller share of the electron density, giving it a partial positive (δ^+) charge.

Ionic bonds transfer an electron from the electropositive element to the electronegative element and occur if the electronegativity difference (> 1.6 D) is significant (see the table below).

Ionic bonds generally occur between a metal (left two columns; groups 1 and 2) and a nonmetal (right side; generally, groups 16 and 17).

Dipole moment

The *dipole moment* measured in debye (D) units is the difference in electronegativity between atoms participating in a bond is the *dipole moment* measured in debye (D) units.

Bonds are covalent, polar covalent, or ionic based on electronegativity.

Type of Bond	Dipole Moment (D)
Covalent	$0 - 0.6$
Polar covalent	$0.6 - 1.6$
Ionic bonds	> 1.6

Notes for active learning

Atomic Size

Electron shells

Electron shells are defined by the principal quantum number n.

With a larger atomic number, elements have more electrons and electron shells.

From top to bottom down a group, the shielding effect increases because the closer shells are between the nucleus and the outermost shell.

Extra shells increase the atom size because of a new orbit (i.e., higher shells are more distant from the nucleus than lower shells).

For elements left to right across a row on the periodic table, the principal quantum number n is the same; therefore, the same shell fills with the additional electrons.

As a shell fills, the effective nuclear charge increases because of the increasing number of protons, while additional electrons fill the same orbit with no increase in shielding.

With the increasing effective nuclear charge, the electrostatic attraction between the nucleus and the electrons increases, making the atom more compact.

Atomic radius

Atomic radius is ½ the distance between the two atomic nuclei and measures atoms' sizes since electron clouds are too ill-defined to measure precisely.

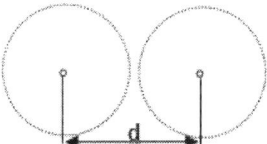

Atomic radii measure the distance between the atomic nuclei of two atoms

The relative atomic radii trends in periods 2 and 3 are illustrated below.

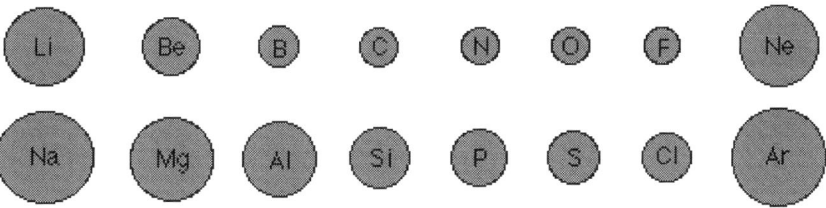

If a nucleus were a ping-pong ball's size on a relative scale, the atom's diameter would be 25 miles.

Therefore, an increase in the nucleus size does not substantially affect an atom's overall size.

The atoms become larger from top to bottom on the periodic table because electrons add to new orbitals farther from the nucleus.

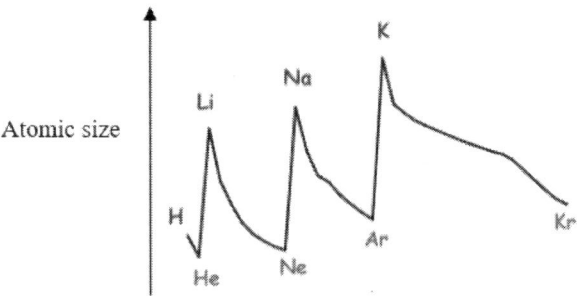

Atomic radii trend increases as principal number n
increases for additional orbitals

The energy *vs.* atomic number graph peaks represent atoms with a single valence electron, while the valleys are atoms with a filled valence shell.

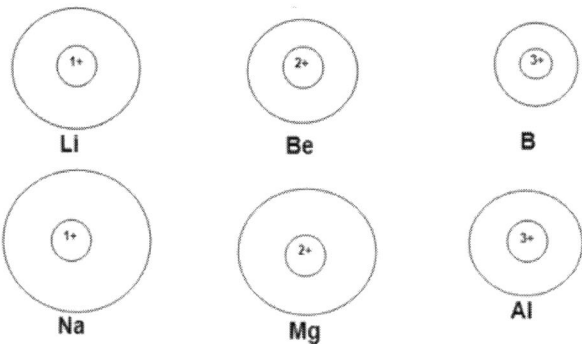

The relative size of the electron shells across a group and down periods The nucleus is shown at the center and the 1s orbital for the ions (Li and Na are 1+; Be and Mg are 2+; B and Al are 3+).

Across the group, the effective *nuclear charge increases with protons* (increasing Z value) in the nucleus.

An atom's *size increases down a column* due to an increasing number of shells (*n* number).

An atom's *size decreases across a row* due to the increased nuclear attraction (more nuclear core protons).

Classification of Elements by Valence Structure

Groups based on valence electrons

The *groups* (or *families*) are vertical columns with recurring trends. For this reason, some of the groups have unique names, sometimes referred to as common names or *unsystematic names* (e.g., alkali metals).

Elements in the same group have the same number of valence (outer shell) electrons and tend to have patterns in physical properties such as ionization energy, atomic radius, and electronegativity.

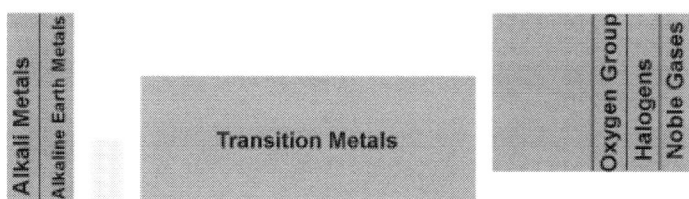

Sample common names for groups on the periodic table

Blocks based on subshells

The periodic table is divided into blocks based on the subshell (*s, p, d, f* orbitals) in which the valence (outermost) electron is located.

The horizontal trends are especially important in some blocks.

The four blocks are the *s*-block, the *p*-block, the *d*-block, and the *f*-block.

These blocks are seen in the figure below, indicating the principal quantum number *n* and the azimuthal quantum number *l* for the subshells *s, p, d,* and *f* are shown.

The *f*-block (i.e., lanthanides and actinides) is displayed separately below the periodic table.

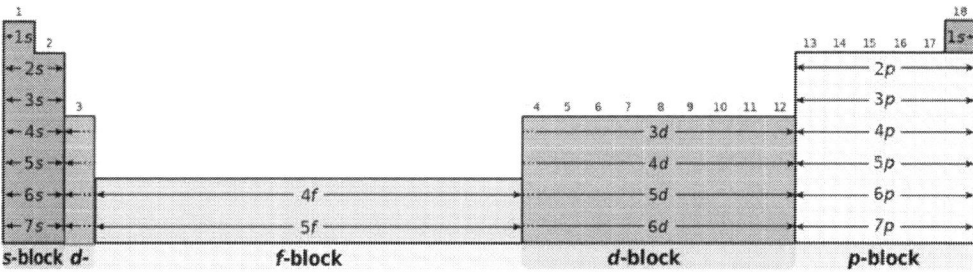

Representative elements

The representative (or main-group) elements (Groups 1, 2 and 13-18) comprise the *s* (1-2) and *p*-blocks (13-18), and properties are as follows:

- No free-flowing (or loosely bound) outer *d* electrons, so the number of valence electrons is the group number

- Valence shell fills from the left (1 electron) to the right (8 electrons)

- Collectively, the essential elements for life on Earth, comprising 80% of the Earth's surface

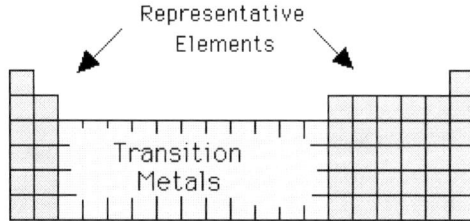

Representative elements have s and p orbitals, while transitions metals have d orbitals

Alkali metals

The alkali metals (Group 1 or formerly 1A) include the elements Li, Na, K, Rb, Cs, and Fr, and the properties of the alkali metals are as follows:

- Single valence electrons are easily ionized; have low ionization energy, and are very reactive

- React to lose one electron and form a cation with +1 oxidation state

- More reactive from top to bottom due to increasing atomic radii (Cs and Fr most reactive)

- Malleable, ductile, good conductor like metals, but usually softer than other metals

- React with oxygen to form strongly alkaline oxides

- React with water to form hydroxides and hydrogen

- React with acids to form salts and release hydrogen

- Commonly found in nature as compounds with halogens

Alkaline earth metals

The alkaline earth metals (Group 2 or formerly 2A) include Be, Mg, Ca, Sr, Ba, and Ra, with the properties:

- 2 valence electrons, so relatively low ionization energy and reactive

- React to lose two electrons and form cation with +2 oxidation state

- Increasingly reactive from top to bottom because of the increasing atomic radii

- React with oxygen to form strongly alkaline oxides

- React with water to form hydroxides and hydrogen

- React with acids to form salts and release hydrogen

- Due to reactivity, generally not found in nature

Notes for active learning

Classification of Elements by Electronic Structure

Oxygen group

The oxygen group (or *chalcogens*) (Group 16 or formerly VIA) includes O, S, Se, Te, Po, and Lv, with properties:

- Oxygen, sulfur, and selenium are nonmetals, while tellurium and polonium are metalloids

- Lighter chalcogens (O, S) are typically non-toxic and critical to life, while heavier chalcogens are toxic

- Form ions with multiple oxidation numbers; sulfur's oxidation states are between –2 to +6

- Exist as polyatomic molecules in the elemental state (e.g., O_2, S_8)

Halogens

Halogens (Group 17 or formerly 7A) include F, Cl, Br, I, and At, with the properties:

- 7 valence electrons (2 from s subshell and 5 from p subshell) - high electron affinity, very reactive (most reactive of the nonmetals)

- Most electronegative of the elements

- React to gain one electron and achieve full valence shell; form anions with –1 oxidation state

- Less reactive from top to bottom because of decreasing atomic radii

- React with alkali metals and alkaline earth metals to form salts (crystalline ionic solids)

- Form diatomic molecules in the elemental state (e.g., Cl_2, Br_2, I_2)

Noble gases

The noble gases (Group 18 or formerly as 8A) include He, Ne, Ar, Kr, Xe, Rn, and Og:

- Full valence shell of 8 have high ionization energy and low electron affinity

- Inert (i.e., unreactive); heavier elements react with a few very electronegative elements (e.g., F).

- Exist in nature as monatomic gases (e.g., He, Ne, Ar)

Transition metals

The transition metals (Groups 3-11 or formerly as 3B-1B) have unique chemical properties due to loosely bound outermost d orbital electrons, with the properties:

- High conductivity and malleability due to loosely bound, free-flowing outer d electrons.

- When bonded with other ions to form metal complexes, the d orbitals become non-degenerate (different energy).

  Electron transitions between non-degenerate d orbitals give transition metal complexes vivid colors.

- Lose electrons from more than one shell to form cations of different oxidation states, indicated using Roman numerals (e.g., Iron (II) Fe^{2+} and Iron (III) Fe^{3+}).

Inner transition elements

The periodic table has two detached rows of elements located under the table (the f-block).

Those elements are collectively known as *inner transition elements* or the *lanthanide and actinide series*. They are in the 6th and 7th periods starting at the 5d block of the periodic table.

Due to natural scarcity, the inner transition elements were historically classified as *rare-earth elements*.

However, even the rarest of the inner transition metals were more common than the platinum-group metals (which are transition metals).

Most inner-transition metals (e.g., plutonium and uranium) only exist in nature as radioisotopes and decay rapidly.

Metals

The elements are divided into metals, nonmetals, and metalloids from shared physical and chemical properties.

Highly ordered, closely-packed atoms characterize metals.

The metals comprise almost three-fourths of known earth elements (e.g., aluminum, iron, calcium, magnesium).

Many metals' chemical and physical properties result from metallic bonds.

The valence electrons of s and p orbitals delocalize and form an aggregate of electrons surrounding the interacting metal atoms' nuclei.

The high degree of freedom for outermost electron movement between metal atoms gives rise to many properties:

- Metallic luster - can shine or reflect light

- Malleable - can be hammered or rolled into thin sheets

- Ductile - can be drawn into wire

- Hardness ranges from hard (iron, chromium) to soft (sodium, lead, copper)

- Conduct heat and electricity

- Crystalline solids at room temperature, except mercury (Hg), which is the only liquid metal

- High melting point

- Chemical reactivity varies from unreactive (e.g., Au, Pt) to very reactive, which burst into flames upon contact with water (e.g., Na, K)

Nonmetals

Nonmetals are substances that do not exhibit chemical or physical traits most associated with metal elements because they bond to form covalently bonded compounds.

Nonmetals are in the upper right-hand corner of the periodic table with the following properties:

- Mostly polyatomic elements, except the monoatomic noble gases

- At room temperature, nonmetal elements are in all phases: gas (H_2, O_2, N_2, F_2, Cl_2), solid (I_2, Se_8, S_8, P_4), and liquid (Br_2)

- Brittle – they pulverize when struck

- Insulators (or extremely poor conductors) of electricity and heat because electrons do not possess the high degree of movement as those in metal atoms

- Chemical reactivity ranges from inert (noble gases) to reactive (F_2, O_2, H_2). Nonmetals react with metals to form ionic compounds

- Some nonmetals have allotropes: different forms in the same phase (carbon as diamond and graphite)

Metals and nonmetals comparison

Chemical Properties	
Metals	**Nonmetals**
Likes to lose electrons to gain a positive (+) oxidation state (good reducing agent)	Likes to gain electrons to form a negative (–) oxidation state (good oxidizing agent)
Lower electronegativity – partially positive in a covalent bond with nonmetal	Higher electronegativity – partially negative in a covalent bond with metal
Forms basic oxides	Forms acidic oxides
Physical Properties	
Good conductor of heat and electricity	Poor conductor of heat and electricity
Malleable, ductile, luster, and solid at room temperature (except Hg)	Solid, liquid, or gas at room temperature. Brittle if solid and without luster.

Metalloids

Metalloids (e.g., B, Si, As) exhibit properties of metals and nonmetals.

The metalloids are in the periodic table's "stair-step" line dividing metals and nonmetals.

Some metalloids are lustrous like metals, while others are brittle like nonmetals.

Metalloids are unique because they do not typically conduct electricity at room temperatures but conduct electricity when heated to higher temperatures.

Practice Questions

1. The attraction of the nucleus on the outermost electron in an atom tends to:

    A. decrease from right to left and bottom to top on the periodic table

    B. decrease from left to right and bottom to top on the periodic table

    C. decrease from left to right and top to bottom of the periodic table

    D. decrease from right to left and top to bottom on the periodic table

    E. increase from right to left and top to bottom on the periodic table

2. Which species shown below has 24 electrons?

    A. $^{52}_{24}Cr$ **B.** $^{55}_{25}Mn$ **C.** $^{24}_{12}Mg$ **D.** $^{45}_{21}Sc$ **E.** $^{51}_{23}V$

3. Which characteristics describe the mass, charge, and location of a proton, respectively?

    A. approximate mass 5×10^{-4} amu; charge +2; inside nucleus

    B. approximate mass 1 amu; charge 0; inside nucleus

    C. approximate mass 1 amu; charge +1; inside nucleus

    D. approximate mass 5×10^{-4} amu; charge –1; outside nucleus

    E. approximate mass 1 amu; charge +2; outside nucleus

4. Which of the following subshell notations for electron occupancy is NOT possible?

    A. $4f^{11}$ **B.** $2p^1$ **C.** $5s^3$ **D.** $4p^5$ **E.** $4d^3$

5. The *f* subshell can hold how many total electrons?

    A. 4   **B.** 8   **C.** 12   **D.** 14   **E.** 18

6. Which of the following is the best description of the Bohr atom?

    A. Sphere with a heavy, dense nucleus containing electrons

    B. Sphere with a heavy, dense nucleus encircled by electrons in orbits

    C. Indivisible, indestructible particle

    D. Homogeneous sphere of protons, neutrons, and electrons

    E. Sphere with a sparse nucleus surrounded by dense electron orbitals

7. Which of the following compounds does NOT exist?

    A. H^-, because H forms only positive ions

    B. PbO_2, because the charge on a Pb ion is only +2

    C. SF_6, because F does not have an empty d orbital to form an expanded octet

    D. OCl_6, because O does not have d orbitals to form an expanded octet

    E. $^-CH_3$, because C lacks d orbitals

8. The elements in Groups 1, 17, and 18 (formerly known as IA, VIIA, and VIIIA) are called, respectively:

    A. alkaline earth metals, transition metals, and halogens

    B. alkali metals, halogens, and noble gases

    C. alkali metals, alkali earth metals, and halogens

    D. alkaline earth metals, halogens, and alkali metals

    E. halogens, alkali earth metals, and noble gases

9. Which of the following is a general characteristic of a metallic element?

    A. Reacts with nonmetals   **C.** Dull, brittle solid

    B. Reacts with metals   **D.** Low melting point

                                  E. None of the above

10. Which of the following elements would be shiny and flexible?

    A. bromine (Br)   **C.** helium (He)

    B. selenium (Se)   **D.** ruthenium (Ru)

                                  E. silicon (Si)

11. The masses on the periodic table are expressed in what units?

    A. picograms   **C.** amu

    B. nanograms   **D.** micrograms

                                  E. gram

12. Isotopes of an element have the same number of [   ] and different [   ] numbers.

    A. protons, electrons   **C.** neutrons, protons

    B. neutrons, electrons   **D.** protons, neutrons

                                  E. electrons, protons

13. Paramagnetism, the ability to be pulled into a magnetic field, is demonstrated by:

  A. any substance containing unpaired electrons

  B. nonmetal elements that have unpaired p orbital electrons

  C. transition elements that have unpaired d orbital electrons

  D. nonmetal elements that have paired p orbital electrons

  E. any substance containing paired electrons

14. Which element has the electron configuration $1s^2 2s^2 2p^6 3s^2 3p^6 4s^2 3d^{10} 4p^6 5s^2 4d^1$?

  A. Y   **B.** La   **C.** Si   **D.** Sc   **E.** Zr

15. Dmitri Mendeleev's chart of elements:

  A. predicted the behavior of unidentified elements

  B. developed the basis of our modern periodic table

  C. predicted the existence of elements undiscovered at his time

  D. placed elements with the same number of valence electrons in the same period

  E. all the above

Notes for active learning

Detailed Explanations

1. D is correct.

The nucleus's attraction on the outermost electrons determines the *ionization energy*, increasing right and up on the periodic table.

2. A is correct.

The number of protons is the element's atomic number (Z) and is shown as a subscript.

Atoms of an element do not necessarily have the same mass (A) because they may have different numbers of neutrons (i.e., isotopes of the element).

For neutral elements, the number of electrons equals the number of protons (i.e., subscript).

An element's atoms may have different numbers of electrons, forming charged ions (i.e., *anions* and *cations*).

For neutral atoms, the number of electrons equals the number of protons.

3. C is correct.

Protons are the positively charged particles located inside the nucleus of an atom.

Like neutrons (also in the nucleus), protons have a mass of approximately 1 amu.

Protons have a +1 charge, while neutrons have a charge of 0 (i.e., neutral).

4. C is correct.

Each orbital can hold two electrons.

Maximum number of electrons in each shell:

The *s* subshell has 1 spherical orbital and can accommodate 2 electrons

The *p* subshell has 3 dumbbell-shaped orbitals and can accommodate 6 electrons

The *d* subshell has 5 lobe-shaped orbitals and can accommodate 10 electrons

The *f* subshell has 7 orbitals and can accommodate 14 electrons

Therefore, the *s* subshell can accommodate only 2 electrons.

5. D is correct.

The number of *orbitals in a subshell* is different from the maximum number of *electrons in the subshell*.

Each orbital can hold two electrons.

Maximum number of electrons in each shell:

The *s* subshell has 1 spherical orbital and can accommodate 2 electrons

The *p* subshell has 3 dumbbell-shaped orbitals and can accommodate 6 electrons

The *d* subshell has 5 lobe-shaped orbitals and can accommodate 10 electrons

The *f* subshell has 7 orbitals and can accommodate 14 electrons

An *f* subshell capacity is 7 orbitals × 2 electrons/orbital = 14 electrons.

6. B is correct.

The *Rutherford-Bohr model* of the hydrogen atom, often referred to as the *Bohr atom*, depicts a sphere with a heavy, dense nucleus encircled by electrons in orbits, held by the electrostatic forces between the positively charged nucleus and the negatively charged electrons.

7. D is correct.

O contains only *s* and *p* orbitals and cannot form 6 bonds due to the lack of *d* orbitals.

Hydrogen forms a negative ion (i.e., hydride) in NaH (sodium hydride), $LiAlH_4$ (lithium aluminum hydride), and $NaBH_4$ (sodium borohydride).

Each of these hydrides (i.e., H^-) is a strong base with a pK_a greater than 32.

Lead (Pb), although not a transition metal, does have two common valences: +2 and +4.

Fluorine in SF_6 does not have an expanded octet because each F is bonded only to S. The sulfur has an expanded octet, but it is in the 3rd row of the periodic table and has empty *d* orbitals.

8. B is correct.

A *group* (or family) is a vertical column, and group elements share similar properties.

Group IA is the alkali metals, which include lithium (Li), potassium (K), sodium (Na), rubidium (Rb), cesium (Cs), and francium (Fr).

Group VIIA is the halogens, which include fluorine (F), chlorine (Cl), bromine (Br), iodine (I), and astatine (As). Halogens gain one electron to become a −1 anion, and the resulting ion has a complete octet of valence electrons.

Group VIIIA is the noble gases that include helium (He), neon (Ne), argon (Ar), krypton (Kr), xenon (Xe), radon (Rn), and Oganesson (Og).

The alkaline earth metals (group IIA) include beryllium (Be), magnesium (Mg), calcium (Ca), strontium (Sr), barium (Ba), and radium (Ra).

9. A is correct.

Most elements on the periodic table (over 100 elements) are metals. There are about four times more metals than nonmetals.

Alkali metals (group IA) include lithium (Li), potassium (K), sodium (Na), rubidium (Rb), cesium (Cs) and francium (Fr).

The alkaline earth metals (group IIA) include beryllium (Be), magnesium (Mg), calcium (Ca), strontium (Sr), barium (Ba), and radium (Ra).

Metals form positive ions (i.e.., cations) by losing electrons during chemical reactions; thus, metals are electropositive elements.

They are *good conductors of heat and electricity* as the metals' molecules are closely packed.

Metals are *not brittle or fragile* (i.e., they do not break easily); to the contrary, they are malleable and can be pressed or hammered without breaking.

They are opaque and not transparent because free moving electrons surround the atoms in a metal; therefore, light striking the metal hits these electrons, which absorb and re-emit it, and light cannot pass through.

Metals, except mercury, are solids under normal conditions. Potassium has the lowest melting point of the solid metals at 146 °F.

10. D is correct.

Transition metals are in groups (vertical columns) 3–12 of the period table, in periods (horizontal rows) 4–7. Transition metal elements include silver, iron, and copper.

11. C is correct.

The masses on the periodic table are the atomic masses of the elements.

The *atomic mass* is a weighted average of the various isotopes of the element (the mass of each isotope multiplied by its relative abundance). Atomic mass is sometimes called atomic weight, but atomic mass is more accurate.

This mass is expressed in *atomic mass units* (amu), called Daltons (D).

> 1 amu equals 1 g/mol

An amu is one-twelfth the mass of a neutral carbon-12 atom.

12. D is correct.

Elements are defined by the number of protons (i.e., atomic number).

The isotopes are neutral atoms: # electrons = # protons.

Isotopes are variants of an element that differ in the number of neutrons; isotopes have the same number of protons and occupy the same position on the periodic table.

The *number of protons* within the atom's nucleus is the *atomic number* (Z) and equals the number of electrons in the neutral (non-ionized) atom.

Atomic number identifies a specific element but not the isotope; an atom of a given element may have a wide range in its number of neutrons.

The number of protons *and* neutrons (i.e., nucleons) in the nucleus is the *mass number* (A), and each isotope of an element has a different mass number.

13. A is correct.

Paramagnetic elements are substances that move into a magnetic field and have one or more unpaired electrons.

Most transition metals and their compounds in oxidation states involving incomplete inner electron subshells are paramagnetic.

An applied magnetic field repels diamagnetic elements because they create an induced magnetic field in a direction opposite to an externally applied magnetic field.

14. A is correct.

Identify an element using the periodic table by using its atomic number.

The *atomic number* equals the number of protons or electrons.

The total number of electrons is determined by summing the electrons in the provided electron configuration:

$$2 + 2 + 6 + 2 + 6 + 2 + 10 + 6 + 2 + 1 = 39$$

Element #39 in the periodic table is yttrium (Y).

15. E is correct.

Russian chemist Dmitri Mendeleev (in 1869) created the prototype version of the periodic table of elements. He placed elements with the same number of valence electrons in the same group (i.e., vertical columns).

Mendeleev's table was the basis of the modern-day periodic table.

However, the modern-day table has an opposite system where elements with the same number of valence electrons are in the same group (or family) rather than periods (i.e., horizontal row).

Mendeleev predicted the existence of undiscovered elements, along with their behavior.

Notes for active learning

CHAPTER 2

Chemical Bonding

- Electrostatic Forces Between Ions

- Electron Pair Sharing

- *Sigma* and *Pi* Bonds

- Hybridization

- Valence Bond (VB) Theory

- Molecular Orbital (MO) Theory

- Lewis Electron Dot Formulas

- Formal Charge

- Resonance

- Partial Ionic Character

- Polarity

- Practice Questions & Detailed Explanations

Electrostatic Forces Between Ions

Ionic bond formation

An *ionic bond* forms when electrons transfer from one atom to another. This electron transfer results in oppositely charged species that attract each other via electrostatic interaction.

Atoms gain and lose electrons and become charged, known as *ions*.

Nonmetals (i.e., right side of the periodic table) tend to gain electrons and form negatively charged anions.

Metals (left side of the periodic table) exhibit weaker nuclear forces on valence electrons and lose electrons forming positively charged cations.

Ionic bonds form between a metal cation and a nonmetal anion so that each atom obtains a complete *valence shell*.

Ionic bonds *transfer an electron* from the electropositive to the electronegative element.

An ionic bond occurs if the *electronegativity difference* (> 1.6 D) is significant (table below).

Ionic bonds generally occur between a metal (left two columns; groups 1 and 2) and a nonmetal (right side; generally, groups 16 and 17).

If a metal has low ionization energy and a nonmetal has a high electron affinity (discussed elsewhere), ionic bonding is more likely.

A familiar ionic compound is sodium chloride (NaCl), or table salt.

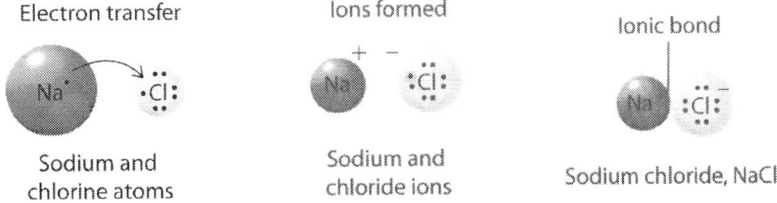

Electron transfer — Sodium and chlorine atoms

Ions formed — Sodium and chloride ions

Ionic bond — Sodium chloride, NaCl

The cations and anions in an ionic compound are like a magnet's North and South poles, which attract.

Similarly, two cations (or anions) repel.

Crystalline structures

Therefore, the molecules arrange into rigid crystalline structures.

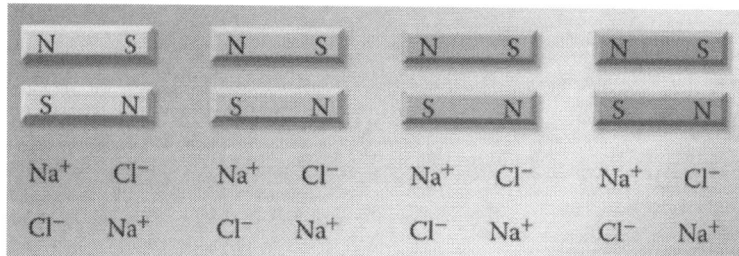

Crystalline Na$^+$ and Cl$^-$ with alternating positive and negative charges

Due to the strong, polarized, intermolecular forces, ionic compounds generally possess high melting and boiling points. Therefore, they are often solid at room temperature.

Ionic crystals are three-dimensional arrangements of ions in an ionic compound, sometimes called a *crystal lattice* (e.g., solid crystalline structure of sodium chloride).

Sodium chloride (NaCl) has a high melting point (800 °C) and dissolves in water to give a conducting solution. Like ionic compounds, sodium chloride is a strong electrolyte, dissociating entirely in water.

Solid ionic compounds are non-conductive; however, molten compounds (i.e., compounds liquefied by heat) or compounds dissolved in water conduct electricity.

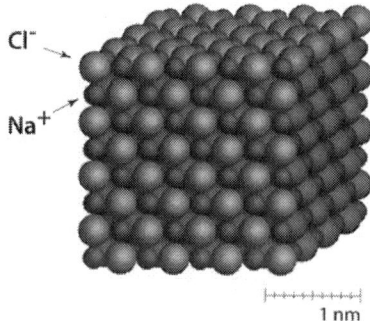

1 nm

The crystalline lattice structure of table salt (NaCl)

Like all types of bonding, the valence (outermost) electrons are involved in ionic bonding.

Transferring the lone 3s electron of a sodium atom to the half-filled 3p orbital of a chlorine atom generates a sodium cation and a chloride anion, forming the compound sodium chloride.

Initially, the sodium and chlorine atoms each had a net charge of 0. After the transfer of one electron took place, they had a charge of +1 and –1, respectively, due to the atom's imbalance between protons and electrons.

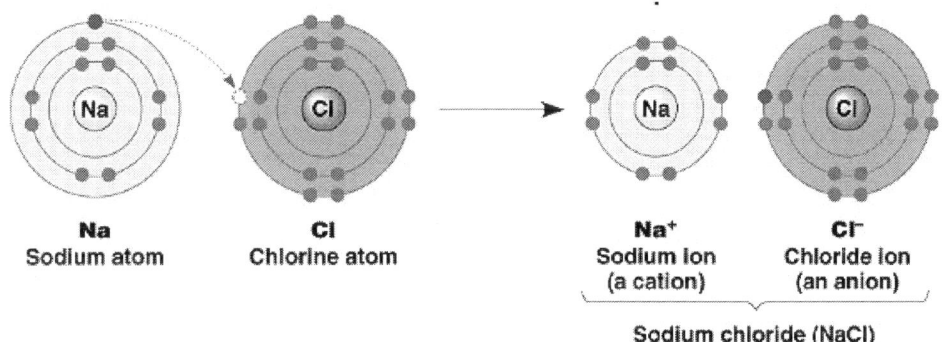

Ionic bonds form by the transfer of a valence electron

Electrostatic potential energy $\propto q_1q_2 / r$

The strength of an ionic bond is related to the amount of *electrostatic potential energy* between opposing charges.

The *electrostatic potential energy* is:

$$U_E = kq_1q_2 / r$$

where U_E is electrostatic potential energy, k is Coulomb's constant (9×10^9 m/F), q_1 and q_2 are the two charges, and r is the distance between them

Electrostatic energy is negative in ionic bonds because q_1 and q_2 have opposing charges.

If q_1 and q_2 did not have opposing charges (i.e., both positive or negative), they repel, and no ionic bond forms.

The negative sign is often dropped, and the magnitude of the electrostatic energy is used.

The higher the magnitude of the electrostatic potential energy, the stronger the ionic bond.

Strong ionic bonds are formed by high charge magnitudes (q values) close together (small r value).

Ions that form strong ionic bonds have a high charge density; the charge-to-size ratio is high.

A cation's high charge density can distort the anion's electron cloud to promote electron sharing.

Lattice energy

Lattice energy is the energy required to break an ionic bond. The energy released by the reaction correlates to the ionic bond strength, proportional to electrostatic energy.

Conversely, it is the energy released when oppositely charged; gaseous ions join to form an ionic crystalline solid.

Electrostatic force $\propto q_1 q_2 / r^2$

The electrostatic force that holds charged particles is defined as *electrostatic energy*.

Coulomb's Law defines electrostatic force F with electrostatic charges and distance.

$$F = k q_1 q_2 / r^2$$

where F is the electrostatic force, k is Coulomb's constant (9×10^9 m/F), q_1 and q_2 are the two charges, and r is the distance between the charges

From the equation, the electrostatic force is directly proportional to the product of opposite charges that attract (negative F) or the same charges that repel (positive F).

The electrostatic force is inversely proportional to the square of the distance between them.

Therefore, larger charge magnitudes closer exhibit a greater electrostatic force.

Coulomb's Law is analogous to the *universal law of gravitation*:

$$F = G m_1 m_2 / r^2$$

where G (gravitational constant) is analogous to k, and m (mass) to q

The major difference is that G is tiny compared to coulomb's constant k (stationary electrically charged particles) because the gravitational force is weak compared to the stronger electrostatic force between charged particles.

There is another way to write the formula for an electrostatic force specific to ionic compounds.

The equation describes the force of attraction between the cation n^+ and the anion n^- at a distance d apart.

$$F = R(n^+ \times e) \cdot (n^- \times e) / d^2$$

where R is coulomb's constant (usually written as k)

$n^+ \times e$ = charge of cation in coulombs = positive charge (n^+) × coulombs per electron (e)

$n^- \times e$ = charge of anion in coulombs = negative charge (n^-) × coulombs per electron (e)

Electron Pair Sharing

Covalent bond formation

A *covalent bond* forms when two atoms share an electron pair.

These electron pairs are shared, or bonding pairs and covalent bonds result from the overlap of electron orbitals.

Covalent bonds typically form between nonmetal elements.

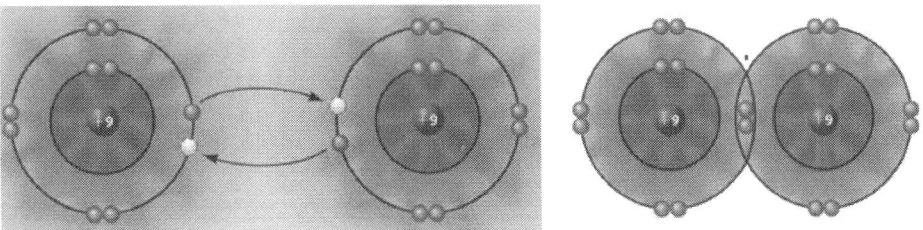

A covalent bond forms between two fluorine atoms. The left image shows the valence electron from each fluorine atom that becomes shared to form the covalent bond in the F_2 molecule

Most atoms with similar electronegativity share electrons to achieve a full valence shell of eight electrons (i.e., octet rule for $2s$ and $6p$ electrons).

Illustrations of covalent bonds following the octet rule are shown below.

Electron dot formulas

Simple atomic notation with dots designating valence electrons:

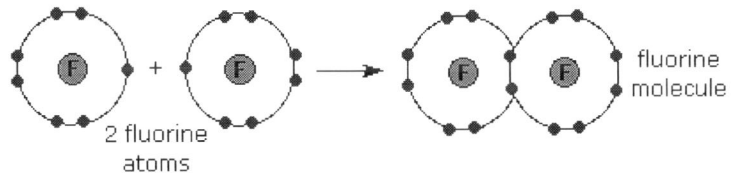

Covalent bond forms between 2 fluorine atoms that share a valence electron

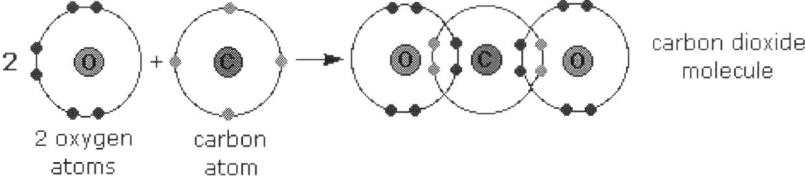

Covalent bond forms between 2 O atoms and one carbon atom
that share two pairs of valence electrons from the carbon atom to form CO_2

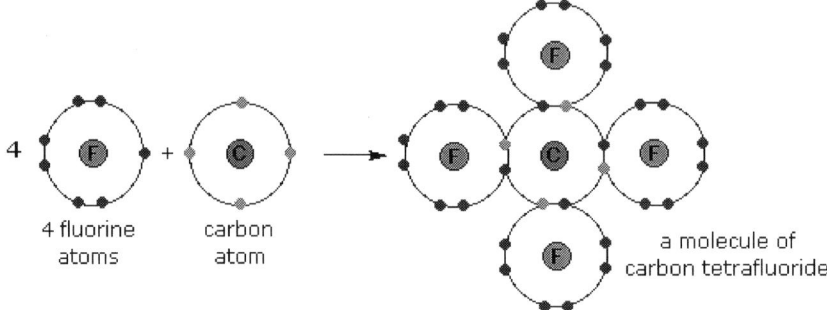

Covalent bond forms between 4 fluorine atoms and one carbon atom

Each atom contributes one valence electron to form carbon tetrafluoride (CF_4)

The hydrogen atom is an exception to the octet rule because it is at its lowest energy when it has two $1s$ electrons in its valence shell.

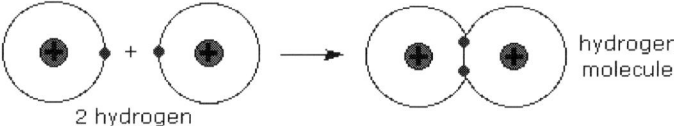

Atoms form covalent bonds to obtain the ideal number of electrons.

Non-bonding lone pairs

The *non-bonding lone pairs* are electrons not involved in bonding.

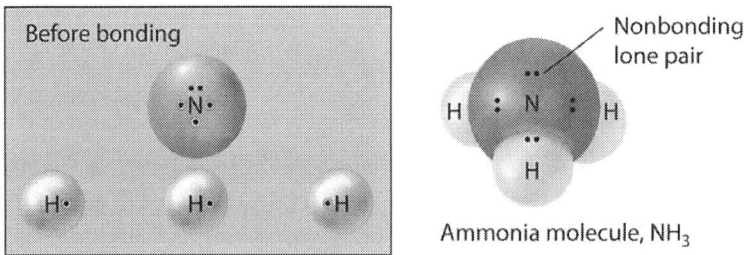

Nitrogen has 5 valence electrons. Nitrogen shares one valence electron with each of three hydrogen atoms. Each hydrogen contributes 1 valence electron. With three covalent bonds, nitrogen has a complete octet, and hydrogen satisfies a complete 1s subshell

　　　　　　　　　　　　　　　　　Copyright © Sterling Education.

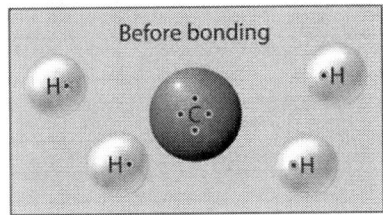

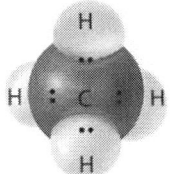

Methane molecule, CH_4

Carbon has 4 valence electrons and shares one electron with four H atoms. Each H contributes 1 valence electron. With four covalent bonds, carbon has a complete octet, and H satisfies a complete 1s subshell

Multiple bonds

Covalent bonds in double- and triple-bonded nonmetal atoms involve more than one pair of electrons.

In double bonds, *two pairs* of electrons (four electrons) are shared.

In triple bonds, *three pairs* of electrons (six electrons) are shared.

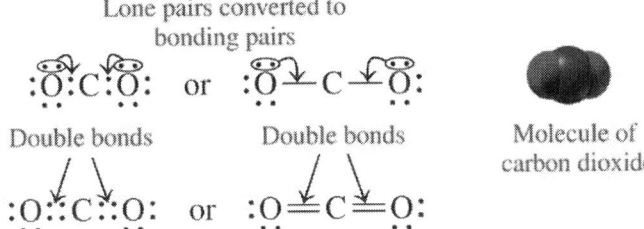

Molecule of carbon dioxide

Carbon dioxide forms from two double bonds on the central carbon. Each double bond results from the oxygen and the carbon contributing a pair of electrons

$$\cdot\ddot{N}{:}\ddot{N}\cdot \longrightarrow \ :N{::}N: \quad :N{\equiv}N:$$

Three shared pairs   *Triple bond* N_2 *molecule*

Coordinate covalent bonds

Coordinate covalent bonds result from one atom donating an electron pair to another atom; both electrons come from one donor atom.

Ozone, O_3, is a molecule with a coordinate covalent bond.

$$:\ddot{O}{::}\ddot{O}: \ + \ \ddot{O}: \ \longrightarrow \ :\ddot{O}{::}\ddot{O}{:}\ddot{O}:$$

nonbonding electron pairs   *coordinate covalent bond: O=O–O*

Notes for active learning

Sigma and *Pi* Bonds

Single bonds

The electron density of a chemical bond can be divided into regions.

Sigma (σ) *bonds* form when one orbital from each atom overlaps and creates a new orbital.

Typically, a *sigma* bond is a single bond and is the first bond involved in double and triple bonds. The sigma bonds lie along the imaginary line joining two nuclei, the *internuclear axis*.

In σ bonds, the electron density is between the two atoms' nuclei and is symmetrical about the internuclear axis.

Atoms on each side of a σ single bond are free to rotate with the bond as their axis.

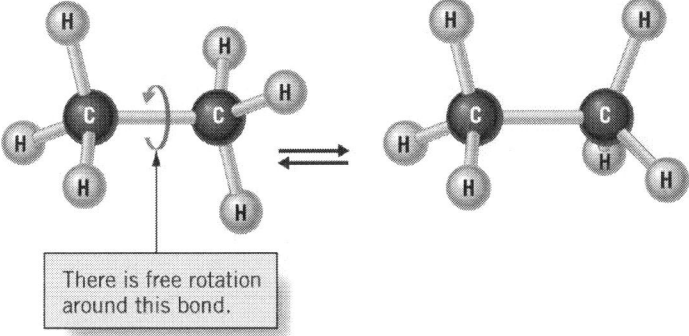

There is free rotation around this bond.

S and *p* orbitals

In the image below, the σ bonds of *s–s* and *p–p* orbitals are shown.

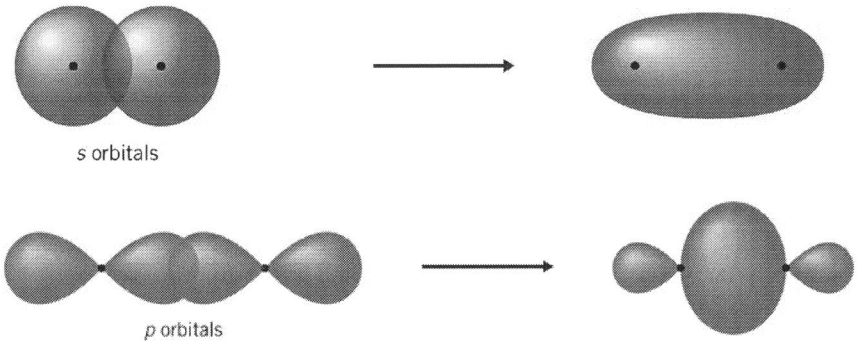

s orbitals

p orbitals

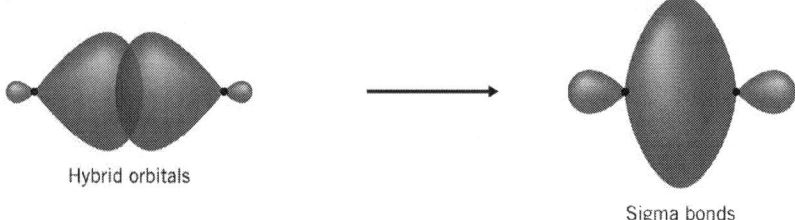

Hybrid orbitals

Sigma bonds

Double bonds

Pi (π) bonds form when unhybridized *p* orbitals on the top and bottom of the internuclear axis overlap.

A single π bond consists of regions above and below the intermolecular axis.

There is no electron density along the π bond intermolecular axis.

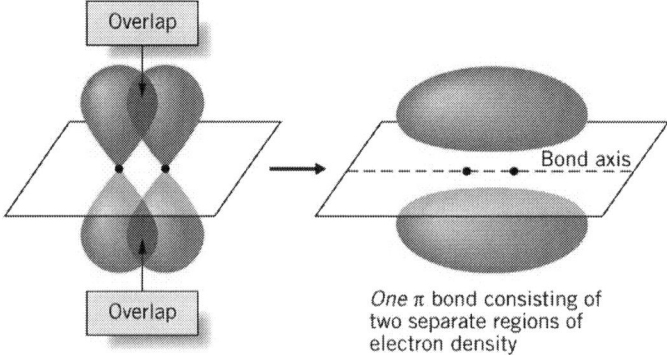

One π bond consisting of two separate regions of electron density

When electron density above and below the σ bonds overlap, a multiple bond (i.e., double or triple) forms. Thus, π bonds form after a σ bond has connected two atoms.

A double bond: one σ bond and one π bond.

A triple bond: one σ bond and two π bonds.

Nitrogen (N_2) gas is connected by a triple bond (N≡N). The first is a σ bond, and the remaining two are π bonds.

The π bond introduces rigidity to molecules; π bonds do not allow for the rotation of atoms.

Molecules that involve π bonds require more energy to break.

Hybridization

Hybrid orbitals

Hybrid orbitals are produced by the hybridization (mixing) of existing electron orbitals.

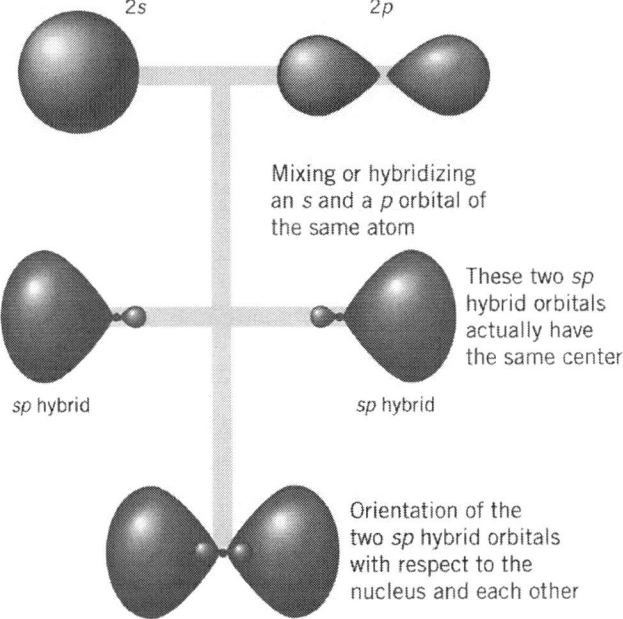

Hybridization and molecular geometry

The possible hybridizations of the carbon atom are as follows.

- sp^3: a hybrid between one s with three p orbitals.

  Tetrahedral geometry.

  Contains single bonds only.

- sp^2: a hybrid between one s with two p orbitals.

  Trigonal planar geometry.

  Contains a double bond.

- sp: a hybrid between one s with one p orbital.

  Linear geometry.

  Contains a triple bond.

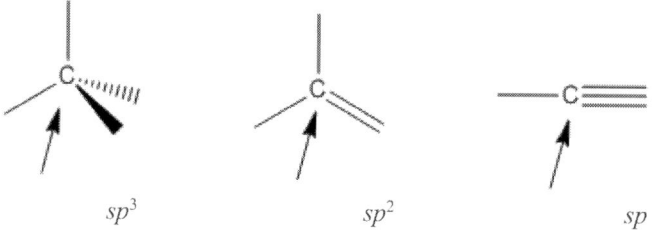

sp^3 sp^2 sp

Hybridization of the carbon atom for 4 single bonds, 1 double or 1 triple bond

Valence shell electron-pair repulsion (VSEPR) theory predicts molecular shapes

Valence Shell Electron Pair Repulsion (VSEPR) theory states that electron pairs surrounding an atom repel each other and have the greatest bond angles separating the bonds and lone pairs around a central atom.

In VSEPR, *electron domains* are electron pairs in the molecule.

Molecules have *bonding* and *non-bonding domains*.

When atoms bond to create molecules, the atoms arrange as far apart to minimize the same-charge repulsion between electrons.

VSEPR theory predicts the *molecular geometry* based on the *number of electron pairs*.

The *electron pair domain geometry* (EDG) indicates bonding and non-bonding electron pairs around the central atom.

The *molecular shape geometry* (MG) indicates atoms' arrangements around the central atom after accounting for electron repulsion.

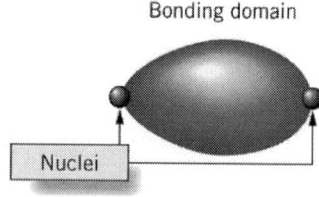

Bonding domain

Nuclei

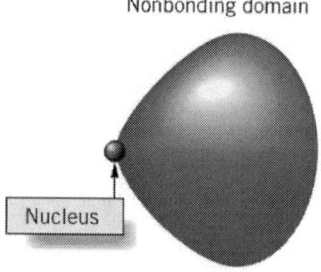

Nonbonding domain

Nucleus

Bonding domains are directed between two adjacent nuclei, while the antibonding domains are directed away from adjacent nuclei

Shape and bond angle

Specific configurations (i.e., specific shapes) maximize the distance between the bonding electron pairs and substituents attached *via* the bond.

Molecules with *two* attached atoms (or substituents) are *linear* (a straight line, with one atom on each side of the central atom).

Bond angle = 180°

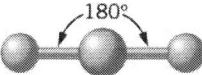

Molecules with *three* attached atoms are *trigonal planar*.

Bond angle = 120°

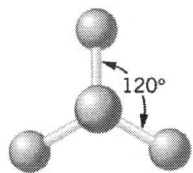

Molecules with *four* attached atoms are a *tetrahedron*.

Bond angle = 109.5°

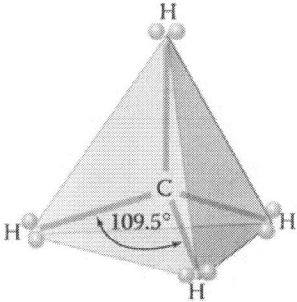

Molecules with *five* attached atoms are *trigonal bipyramidal*.

Bond angles = 120° and 90°

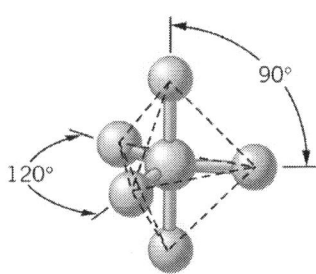

Molecules with *six* attached atoms are an *octahedron*.

Bond angle = 90°

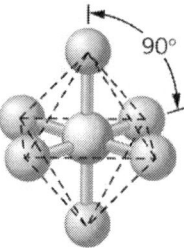

The molecular shapes above apply to molecules containing only bonding domains (i.e., central atoms do not have non-bonding domains or lone pairs of electrons).

Non-bonding domains

When the central atom has non-bonding domains, the lone pair occupies the space where a bond usually forms.

These non-bonding electron domains are relatively dispersed in space and exert electrostatic repulsions that cause bonding electrons to bend away.

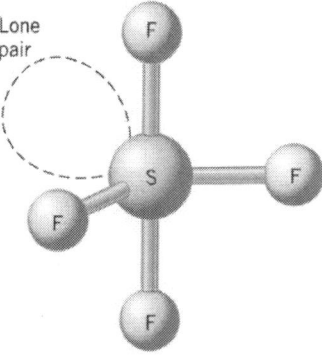

Lone pairs must be closer to the atom whose orbital they occupy than the bonding pairs shared along the two bonding atoms' internuclear axis.

Thus, lone pairs are not restricted to the internucleus axis occupy more available space within the orbital (i.e., more electrostatic repulsion).

An example is SF_4, which has one lone pair:

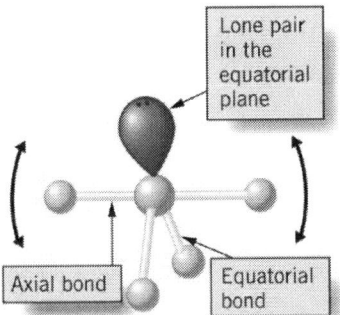

Electrons in sigma bonds occupy a restricted space while nonbonding lone pairs are more dispersed and therefore contribute electrostatic repulsion to the bonding electrons

Hybridization, bond angles, and molecular geometry rules

Hybrid ization	Electron geometry	Electron groups	Bonding groups	Lone pairs	Approx. bond angles	Molecular geometry
sp	Linear	2	2	0	180°	Linear
sp^2	Trigonal planar	3	3	0	120°	Trigonal planar
		3	2	1	< 120°	Bent
sp^3	Tetrahedral	4	4	0	109.5°	Tetrahedral
		4	3	1	< 109.5°	Trigonal pyramidal
		4	2	2	<< 109.5°	Bent
sp^3d	Trigonal bipyramidal	5	5	0	120° (equatorial) 90° (axial)	Trigonal bipyramidal
		5	4	1	< 120° (equatorial) < 90° (axial)	Seesaw
		5	3	2	< 90°	T-shaped
		5	2	3	180°	Linear
sp^3d^2	Octahedral	6	6	0	90°	Octahedral
		6	5	1	< 90°	Square pyramidal
		6	4	2	90°	Square planar

Determining the molecule's geometric shape

1. Start by counting the domains (bonding and non-bonding) around the central atom.

   The number of domains determines the molecule's general shape (discussed above).

   For every non-bonding domain, remove a surrounding atom.

   For example, a four-domain molecule has a tetrahedral shape.

   If one of the domains is non-bonding, one of the surrounding atoms is removed, leaving a central atom with three surrounding atoms; the molecule has a trigonal pyramid shape.

   If there are two non-bonding domains, remove two surrounding atoms; the molecule is a bent shape.

2. For molecules with five domains, start by removing atoms above and below the pyramid before the 3 atoms surrounding it.

   The molecule is more stable when the atoms adopt a configuration with the least repulsion.

   The angles between the top or bottom atoms and the three side atoms are 90°.

   The angles between the side atoms are 120°.

   Thus, the molecule sheds the more repulsive atoms first (i.e., ones with smaller angles).

   The same approach applies to molecules with six domains.

   The domains above and below are removed before the central atom domains.

Non-bonding domains affect the angle of bonds in the molecule due to occupying more significant space (increased electrostatic repulsion) than bonded electrons.

The following diagram summarizes the shapes of atoms based on the number of bonding and non-bonding electron pairs (also called nonbonded electron pairs):

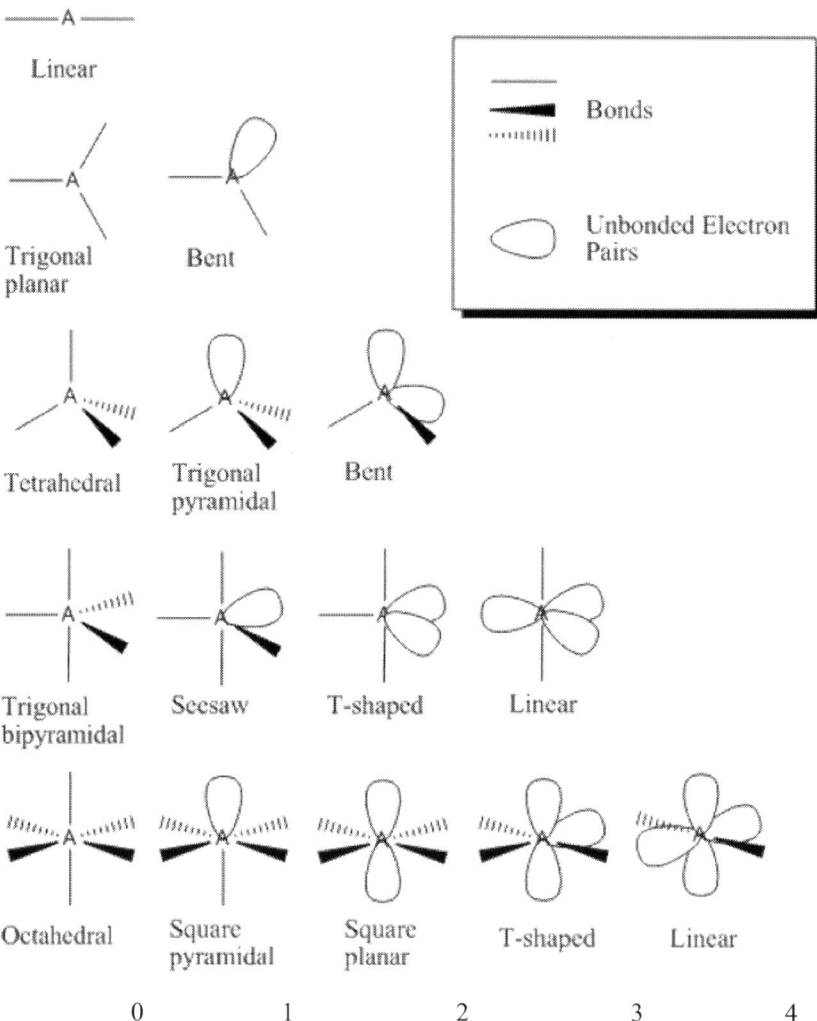

Numbers along the bottom indicate the # of nonbonded electron pairs

The oxygen of water (H_2O, shown below) has two bonding and two non-bonding domains. The water molecule experiences electron repulsion from the two lone pairs on its bonds.

The water molecule adopts a bent shape (see the previous chart) to minimize the van der Waals repulsion of the negatively charged electrons.

Water's bent shape comes from a tetrahedral configuration, with two vertices removed.

Bonds in a tetrahedron are 109.5°, which means bonds in a bent molecule should be 109.5°.

However, because of the repulsion between lone electron pairs and the bonding electrons, the reported bond angle is slightly decreased to approximately 105°.

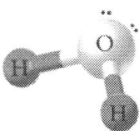

A water molecule (H_2O) is bent due to the two lone pairs of e^- on oxygen

VSEPR geometric shapes classify double bonds as a single domain.

Formaldehyde (CH_2O) has one double and two single bonds, adding to three bonding domains.

The molecular geometry of formaldehyde is trigonal planar (i.e., flat shape with ~120 bonds).

Trigonal plane geometry for formaldehyde (CH_2O) with 1 double (π) and 2 single (σ) bonds

Two major theories explain how molecules bond: valence bond theory and molecular orbital theory. Both theories are used to predict the structure of a molecule.

Notes for active learning

Valence Bond (VB) Theory

Atomic orbital (AO) theory

The *Valence bond theory* (VB theory) is primarily based on valence electrons and states that each atom has an orbital, and *orbitals overlap to form bonds*.

The extent of overlap of atomic orbitals (AO) is related to bond strength; more overlap results in a stronger bond.

The VB theory explains the bonding of F_2. F_2 bonds form from atomic valence orbitals overlap (e.g., $2p$ orbitals), one from each fluorine.

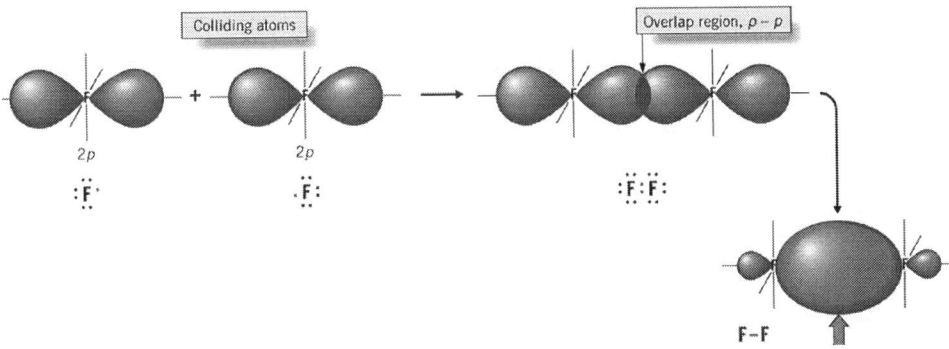

The overlap of p–p orbitals form the F_2 bond

The orbital overlap is not limited to identical orbitals.

Bonding in HF molecules uses the overlap between the $1s$ orbital of H and the $2p$ orbital of F.

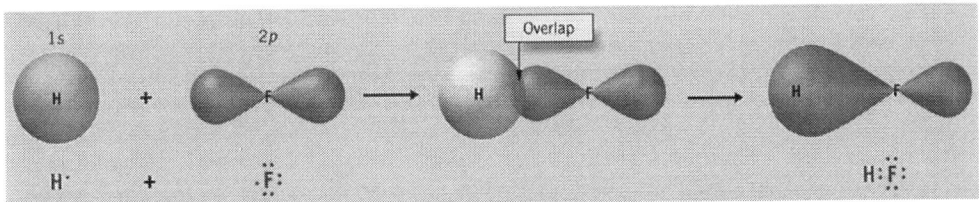

The bonding of H_2S can be described according to VB theory. Sulfur has one filled orbital (contains two electrons) and two partially filled orbitals (contains one electron).

The partially filled p orbitals overlap with hydrogen's s orbital.

The predicted 90° bond angle is close to the measured value of 92°.

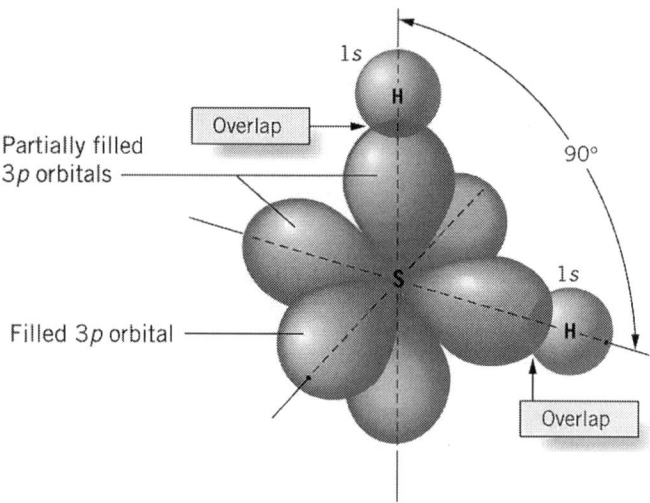

VB theory describes many molecules' bonding, but it cannot explain existing molecular geometries.

Methane (CH_4) has a tetrahedral shape, seen in the image below.

Carbon's electron configuration ($1s^2 2s^2 2p^2$) indicates that its electrons are paired, except for two electrons that occupy p orbitals.

VB theory predicts that carbon can only bond with two hydrogen molecules, resulting in a CH_2 molecule and that the bond angle is close to 90° (like H_2S above).

Observations of methane reveal four equal bonds in the shape of a tetrahedral with bond angles of 109.5°.

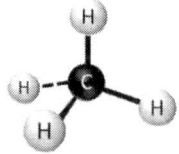

Methane (CH_4) adopts a tetrahedral shape to maximize repulsion from the 4 H substituents

Hybridization mixes atomic orbitals

The *hybridization* concept improved VB theory to resolve theoretical and experimental values differences.

Hybridization is mixing *atomic orbitals* (AO) to form bonds with unexpected bond angles.

Atomic orbitals have the highest probabilities of finding electrons, and *hybridization* is a rearrangement of orbitals.

Hybrid orbitals have new shapes, directional properties, and combined constituent orbitals (shown below).

Hybrid orbitals combine the symbols of the initial orbitals.

The sum of exponents in hybrid orbital notation equals the number of atomic orbitals used.

- One *s* and one *p* orbital form two *sp* hybrid orbitals

- One *s* and two *p* orbitals form three *sp²* hybrid orbitals

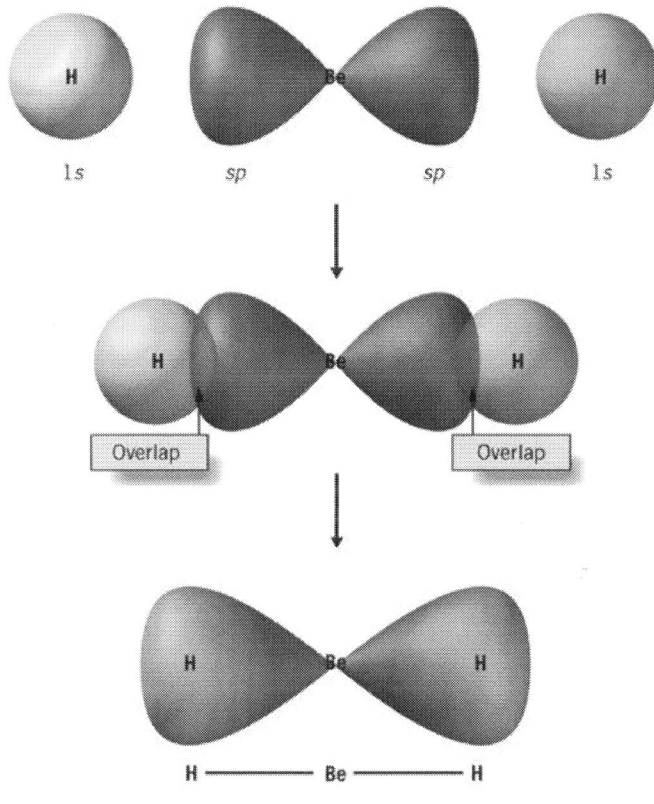

Bonding of BeH_2 Be: $1s^2 2s^2$ H: $1s^1$

Beryllium (Be) has two electrons in the 2*s* orbital. When bonding with two hydrogens, it splits the electrons into two orbitals. The Be atom uses the next available orbital (2*p*) combined with 2*s* orbital, resulting in two *sp* orbitals.

Each of those *sp* orbitals overlaps with hydrogen's 1*s* orbital, creating a covalent bond.

VSEPR theory predicts this molecule has a linear shape confirmed by experimental observations.

Most atoms form up to four bonds (i.e., octet rule); four orbitals are needed (1 *s* and 3 *p*).

Bonding and non-bonding pairs

For molecules with more than four bonds/lone pairs around them, *d* orbitals are included in the hybridization to accommodate the extra bonds, known as *expanded octet hybridization.*

Bond angle experiments suggest that NH_3 and H_2O use *sp³* hybrid orbitals in bonding (a total of four orbitals) even though each has two bonds.

This evidence for *sp³* hybrid orbitals concluded that hybrid orbitals are not exclusively used for bonding; they accommodate the non-bonding electron pairs.

Bonding and non-bonding pairs contribute to the molecule's geometry.

For example, the orbital diagram of hybridized nitrogen in NH_3:

Hybridization accommodates 3 bonding and 1 nonbonding electron pairs

Diagram showing the hybridization of oxygen in H_2O:

Hybridization accommodates 2 bonding and 2 nonbonding electron pairs

Molecular Orbital (MO) Theory

Wave interference

Molecular orbital theory (MO theory), the second dominant theory, views molecules as a collection of positively charged nuclei having a set of molecular orbitals filled with electrons (like filling atomic orbitals with electrons).

It does not focus on how individual atoms join to form molecules.

The MO theory is more difficult to visualize than the VB theory.

MO theory is advantageous because it accurately predicts a molecules' *magnetic properties* and other characteristics.

The energies of molecular orbitals (MO) are determined by combining *atomic orbitals'* (AO) *electron waves*.

For H_2 (below), two $1s$ wave functions, one from each atom, combine to make two MO wave functions.

Two MO produce *constructive interference* of the waves.

Bonding and antibonding molecular orbitals

For H_2, the bonding MO's energy is lower than the atomic orbitals' energy (AO), and the molecule is collectively more stable than the individual atoms (atoms forming the molecule).

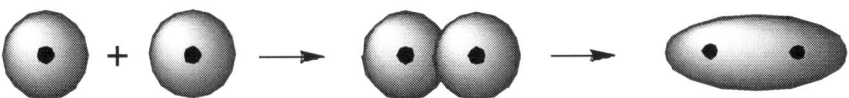

As there are σ orbitals, there are corresponding σ^* antibonding MOs.

The possible combination of two $1s$ orbitals involves destructive interference of the $1s$ waves.

In the example below, the antibonding MO's energy is higher than the energy of parent atomic orbitals. The molecule *does not form*, and the atoms remain separate.

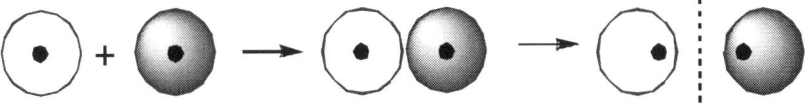

Antibonding MOs are high in energy and do not contribute to bond formation

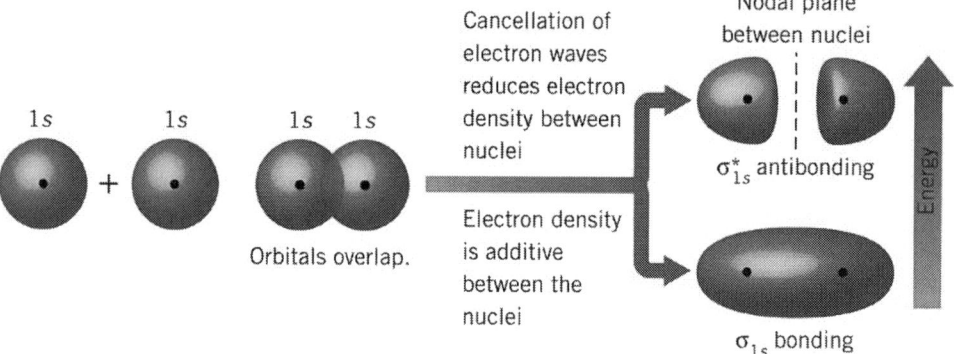

Bonding and antibonding MOs with the relative energy of each

In *bonding MOs*, electron density is between nuclei; the electrons tend to stabilize the molecule.

In the *antibonding MO*, the cancellation of electron waves reduces the nuclei's electron density.

The electrons in antibonding MOs tend to destabilize the molecule.

Molecular orbital theory diagrams

MO energy diagrams (shown below) represent the interactions between atomic orbitals.

A MO energy diagram displays the orbitals arranged vertically, from highest to lowest energy.

The atomic orbitals (AO) for the atoms are on the left and right sides of the diagram.

The molecular orbitals (MO) are in the central column of the diagram.

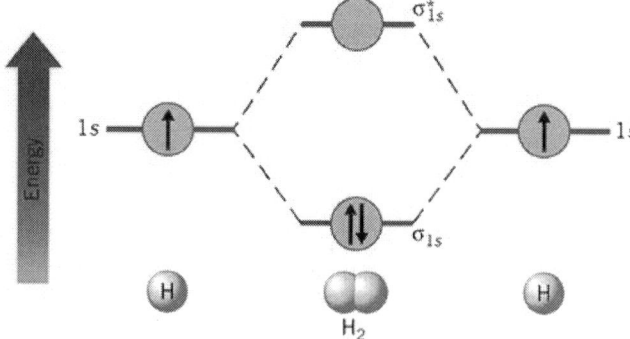

The MO energy diagram for H_2: atomic orbitals are designated 1s and have single arrows up, while the lower energy bonding molecular orbital has 2 electrons of opposite spin.

The high-energy antibonding MO is *vacant*. The molecule forms from separate atoms because the bonded molecule is *more stable* than the separate atoms (relative stability with energy on the y-axis).

Three rules for filling MO energy diagrams:

1. Electrons fill the lowest-energy orbital (i.e., Aufbau's principle).

2. No more than two electrons (with opposite spins) can occupy any orbital (i.e., Pauli exclusion principle).

3. Electrons with unpaired spins spread over the same energy orbitals (i.e., Hund's rule).

Bond order

Bond order indicates the number of electron pairs shared between atoms (i.e., number of bonds between two atoms).

A bond order of one corresponds to a single bond.

For example, the H_2 bond order and the C−H bond order are each one, while the N≡N bond order in diatomic nitrogen is three.

The bond order is calculated by:

$$\text{Bond order} = \frac{(\text{number of bonding } e^-) - (\text{number of antibonding } e^-)}{2 \text{ electrons / bond}}$$

Valence bonding and molecular orbital theory limitations

Neither the valence bonding (VB) nor the molecular orbital theory (MO theory) is entirely correct because neither theory explains all aspects of bonding; each theory has its strengths and deficiencies.

VB theory is based on Lewis structures (discussed later) and the related geometric shapes.

MO theory correctly predicts the unpaired electrons in O_2, while the Lewis structures do not. MO theory is complicated because even simple molecules have complex energy level diagrams, and molecules with three or more atoms require extensive calculations.

VB theory uses three-dimensional structures based on electron domains without extensive calculations.

Simple hybrid orbitals are invoked where experimental evidence shows the need, and the integer bond orders are often correct.

Structural formulas for heteroatom molecules

Lewis structures (explained later) are used to represent atoms and molecules visually.

The Lewis structures for elements in the same column (group or family) are similar.

For example, sulfur can be substituted for oxygen in Lewis structures of oxygen.

Below are structural formulas for some important molecules.

- **Hydrogen** Lewis structures

  hydrogen proton: H^+

  hydride ion: H^-

- **Boron** – Group 13 Lewis structures

Since group 13 elements have three valence electrons, they often have six total electrons in a molecule (an exception to the octet rule).

Borane:

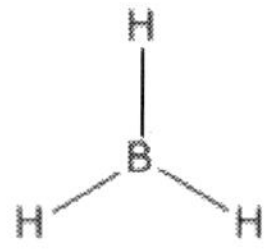

Borohydride ion:

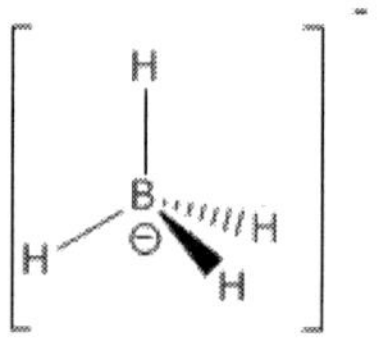

Borohydride is expressed as BH_4^- with a formal charge on boron (B), shown with brackets and a negative charge on the top right side.

- **Carbon** – Group 14 Lewis structures

Methane:

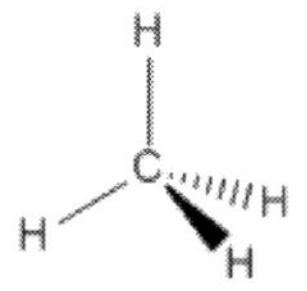

Carbocation:

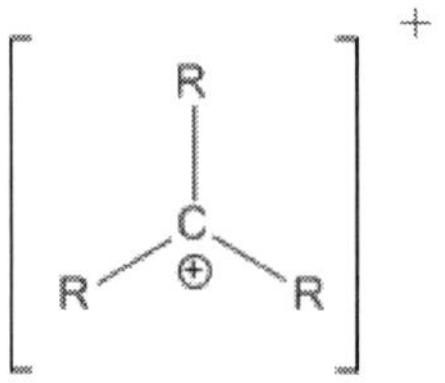

Carbanion: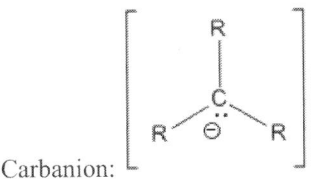

Note: R in the figures is carbon or hydrogen.

- **Nitrogen** – Group 15 Lewis structures

Some group 15 elements have more than eight electrons after bonding, an exception to the octet rule (e.g., PCl_5).

The presence of *d* orbitals permits the atom to accommodate these additional (> than 4) bonds.

Amine / Ammonia:

Ammonium:

Imine:

- **Oxygen** – Group 16 Lewis structures

Some group 16 elements have more than eight electrons after bonding, an exception to the octet rule (e.g., SF_6).

Molecular oxygen:

Water, alcohol, and ethers:

Ozone:

- **Halogen** – Group 17 Lewis structures

Hydrogen fluoride: H—F̈:

Chloromethane:

$$
\begin{array}{c}
\text{H} \\
| \\
\text{H—C—Cl:} \\
| \\
\text{H}
\end{array}
$$

Bromide ion: [:Br̈:]⁻

Lewis Electron Dot Formulas

Lewis structures

Lewis dot formulas, or *Lewis structures*, are notations used to describe bonds and electrons in a molecule.

For Lewis structures, each *dot represents one electron*, and each line represents one bond (two electrons); two dots represent a lone pair.

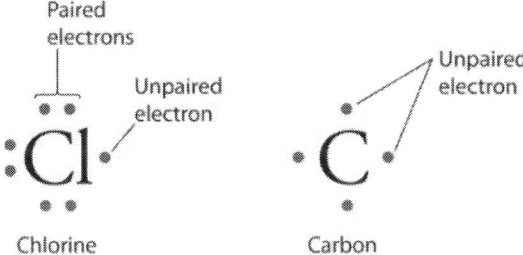

Lone pairs as two dots while an unpaired electron has no partner

$$Cl:P:Cl \quad \text{or} \quad Cl—P—Cl$$
$$\ddot{C}l \qquad\qquad\quad | $$
$$\qquad\qquad\qquad Cl$$

Lone pairs are usually omitted from the molecule's final structure:

$$:\ddot{F}—\ddot{F}: \qquad F–F$$

Recall the *octet rule*: most atoms need a total of eight valence electrons to achieve stable electron configurations.

Electrons in a bond are shared and can satisfy the bonded atoms' octet.

This electron configuration principle includes coordinate covalent bonds, where both bonding electrons are contributed by one atom.

However, the electron pair counts towards the other atom's octet rule.

Exceptions for the octet rule include atoms with other than 8 e⁻:

> the boron column (they form three bonds and have a sextet),

> large elements with periods of 3 or higher where these elements have *d* orbitals (e.g., the dectet P in PO_4^{3-} and the duodectet S in SO_4^{2-}),

> radicals (compounds with an odd number of total electrons possess a single, unpaired electron).

General rules for Lewis structures

- From the octet rule, most atoms have up to eight electrons.

- For atoms with up to four valence electrons, separate them evenly around the molecule.

- If the atom has five or more valence electrons, divide them into four sections with a maximum of two electrons on each side. This is not required but makes it easier to determine if an atom has achieved stability (e.g., eight electrons).

When sketching Lewis structures, use different symbols for electron sources.

For example, consider the Lewis structure of PCl_3 below; it is unclear which electrons on P originated from P or Cl.

P has five valence electrons; deduce that the remaining three come from Cl.

However, it can be confusing when the bonds get more complicated (e.g., SO_3).

Use clear designations to differentiate electrons by source.

In $BeCl_2$ below, "x" represents electrons from Be, and dots represent electrons from Cl.

Place square brackets around complete Lewis structures and indicate the charge on the top of the right bracket.

Drawing electron dot formulas

1) Determine the arrangement of the atoms.

2) Determine the total number of valence electrons.

3) Attach bonded atom to the central atom with a pair of electrons.

4) Place the remaining electrons using single or multiple bonds to complete the octet.

Electron-dot formulae using valence electrons

Molecule or Polyatomic Ion	Total Valence Electrons	Form Single Bonds to Attach Atoms (electrons used)	Electrons Remaining	Completed Octets (or H:)
Cl_2	$2(7) = 14$	$Cl—Cl \ (2\,e^-)$	$14 - 2 = 12$	$:\ddot{C}l—\ddot{C}l:$
HCl	$1 + 7 = 8$	$H—Cl \ (2\,e^-)$	$8 - 2 = 6$	$H—\ddot{C}l:$
H_2O	$2(1) + 6 = 8$	$H—O—H \ (4\,e^-)$	$8 - 4 = 4$	$H—\ddot{O}—H$
PCl_3	$5 + 3(7) = 26$	$\begin{array}{c} Cl \\ | \\ Cl—P—Cl \end{array} (6\,e^-)$	$26 - 6 = 20$	$\begin{array}{c} :\ddot{C}l: \\ | \\ :\ddot{C}l—P—\ddot{C}l: \end{array}$
ClO_3^-	$7 + 3(6) + 1 = 26$	$\left[\begin{array}{c} O \\ | \\ O—Cl—O \end{array}\right]^- (6\,e^-)$	$26 - 6 = 20$	$\left[\begin{array}{c} :\ddot{O}: \\ | \\ :\ddot{O}—Cl—\ddot{O}: \end{array}\right]^-$
NO_2^-	$5 + 2(6) + 1 = 18$	$\left[O—N—O \right]^- (4\,e^-)$	$18 - 4 = 14$	$\left[:\ddot{O}—\ddot{N}=\ddot{O}:\right]^-$ $\updownarrow$ $\left[:\ddot{O}=\ddot{N}—\ddot{O}:\right]^-$

Lewis structure rules for common elements

- **Carbon**: four bonds, zero lone pairs. (e.g., CH_4, CO_2)

- **Oxygen**:

  O: two bonds total, 2 lone pairs (e.g., H_2O, O_2)

  O^{1-}: one bond, 3 lone pairs, a formal charge of -1

  O^{1+}: three bonds, 1 lone pair, a formal charge of $+1$

- **Nitrogen**:

  N: three bonds total, one lone pair (e.g., amines, ammonia NH_3)

  N^+: four bonds, zero lone pairs, a formal charge of $+1$ (e.g., ammonium NH_4^+)

- **Halogens**: one bond, three lone pairs (e.g., CCl_4)

- **Hydrogen:** one bond, zero lone pairs (octet rule exception)

- **Carbocation**: C^+ has three bonds, no lone pairs

- **Carbanion**: C^- has three bonds, one lone pair

- **Boron**: three bonds, zero lone pairs (an exception to the octet rule) (e.g., BH_3)

Formal Charge

Formal charge for resonance structures

The *formal charge* is assigned to each atom in a Lewis structure from equal sharing of bonded electron pairs.

Formal charges are used in drawing resonance structures (multiple Lewis structures that collectively describe a single molecule) of covalently bonded molecules.

This technique describes and compares resonance structures, like how oxidation numbers balance chemical equations in oxidation-reduction reactions.

Formal charge is calculated by:

Formal charge = valence e^- in neutral atom – (unshared valence e^- + ½ shared e^-)

where e^- represents electrons

Every atom in a molecule has a formal charge.

In a neutral molecule, the sum of atomic charges is zero. That does not mean the atoms are not charged; they cancel to create a neutral molecule.

The charge is *not evenly distributed* in charged molecules, and formal charges can help determine where the charge rests.

Calculating formal charge

Formal charge = valence e^- in a neutral atom – (unshared valence e^- + ½ shared e^-)

For Lewis structures:

Formal charge = [# of valence e^-] – [dots around atom + lines connected to atom]

The number of valence electrons is generally equal to the atom's group number on the periodic table (e.g., N has five valence electrons, O has six valence electrons, and F has seven valence electrons).

Dots around the atom represent electrons held entirely by the atom.

The lines connecting the atom represent bonding electron pairs; the atom gets one of the two electrons.

Label the atom with the formal charge for formal charges other than zero.

Examples of atoms with formal charges:

Oxygen with a single bond: −1

Oxygen with no bonds but with an octet: −2

Carbon with three bonds:

+1 as a carbocation

−1 as a carbanion

Nitrogen with four bonds: +1

Nitrogen with three bonds: −1

Halogen with no bonds but has an octet: −1

Boron with four bonds: −1. (e.g., ⁻BH₄)

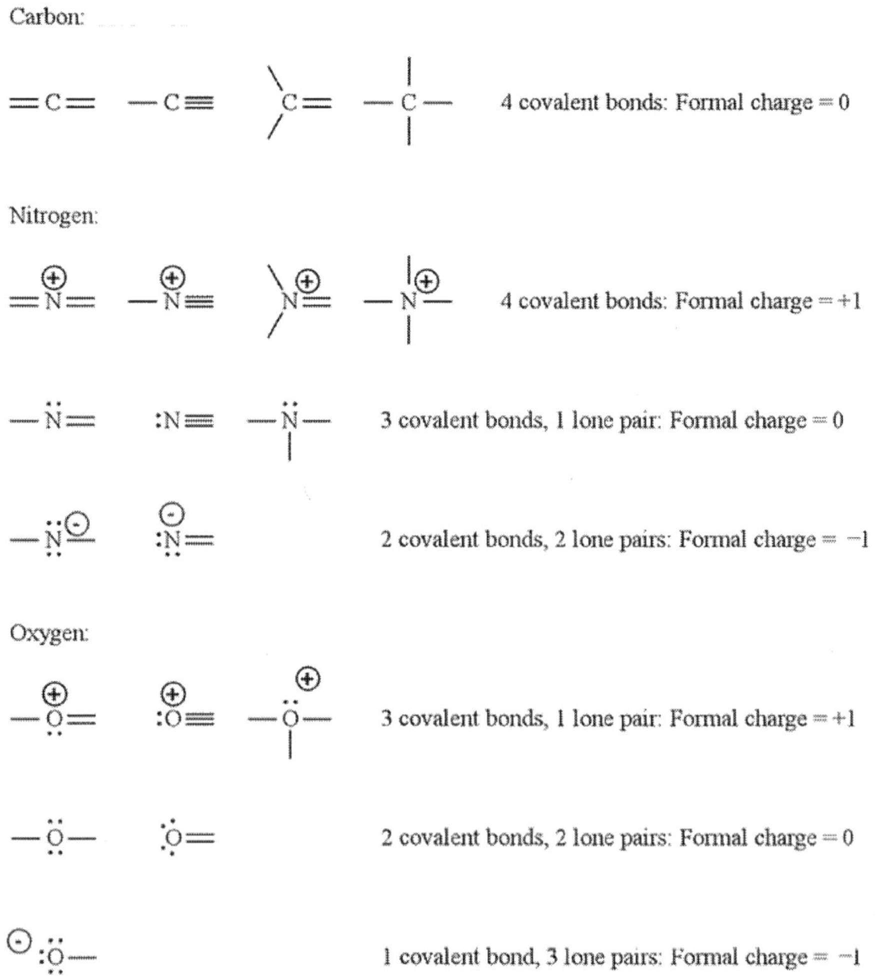

Carbon:

$=C=$ $-C\equiv$ $>C=$ $-C-$ 4 covalent bonds: Formal charge = 0

Nitrogen:

$=\overset{\oplus}{N}=$ $-\overset{\oplus}{N}\equiv$ $>\overset{\oplus}{N}=$ $-\overset{\oplus}{N}-$ 4 covalent bonds: Formal charge = +1

$-\overset{..}{N}=$ $:N\equiv$ $-\overset{..}{N}-$ 3 covalent bonds, 1 lone pair: Formal charge = 0

$-\overset{..}{N}\overset{\ominus}{\underset{..}{\cdot}}$ $\overset{\ominus}{:N}=$ 2 covalent bonds, 2 lone pairs: Formal charge = −1

Oxygen:

$-\overset{\oplus}{\underset{..}{O}}=$ $\overset{\oplus}{:O}\equiv$ $-\overset{\oplus}{O}-$ 3 covalent bonds, 1 lone pair: Formal charge = +1

$-\overset{..}{\underset{..}{O}}-$ $\overset{..}{:O}=$ 2 covalent bonds, 2 lone pairs: Formal charge = 0

$\overset{\ominus}{:}\overset{..}{\underset{..}{O}}-$ 1 covalent bond, 3 lone pairs: Formal charge = −1

The valence number is the number of outer shell (valence) electrons an atom needs to gain or lose to achieve a complete octet.

In covalent compounds, the number of bonds characteristically formed by an atom equals the valence number.

Atom	H	C	N	O	F	Cl	Br	I
Valence	1	4	3	2	1	1	1	1

Determining formal charge

The valences in the chart above represent the most common form of these elements in organic compounds.

Many elements (e.g., chlorine, bromine, iodine) have several valences in different inorganic compounds.

If the number of covalent bonds to an atom is higher than its typical valence, it carries a positive formal charge.

If the number of covalent bonds to an atom is lower than its typical valence, it carries a negative formal charge.

A *step-by-step approach* to determine the formal charges of each atom in a molecule of nitric acid (HNO_3), using the structure displayed below.

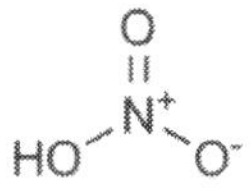

The formal charge for nitric acid (HNO_3) with formal charge shown on the atoms

Formal charge on **H**

- Hydrogen shares two electrons with oxygen.

- Assign one electron to H and one to O.

- Hydrogen has one valence electron, and $1 - 1 = 0$.

- Therefore, the formal charge of H in nitric acid is zero.

Formal charge on the **O** bonded to N and H

- Oxygen has four electrons in covalent bonds (2 with N, 2 with H).

- Assign two of these four electrons to O.

- Oxygen has two unshared pairs; assign four electrons to O.

- The total number of electrons assigned to O is $2 + 4 = 6$.

- Oxygen has six valence electrons, and $6 - 6 = 0$.

- Formal charge of O bonded to H and N in nitric acid is zero.

Formal charge on the **double-bonded O**

- Oxygen has four electrons in covalent bonds with N.

- Assign two of these four electrons to O.

- Oxygen has two unshared pairs; assign four electrons to O.

- The total number of electrons assigned to O is $2 + 4 = 6$.

- Oxygen has six valence electrons, and $6 - 6 = 0$.

- Formal charge of the double-bonded O in nitric acid is 0.

Formal charge on the **single-bonded O**

- O has two electrons in a covalent bond.

- Assign one of those electrons to O.

- O has three unshared pairs. Assign six electrons to O.

- The total number of electrons assigned to O is $1 + 6 = 7$.

- Oxygen has six valence electrons, $6 - 7 = -1$.

- Typical charge of single-bonded O in nitric acid is -1.

Formal charge of **N**

- N has eight electrons in covalent bonds.

- Assign four of those electrons to N.

- Nitrogen has five valence electrons, $5 - 4 = 1$.

- Therefore, the formal charge of the N in nitric acid is $+1$.

If atoms in a molecule have a formal charge, it does not necessarily mean that the entire molecule has a charge.

In the ozone structure (O_3) below, the central oxygen atom has three bonds and is positively charged.

The right-hand oxygen has a single bond and is negatively charged.

The overall charge of the ozone molecule is, therefore, zero.

ozone   nitromethane   azide ion

Nitromethane (CH_3NO_2) has positively charged nitrogen and negatively charged oxygen; the total molecular charge is again zero.

The azide (N_3^-) anion has two negatively charged nitrogens (2 bonds each) and one positively charged nitrogen (four bonds), and the total charge is minus one.

Notes for active learning

Resonance

Resonance structures

When more than one relatively stable structure exists, it forms *resonance structures*.

Resonance is the averaging of electron distribution over two or more hypothetical contributing structures to produce a hybrid electronic structure.

Resonance structures describe molecules that may contain fractional bonds and charges.

The resonance structures must have the same number of paired and unpaired electrons.

No atoms change position within the common structural framework, not breaking existing chemical bonds.

Electrons (not atoms, as is the case for isomers) move. Visualize the molecule as quickly "shifting" between its resonance structures and intermittently existing in several configurations.

However, the molecule spends more time in the most stable resonance form.

More accurately, the molecule structure is a "combination" (or a weighted average) of its resonance structures, favoring the most stable resonance structures.

In a molecule with a single-and a double-bond resonance structure, the bond length is between a single- and a double-bond length.

Resonance structures for SO_3^{2-} (single-bonded O with negative charge). The images above do not include the formal charge or the double-headed arrows to show the individual contributing resonance structures.

Three contributing resonance structures for CO_3^{2-}

The double-headed arrow seen in the image above for CO_3^{2-} acts as a notation symbol for resonance structures contributing to the hybrid structure. The principle of resonance is useful in rationalizing the chemical behavior of compounds.

The electron delocalization provided by resonance significantly enhances the molecules' stability (lowers PE).

Determine the most stable resonance structure of a molecule when a stable molecule's properties are known.

In stable molecules, the octet is satisfied in each atom (aside from hydrogen and exceptions, like boron).

Neutral molecules are more stable if the formal charges on atoms are spread accordingly (i.e., + charge on a positive atom, – charge on electronegative atoms).

Resonance structure for formaldehyde (CH₂O)

In the example above, the neutral molecule is the most stable.

The preferred charge distribution has a positive charge on the less electronegative atom (carbon) charged species, and a negative charge is held by the more electronegative atom (oxygen).

Therefore, the middle formula represents a more reasonable and stable structure than the right.

Resonance hybrids

The stable structure(s) indicates that the molecule spends most time in resonance structure(s) over the others.

If the double bond is broken heterolytically (both electrons going to one of the bonding atoms), formal charge pairs result, as shown in the other two structures.

Since the middle, the charge-separated contributor has an electron-deficient carbon atom; this explains the tendency of electron donors (nucleophiles) to bond at this site.

The application of resonance to this example requires a weighted average of these structures.

Consequently, if one structure has greater stability than the others, the hybrid closely resembles it electronically and energetically.

The resonance hybrid is exceptionally stable if two or more forms have identical low-energy structures.

Examples of low-energy molecules stabilized by resonance hybrid structures include sulfur dioxide (SO_2) and nitric acid (HNO_3).

Three resonance forms of sulfur dioxide (SO_2); the central structure is neutral and therefore the greatest contributor (most molecules exist in this state)

Two resonance forms of nitric acid (HNO_3); structures are equal in energy with nitrogen positive, and the negative charge shared by two oxygens

Three considerations for drawing resonance structures

1. The number of covalent bonds in a structure - the greater the bonding, the more essential and stable the contributing structure.

2. Formal charge separation: *charge separation* decreases the contributing structure's stability and importance.

3. Evaluate electronegativity of charge-bearing atoms and charge density.

    The *positive charge* is accommodated on atoms of low electronegativity.

    The *negative charge* is on highly electronegative atoms.

Lewis acids and bases

Molecules can be categorized by the number of free electron pairs:

Lewis *acids accept* electron pairs.

Lewis *bases donate* electron pairs.

Acids have vacant orbitals and do not have lone pairs on the central atom (e.g., BF_3), while Lewis bases have lone pair electrons (e.g., NH_3).

Lewis acid (A): electron acceptor   Lewis base (B): electron donor

$$F-B \overset{\displaystyle F}{\underset{\displaystyle F}{|}} \ + \ :N \overset{\displaystyle H}{\underset{\displaystyle H}{|}} -H \ \longrightarrow \ F-B \overset{\displaystyle F}{\underset{\displaystyle F}{|}} -N \overset{\displaystyle H}{\underset{\displaystyle H}{|}} -H$$

Acid   *Base*

The product would have a formal negative charge on boron and a positive charge on nitrogen

Acids and bases are discussed in the acids and bases chapter.

Partial Ionic Character

Electron sharing

Covalent bonds can be categorized into nonpolar covalent bonds and polar covalent bonds.

Nonpolar covalent bonds occur between atoms of the same element or atoms with similar electronegativity sharing electrons equally to achieve a more stable electron configuration.

Polar covalent bonds form between atoms with differences in electronegativity. These bonds have *partial ionic character* due to the unequal distribution of the electrons. One atom has a stronger attraction for the shared electrons resulting in one end of the bond being partially negative (δ^-), while the other is partially positive (δ^+).

 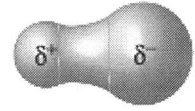 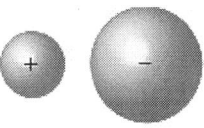

Three types of bonds: covalent (left), polar covalent (center), and ionic (right)

The left structure above shows a covalent bond for identical atoms (i.e., same electronegativity).

The center structure is a polar covalent bond between elements of moderately different electronegativities with the notation for the molecule's resulting partial positive and partial negative ends.

The right figure shows an ionic bond between elements with significant differences in electronegativities, usually a metal and a nonmetal. In ionic bonds electrons from one atom are transferred completely to another, resulting in ions with opposite charges that are attracted to each other by electrostatic forces (e.g., in NaCl sodium donates an electron to chlorine).

Nonpolar covalent bonds involve equal sharing of electrons, polar covalent bonds involve unequal sharing, and ionic bonds result from the complete transfer of electrons from one atom to the other. The difference in electronegativity between the atoms determines the type of bond.

Electronegativity for determining charge distribution

Electronegativity (EN) is an element's ability to attract electrons through bonding due to differing nuclear charges and shielding by inner electron shells; each element has different electronegativity.

The molecule's polarity is the difference in EN values of its atoms.

Linus Pauling (Nobel laureate in chemistry, 1954) elucidated the chemical bond's nature and application to complex molecules' structure. Pauling established a quantitative scale of electronegativity values.

A larger number (e.g., F = 3.98) signifies a higher affinity for electrons.

H 2.20	Electronegativity Values for Some Elements					
Li 0.98	**Be** 1.57	**B** 2.04	**C** 2.55	**N** 3.04	**O** 3.44	**F** 3.98
Na 0.90	**Mg** 1.31	**Al** 1.61	**Si** 1.90	**P** 2.19	**S** 2.58	**Cl** 3.16
K 0.82	**Ca** 1.00	**Ga** 1.81	**Ge** 2.01	**As** 2.18	**Se** 2.55	**Br** 2.96

For the periodic table, EN values increase from left to right, from bottom to top

Higher electronegativity values are located towards the nonmetal side (right side) on the periodic table.

Fluorine has the highest electronegativity (3.98). The heavier alkali metals (e.g., potassium, rubidium, cesium) have lower electronegativity values.

Carbon has an electronegativity of 2.55. This value is mid-range for electronegativity and is slightly more electronegative than hydrogen (2.20).

Of an atom's subatomic particles (protons, neutrons, electrons), only the electrons move within the atom.

When two atoms bond, most electrons congregate between the two nuclei.

The distribution depends on the electronegativity of the atoms.

Shared electrons in a bond are attracted to the more electronegative atom, shifting electron density toward the electronegative atom.

Nonpolar covalent bonds form between identical atoms as *homonuclear diatomic molecules*.

For example, in diatomic hydrogen (H_2), both atoms are identical; each has the identical EN value.

The electrons, therefore, are evenly distributed between the two hydrogens.

H : H   H H

Polarity based on electronegativity

Hydrogen fluoride (HF) has atoms of different EN values – fluorine (3.98) is more electronegative than hydrogen (2.20). The electrons are more attracted to the fluorine nucleus and are closer to the F than H.

H ∶ F   H F

The electron density for the H–F bond with the greatest density on the electronegative F

For example, suppose one atom in a molecule is more electronegative than another. This causes a higher concentration of electrons to locate on one side of the bond, and the molecule is polar (i.e., dipole present).

The molecule's *polarity* is proportional to the difference in the bonded atoms' electronegativity, shown below.

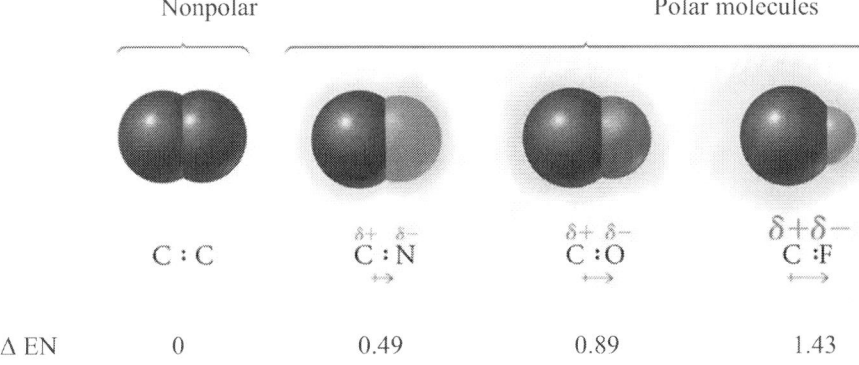

Nonpolar		Polar molecules	
C ∶ C	C ∶ N	C ∶ O	C ∶ F
ΔEN			
0	0.49	0.89	1.43

The lowercase Greek letter delta δ denotes partial charges on a polar molecule.

A plus or minus sign (i.e., (δ^+ or δ^-) indicates the partial positive or negative charges.

δ^+ H–Cl δ^-

Arrows show the direction of partial charges. The arrow points in the direction of electron movement, originating at the positively charged atom and pointing toward the negatively charged atom.

	δ^- δ^+	δ^- δ^+	δ^+ δ^-	δ^- δ^+
	C——H	O——H	C——Cl	C——Li

The partial positively δ^+ and negatively δ^- charged ends are *dipoles*.

If the difference in electronegativity is high enough in a direction, the molecule acquires a net dipole moment (geometric polarity).

Dipole moment

The *dipole moment* measures polarity in a molecule; it is the difference in electronegativity values of participating or bonded atoms.

The geometric shape of a molecule affects its overall dipole moment.

A molecule may have polar bonds, but if those bonds arrange to cancel the dipole moments, it has a zero net dipole and is nonpolar.

A molecule is nonpolar if:

1) It is symmetrical.

2) It has identical atoms surrounding the central atom.

3) Non-bonding electron pairs are evenly distributed and cancel (illustrated below).

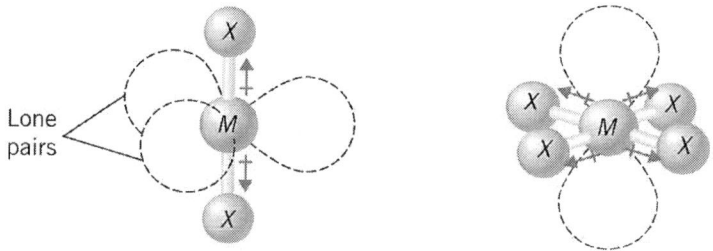

Symmetrical molecules have no net dipole even with individual bond polarity

Dipole based on shape

In the right-hand linear configuration of the molecule below (bond angle 180°), the bond dipoles cancel, and the molecular dipole is zero.

The molecular dipole varies in size for bond angles 90° to 120°.

The molecular dipole is the largest at the 90° configuration.

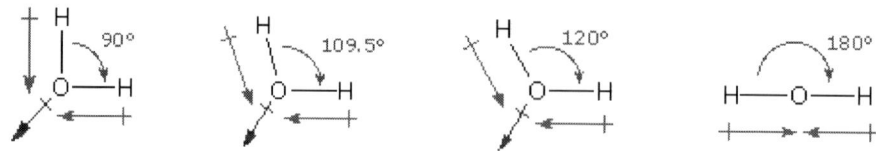

For example, boron trichloride (BCl_3) has three B−Cl bonds. B−Cl is a polar bond, but because the molecule is symmetrical and charges evenly distribute in a single plane, the charges cancel (shown below).

BCl_3 does not have non-bonding electron pairs, and the symmetrical BCl_3 molecule has a zero dipole moment.

$$Cl - B \substack{Cl \\ \diagdown Cl}$$

Boron trichloride BCl_3 is a symmetric molecule with a zero dipole moment

The configurations of methane (CH_4) and carbon dioxide (CO_2) may be deduced from their zero dipole moments.

Since the dipoles cancel, these molecules' configurations must be tetrahedral and linear, respectively.

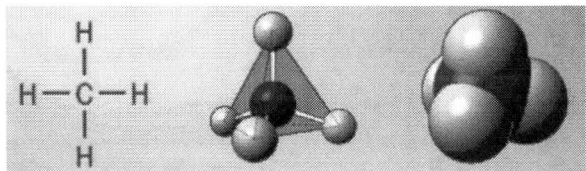

Methane (l to r) as simple drawing, ball-and-stick, and space-filling models

Carbon dioxide (CO_2) is a linear molecule and exhibits a net dipole of zero

Notes for active learning

Polarity

Molecular polarity

The methane CH_4 molecule (below) provides insight confirming its tetrahedral configuration.

Substituting one hydrogen by a chlorine atom gives the methyl chloride CH_3Cl molecule below.

Since the tetrahedral, square-planar, and square-pyramidal configurations have structurally equivalent hydrogen atoms, each give a single substitution product.

One chlorine atom at the apex has a different electronegativity in the trigonal-pyramidal configuration than the three hydrogens at the pyramid base.

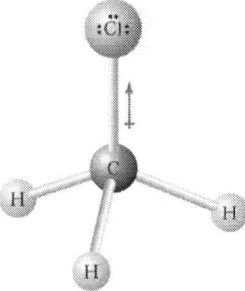

Methyl chloride (CH_3Cl; or chloromethane) with dipole pointing towards the single chlorine

Substitution should give two CH_3Cl compounds when hydrogen reacts.

The tetrahedral configuration of methane leads to a single dichloromethane CH_3Cl product below.

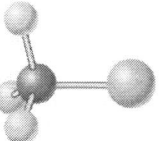

Methyl chloride CH_3Cl (above) is a refrigerant and used to manufacture synthetic rubber and silicones.

The *dipole moment* points toward the single electronegative chlorine atom.

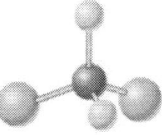

Dichloromethane (methyl chloride; above) CH_2Cl_2 is an industrial solvent used to remove paint and varnish.

The *dipole moment* points toward the two electronegative chlorine atoms.

The vector is resolved between the two chlorines and away from the two hydrogens.

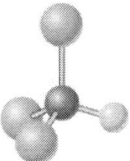

Trichloromethane (chloroform; above) $CHCl_3$ is a solvent to remove resins and adhesives.

The *dipole moment* points toward the three electronegative chlorine atoms.

The vector is resolved mathematically between the three Cl and away from H.

Most covalent compounds show some degree of local charge separation, resulting from differences in electronegativity between the atoms.

If a molecule has polar bonds arranged so that the bonds are in a specific direction, the molecule has a net dipole in that same direction. Some examples of polar molecules are PCl_3 and HCN.

$$Cl^{\cdots\cdots}\overset{\overset{\bullet\bullet}{P}}{\underset{Cl}{|}}\diagdown Cl$$

Phosphorus trichloride (PCl_3) has a trigonal pyramidal molecular shape

The phosphorus trichloride (PCl_3) molecule has three chlorine atoms positioned on one side.

Since chlorine is very electronegative, the electrons are attracted to the chlorine side, creating a dipole moment.

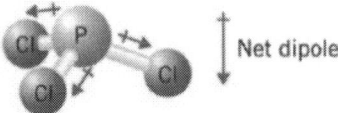
Net dipole

Phosphorus trichloride exhibits a net dipole because of the asymmetric pyramidal shape

Hydrogen cyanide (HCN) is a symmetrical molecule with one atom on each side of carbon.

Nitrogen is more electronegative than hydrogen; electrons concentrate near the nitrogen.

Bond dipoles
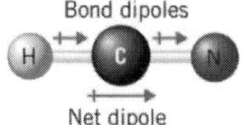
Net dipole

Polarity affects physical properties

Some physical properties (e.g., melting, boiling) are affected by molecular polarity.

Polar molecules have stronger intermolecular forces, resulting in higher boiling and melting points (below).

Substance	Boiling Point (°C)
Polar	
Hydrogen fluoride, HF	20
Water, H_2O	100
Ammonia, NH_3	−33
Nonpolar	
Hydrogen, H_2	−253
Oxygen, O_2	−183
Nitrogen, N_2	−196
Boron trifluoride, BF_3	−100
Carbon dioxide, CO_2	−79

Boiling points of polar (high BP) and nonpolar (low BP) substances

A critical example of polarity is the water (H_2O) molecule (shown below).

The center oxygen of water has two pairs of non-bonding electrons, two hydrogens (one on each side), and is theoretically symmetrical if the two lone pairs are ignored.

VSEPR theory predicts water has a bent shape, and the electron pairs are grouped (below).

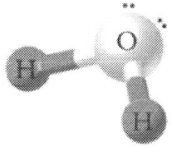

Water with the two lone pairs. The dipole moment is indicated in the central structure.

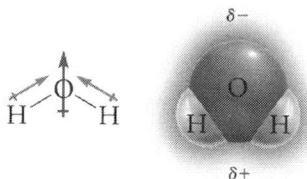

The space-filling water model with the molecule's partial negative and partial positive ends.

For water, the electron pairs are not evenly distributed; they are shifted downwards and concentrated on one side of the oxygen atom. That side has a partial negative charge, while the opposing side has a partial positive charge.

Therefore, the water molecule has a net dipole moment and is a polar molecule.

The polar water molecules exert strong intermolecular forces, known as *hydrogen bonds*.

Hydrogen bonds are a particular class of strong dipole-dipole interactions when a hydrogen atom is bonded *directly* to one of the three most electronegative atoms (F, Cl, N).

Practice Questions

1. Unhybridized p orbitals participate in π bonds as double and triple bonds. How many distinct and degenerate p orbitals exist in the second electron shell, where n = 2?

    A. 3   **B.** 2   **C.** 1   **D.** 0   **E.** 4

2. Which type of attractive forces occurs in molecules regardless of the atoms they possess?

    A. Dipole–ion interactions   **C.** Dipole–dipole attractions

    B. London dispersion forces   **D.** Hydrogen bonding

                                            E. Ion–ion interactions

3. Ignoring other factors, it is possible to approximate the energy released when an ionic bond forms by considering how much the electrostatic potential energy changes while forming the bond (i.e., bringing the two ions together from infinity). Using approximation, what is the amount of energy released when 1 mole of NaCl forms?

For two ions with charges q_1 and q_2 separated by distance r, the electrostatic potential energy is given by $U_e = k \cdot q_1 q_2 / r$, where electrostatic constant $k = 9 \times 10^9$ J·m / C^2. (The charge on $Na^+ = +e$, the charge on $Cl^- = -e$, the charge of electron $e = 1.602 \times 10^{-19}$ C, and the separation distance $r = 282$ pm)

    A. −493 kJ/mol   **C.** +91.2 kJ/mol

    B. −288 kJ/mol   **D.** +421 kJ/mol

                                            E. −464 kJ/mol

4. During strenuous exercise, why does perspiration on a person's skin forms droplets?

    A. The ability of H_2O to dissipate heat   **C.** Adhesive properties of H_2O

    B. High specific heat of H_2O **D.** Cohesive properties of H_2O

                                            E. High NaCl content of perspiration

5. Based on the Lewis structure, how many polar and nonpolar bonds are present in H_2CO?

    A. 1 polar bond and 2 nonpolar bonds   **C.** 3 polar bonds and 0 nonpolar bonds

    B. 2 polar bonds and 1 nonpolar bond   **D.** 0 polar bonds and 3 nonpolar bonds

                                            E. 2 polar bonds and 2 nonpolar bonds

6. Which of the following occur(s) naturally as nonpolar diatomic molecules?

     I. sulfur   II. chlorine   III. argon

A. I only   **C.** I and III only

B. II only   **D.** I and II only

                                        E. I, II and III

7. In the nitrogen monoxide molecule, the dipole moment is 0.16 D, and the bond length is 115 pm. What is the sign and magnitude of the charge on the oxygen atom? (Use the conversion factor of $1 \text{ D} = 3.34 \times 10^{-30}$ C·m and the charge of 1 electron $= 1.602 \times 10^{-19}$ C)

    A. $-0.098\ e$ **B.** $-0.71\ e$ **C.** $-1.3\ e$ **D.** $-0.029\ e$ **E.** $+1.3\ e$

8. From the electronegativity, which single covalent bond is the most polar?

Element:	H	C	N	O
Electronegativity	2.1	2.5	3.0	3.5

    A. O–C   **B.** O–N   **C.** N–C   **D.** C–H   **E.** C–C

9. Under what conditions is graphite converted to diamond?

A. low temperature, high-pressure   **C.** high temperature, high-pressure

B. high temperature, low-pressure   **D.** low temperature, low-pressure

                                         E. none of the above

10. An ion with an atomic number of 34 and 36 electrons has what charge?

    A. +2   **B.** –36   **C.** +34   **D.** –2   **E.** neutral

11. Which of the following pairings of ions is NOT consistent with the formula?

A. Co_2S_3 (Co^{3+} and S^{2-})   **C.** Na_3P (Na^+ and P^{3-})

B. K_2O (K^+ and O^-)   **D.** BaF_2 (Ba^{2+} and F^-)

                                         E. KCl (K^+ and Cl^-)

12. The ability of an atom in a molecule to attract electrons to itself is:

A. ionization energy

B. paramagnetism

C. electronegativity

D. electron affinity

E. hyperconjugation

13. All of the following are examples of polar molecules, EXCEPT:

A. H_2O

B. CCl_4

C. CH_2Cl_2

D. HF

E. CO

14. When NaCl dissolves in water, what is the force of attraction between Na^+ and H_2O?

A. ion–dipole

B. hydrogen bonding

C. ion–ion

D. dipole–dipole

E. van der Waals

15. Which element likely forms a cation with a +2 charge?

A. Na   **B.** S   **C.** Si   **D.** Mg   **E.** Br

Notes for active learning

Detailed Explanations

1. A is correct.

Three degenerate p orbitals exist with an electron configuration in the second shell or higher.

The first shell only has access to s orbitals.

The d orbitals become available from $n = 3$ (third shell).

2. B is correct.

London dispersion forces result from the momentary flux of valence electrons and are present in all compounds; they are the attractive forces that hold molecules.

They are the weakest intermolecular forces, and strength increases with increasing size (i.e., surface area contact) and the molecules' polarity.

3. A is correct.

Calculate the potential energy for each molecule of NaCl:

$$U_e = k\, q_1 q_2 / r$$

$$U_e = [(9 \times 10^9 \text{ J·m·C}^{-2}) \cdot (1.602 \times 10^{-19} \text{ C}) \cdot (-1.602 \times 10^{-19} \text{ C})] / (282 \times 10^{-12} \text{ m})]$$

$$U_e = -8.19 \times 10^{-19} \text{ J per molecule of NaCl}$$

Calculate the energy released for one mole of NaCl:

$$[(-8.19 \times 10^{-19} \text{ J}) \times (6.02 \times 10^{23} \text{ mol}^{-1})] = -493{,}077 \text{ J}$$

$$-493{,}077 \text{ J} \approx -493 \text{ kJ/mol}$$

This approximation disregards a few important effects.

For example, the crystal lattice orientation affects the energy, and this model does not consider quantum mechanical effects.

4. D is correct.

Water molecules stick to each other (i.e., cohesion) due to hydrogen bonds' collective action between individual polarized water molecules.

These hydrogen bonds are continually breaking and reforming; these bonds hold many molecules.

Water sticks to surfaces (i.e., *adhesion*) because of water's polarity.

continued...

Water may form a thin film on a highly smooth surface (e.g., glass) because the molecular forces between glass and water molecules (adhesive forces) are stronger than the water molecules' *cohesive* forces.

5. A is correct.

The *electronegative oxygen* pulls electron density away from the carbon atom and creates a net dipole towards the oxygen, resulting in a polar covalent bond.

The electronegativity values of carbon and hydrogen are similar and result in a covalent bond (i.e., about equal sharing of bonded electrons).

$$\overset{\displaystyle O}{\underset{\displaystyle H\diagup C\diagdown H}{\|}}$$

6. B is correct.

The *octet rule* states that atoms of main-group elements combine so that each atom has eight electrons in its valence shell.

7. D is correct.

Formula to calculate *dipole moment*:

$$\mu = qr$$

where μ is dipole moment (coulomb meter or C·m), q is the charge (coulomb or C), and r is the radius (meter or m)

Convert the unit of dipole moment from Debye to C·m;

$$0.16 \times 3.34 \times 10^{-30} = 5.34 \times 10^{-31}\,\text{C·m}$$

Convert the unit of radius to meter:

$$115\text{ pm} \times 1 \times 10^{-12}\text{ m/pm} = 115 \times 10^{-12}\text{ m}$$

Rearrange the dipole moment equation to solve for q:

$$q = \mu / r$$

$$q = (5.34 \times 10^{-31}\,\text{C·m}) / (115 \times 10^{-12}\,\text{m})$$

$$q = 4.65 \times 10^{-21}\,\text{C}$$

Express the charge in terms of electron charge (e).

$$(4.65 \times 10^{-21}\,\text{C}) / (1.602 \times 10^{-19}\,\text{C/e}) = 0.029\text{ e}$$

In this NO molecule, oxygen is the more electronegative atom. Therefore, the charge experienced by the oxygen atom is negative: –0.029e.

8. A is correct.

The greater the difference in *electronegativity* between two atoms in a compound, the more polar a bond these atoms form.

The atom with the higher electronegativity is the partial (delta) negative end of the dipole.

9. C is correct.

The substantial difference between the two forms of carbon (i.e., graphite and diamonds) is mainly due to their crystal structure, hexagonal for graphite and cubic for diamond.

The conditions to convert graphite into diamond are high pressure and high temperature.

Creating synthetic diamonds is time-consuming, energy-intensive, and expensive since carbon is forced to change its bonding structure.

10. D is correct.

The *valence shell* is an atom's outermost shell (i.e., highest principal quantum number, n).

Valence electrons are those electrons of the outermost electron shell that can participate in a chemical bond.

11. B is correct.

If O's charge is -1, the proper formula of its compound with K ($+1$) should be KO.

12. C is correct.

Electronegativity is a chemical property that describes an atom's tendency to attract electrons.

The most common use of electronegativity pertains to polarity along the *sigma* (single) bond.

The most electronegative atom is F, while the least electronegative atom is Fr.

The trend for increasing electronegativity on the periodic table is up and toward the right (i.e., fluorine).

13. B is correct.

The greater the difference in electronegativity between two atoms, the more polar bond these atoms form.

The atom with the higher electronegativity is the partial (delta) negative end of the dipole.

Although each C–Cl bond is polar, the dipole moments of each of the four bonds in CCl_4 (carbon tetrachloride) cancel because the molecule is a symmetric tetrahedron.

14. A is correct.

When NaCl dissolves in water, ion–dipole is the force of attraction between Na^+ and H_2O.

15. D is correct.

Group IA elements (e.g., Li, Na, and K) *lose* 1 electron to achieve a complete octet as +1 cations.

Group IIA elements (e.g., Mg and Ca) *lose* 2 electrons to achieve a complete octet as +2 cations.

Group VIIA elements (halogens such as F, Cl, Br, and I) *gain* 1 electron for a complete octet as a –1 anion.

Notes for active learning

Notes for active learning

CHAPTER 3

Phases
&
Phase Equilibria

- Phase Equilibria
- Phase Diagrams
- Molality
- Colligative Properties
- Gas Phase
- Gas Pressure
- Kinetic Molecular Theory of Gases
- Gas Laws
- Partial Pressure
- Intermolecular Forces
- Practice Questions & Detailed Explanations

Phase Equilibria

Phases of substances

Most substances exist in solid, liquid, or gas form.

A *solid* has a definite shape and volume (e.g., ice). The molecules in a solid vibrate about a fixed position, and a solid *cannot be compressed*.

A *liquid* has a definite volume but takes its container shape (e.g., water). The molecules move about but are close and *bound by intermolecular forces*.

A *gas* has neither a definite volume nor a definite shape; it takes the container's volume and shape (e.g., steam or water vapor). The molecules *move independently* and are not held together by intermolecular forces, and gas is *easily compressible*.

Freezing and melting point

The *freezing point* is the temperature at which a liquid changes state to become a solid (i.e., solidification).

The *melting point* is when a solid changes its state and becomes a liquid.

The melting point of a substance is the temperature at which it changes from a solid to a liquid, and it has the same temperature range as the *freezing point*; the reverse reaction occurs at the same temperature.

Theoretically, a substance's melting point and freezing point should be identical.

Experimentally, a difference in the respective processes' quantities may be observed.

Vaporization

Vaporization is the phase transition at which a liquid becomes a gas.

There are two types of vaporization: *boiling* and *evaporation*.

The *boiling point* is when the liquid is heated to a temperature (i.e., the kinetic energy of the molecules) where the vapor pressure is high enough for bubbles to form inside the liquid. It is in the same temperature range as the *condensation point*.

As a liquid is boiling, the temperature remains constant until the liquid has converted into gas.

Since the boiling point is based on pressure, the precise value depends on the environment (e.g., elevation relative to sea level).

For example, at high-altitude and low-pressure, a substance's boiling point is lower, requiring less heat energy to increase the vaporizing molecule's kinetic energy to break intermolecular attraction to adjacent molecules.

Evaporation occurs at temperatures below the boiling point. It occurs at the liquid's surface when the gaseous phase above the liquid is not saturated with the evaporating substance.

Boiling occurs from the bulk of the liquid, not just the surface.

Evaporation tends to occur more quickly in liquids with higher vapor pressure.

Condensation and sublimation

Condensation (reverse of evaporation) is when the substance changes from the gas phase to the liquid phase. Condensation commonly occurs when a vapor is cooled or compressed to its saturation limit.

The *condensation po*int is the temperature at which condensation occurs.

For example, condensation is the formation of water droplets in atmospheric clouds.

Sublimation (reverse of deposition) is the phase transition from a solid to a gas, essentially evaporation that occurs directly from the solid phase below the melting point.

Deposition (reverse of sublimation) is the phase transition from a solid directly into a gas (i.e., the matter is not present as an intermediate liquid). This process is observed in the winter when the temperature rises above freezing, and the snow does not thaw to liquid, yet the snow melts.

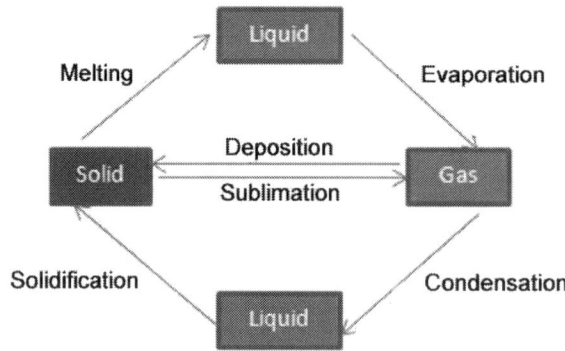

Interconversion of states of matter

Phase Diagrams

Phase change diagram

When an ice block is placed in the Sun, it slowly absorbs the thermal energy through radiation and undergoes a series of phase changes. The ice cube changes from a solid to a liquid to a gas.

The *melting point* of a substance is the temperature at which it changes from a solid to a liquid, and it has the same temperature range as the *freezing point*; the reverse reaction occurs at the same temperature.

The *boiling point* (vaporization) of a substance is the temperature at which it changes phase from a liquid to a gas, and it is the same temperature range as the *condensation point*.

From the graph below, once the ice cube absorbs enough heat to melt (i.e., reaches 0 °C), its temperature remains constant.

Melting is points B to C on the graph where the molecules have reached a temperature of 0 °C. The ice melts only when this occurs, as illustrated by the graph's horizontal plateau from B to C.

When the liquid water changes to a gas, the liquid absorbs enough heat to vaporize (i.e., at 100 °C). The liquid temperature remains constant from point D to E until all molecules reach a temperature of 100 °C. The horizontal plateau from D to E signifies the phase transition of vaporization.

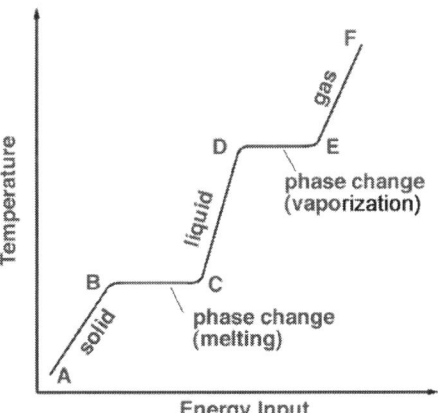

Heat during phase changes

Although the energy (heat) input increases with temperature, it does not remain constant as the temperature does during a phase change.

Technically, heat converts the potential energy (PE) stored within each atom to kinetic energy (KE) before changing phase.

As energy is continuously added to the process, the substance uses the added heat to transition each molecule before increasing the temperature.

Pressure *vs.* temperature relationship

A *phase diagram* displays the phases of a substance. A phase diagram is a graph plotted as pressure *vs.* temperature.

The graph is split into three disproportionate parts, representing the substance's three phases.

Each point on the graph represents a possible combination of pressure and temperature for the system.

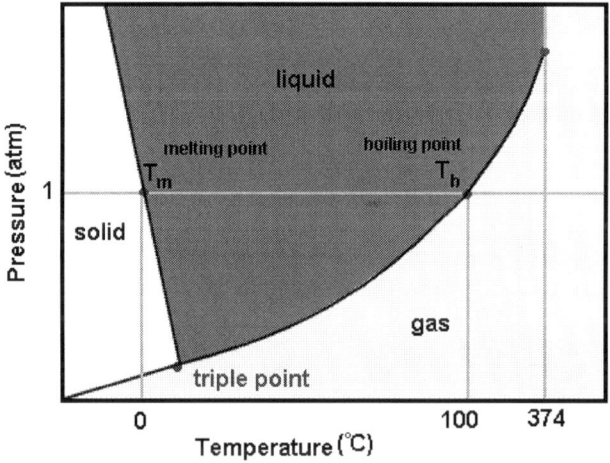

The horizontal line intersects on the phase diagram the temperature where the substance changes phases.

The *melting point* and *boiling point* are found by drawing a horizontal line at 1 atm of pressure, atmospheric pressure.

The *triple point* is shown when all three lines intersect and three phases exist simultaneously.

The *critical point* is the temperature above which the substance is a gas, regardless of pressure.

Molality

Molal concentration

Molality (or *molal concentration*) measures the moles (i.e., the concentration of solutes) in a solvent in terms of the mass (on kg) of the solvent.

The SI unit for molality is mol/kg.

Molality *m* is different from the uppercase *M* for molarity (i.e., M = moles/liter).

> *Molality* = (moles of solute) / (mass in kg of solvent)

Calculating molality

For example, if 3.00 moles of sucrose are mixed into 1.00 L water until the sucrose is entirely dissolved (i.e., saturated solution). What is the molality of the solution? (Use the density of water = 1.00 g/mL or 1.00 kg/L).

Solution, the mass of the liter of water is 1.00 kg.

Use the solvent mass of 1. Kg/L and solve for molality.

> m = 3.00 mol / 1.00 kg

> m = 3.00 mol/kg of sucrose

> m = 3.00

The molality of the solution is 3.00 m.

Instead of expressing the number of moles, the solute's mass is given in grams.

When calculating molality, the solvent's mass is the pure solvent, excluding mass of the solute.

Molarity is the volume of solution, including the mass of the solute *and* the solvent.

> *Molality m* is expressed as mol solute/kg solvent.

> *Molarity M* is expressed as mol solute/L solution

For example, suppose 16.88 grams of sodium chloride (NaCl) are dissolved into 4.00 kg of water. What is the molality of the solution?

Convert grams of NaCl to moles (molar mass of NaCl = 58.44 g/mol).

> moles = (mass) / (molar mass)

> moles = (116.88 g) / (58.44 g/mol)

> moles = 2.00 mol NaCl

Solve for molality:

m = (moles of solute) / (mass in kg of solvent)

m = (2.00 mol) / (4.00 kg of solvent)

m = 0.500 mol/kg of the solution

m = 0.500

The molality of the solution is 0.500 mol/kg.

.

Colligative Properties

Solute particles

Colligative properties depend on the ratio of solute particles to solvent but *not* on the type of chemical species.

Solute particles in mixtures increase the strength of intermolecular (e.g., dipole-dipole) bonds. It is more difficult to boil the solution (i.e., boiling point elevation) or freeze the solution (i.e., freezing point depression).

Colligative properties include:

> 1) the relative lowering of vapor pressure,
>
> 2) the elevation of boiling point,
>
> 3) the depression of freezing point, and
>
> 4) osmotic pressure.

Van 't Hoff factor

The colligative properties calculations incorporate the van 't Hoff factor (i), which measures a solute's effect on the solution's colligative properties.

The van 't Hoff factor is:

$$\frac{[\text{concentration of particles produced when a substance dissolves}]}{[\text{concentration of a substance calculated from its mass}]}$$

Concentration (represented in [ ]) is converted to reflect the total number of particles in the solution. (e.g., 5 g of NaCl dissociates into Na^+ and Cl^-).

For example, glucose ($C_6H_{12}O_6$) has an *i* value of 1 because it does not dissociate in solution.

Nonelectrolytes do not dissociate in solution (e.g., CH_4 and O_2).

The salt NaCl has an *i* value of 2 because it dissociates into 1 Na^+ and 1 Cl^- (2 resulting ions per one reactant molecule).

Vapor pressure depression (Raoult's Law)

In 1888, French chemist Francois-Marie Raoult (1830-1901) proved that the vapor pressure of a solution equals the mole fraction of the solvent times the vapor pressure of the pure solvent.

Raoult's Law is expressed by:

$$P = \chi_{solvent} \cdot P°_{solvent}$$

where P is the vapor pressure, $\chi_{solvent}$ is the mole fraction of solvent (moles of solvent / total moles of solute *and* solvent), and $P°_{solvent}$ is the vapor pressure of the pure solvent

Calculate the χ_{solute} using the van 't Hoff factor (i.e., 1 mol of NaCl yields 2 mol of ions in the solution).

For example, suppose 2.00 moles of sucrose is added to a pitcher containing 2.00 liters of water. What is the vapor pressure of the sucrose solution? (Use the value 1.00 L water = 1.00 kg, the vapor pressure of pure water = 23.8 mmHg and the molecular mass of H_2O = 18.02 g/mol)

Convert the liters of water into mass:

2.00 L water = 2 Kg

2 Kg = 2,000 g

Convert the mass of the water into moles:

molecular mass = mass / moles

moles = mass / molecular mass

moles = (2,000 g) / (18.02 g/mol)

moles = 110.9 moles H_2O

Solve for the mole fraction, $\chi_{solvent}$:

$\chi_{solvent}$ = (moles of solvent) / (moles of the solute *and* solvent)

$\chi_{solvent}$ = (110.9 moles H_2O) / (moles of the solute *and* solvent)

$\chi_{solvent}$ = 110.9 moles H_2O / (110.9 moles solvent + 2 moles solute)

$\chi_{solvent}$ = 0.98

Use Raoult's Law to determine the pressure:

$P = \chi_{solvent} \cdot P°_{solvent}$

$P = (0.98) \cdot (23.8 \text{ mmHg})$

$P = 23.4$ mmHg

The addition of a solute caused the vapor pressure to decrease.

Boiling point elevation ($\Delta T_b = K_b m$)

Adding a solute to a solvent stabilizes the solvent in the liquid phase, thus lowering the solvent molecules' tendency to move to the gas or solid phases. Therefore, the boiling point increases.

The *boiling point elevation* is proportional to the decrease of vapor pressure in a dilute solution:

$\Delta T_b = k_b \cdot m \cdot i$

where ΔT_b is the increase in boiling point, k_b is the molal boiling point constant (a given value determined experimentally), m is the molality (mol solute/kg solvent), and i is the van 't Hoff factor

For example, what is the boiling point elevation if 6.4 g of ammonia is dissolved in 0.3 kg of water? (Use the k_b for water = 0.52 °C/m and the molar mass of ammonia = 17.031 g/mol)

Convert the mass of ammonia (NH_3) into moles:

molecular mass = mass / moles

moles = mass / molecular mass

moles = (6.4 g) / (17.031 g/mol)

moles = 0.38 mol NH_3

Calculate the molality of the solution:

molality = (moles of solute) / (mass (in kg) of solvent)

m = (0.38 mol NH_3) / (0.3 kg of water)

m = 1.27 m

Calculate the boiling point elevation:

$\Delta T_b = k_b \cdot m \cdot i$

ΔT_b = (0.52 °C/m)·(1.27 m)·(1)

ΔT_b = 0.66 °C

The boiling point increased by 0.66 °C due to the addition of 6.4 g of NH_3.

Freezing point depression ($\Delta T_f = K_f m$)

The *freezing point depression* (i.e., lowering the freezing point) is calculated by the following equation: (note the negative sign indicates that the change is decreased).

The freezing point depression is expressed as:

$$\Delta T_f = -k_f \, m \cdot i$$

where ΔT_f is the decrease in freezing point, k_f is the molal freezing point constant (a given value determined experimentally), m is the molality (mol solute/kg solvent), and i is the van 't Hoff factor (given since it is determined experimentally)

For example, a 48.0 g sample of a nonelectrolyte (does not dissociate in solution) is dissolved in 500.0 g of water to produce a solution with a freezing point of –3.5 °C. What is the molar mass of the compound? (Use k_f of water = 1.86 °C/m, the freezing point of pure water = 0 °C)

The freezing point of pure water = 0 °C, and the freezing point depression, ΔT_f = 3.5 °C.

Calculate the moles using the freezing point depression expression:

$$\Delta T_f = k_f \cdot m \cdot i$$

$$3.5 \text{ °C} = (1.86 \text{ °C/m}) \cdot (x \, / \, 0.5 \text{ kg}) \cdot (1)$$

$$3.5 \text{ °C} = (3.72 \text{ °C/m}) \cdot (x)$$

$$3.5 \text{ °C} \, / \, (3.72 \text{ °C/m}) = x$$

$$x = 0.94 \text{ mol nonelectrolyte}$$

Calculate molar mass by dividing the sample's mass by moles in the sample.

Molar mass = mass / moles

Molar mass = 48.0 g / 0.94 mol

Molar mass = 51.1 g/mol

The molar mass of the unknown nonelectrolyte is 51.1 g/mol.

Osmotic pressure

Osmotic pressure is the *minimum pressure that needs to be applied to a solution to prevent the inward flow of water across a semipermeable membrane.*

The semipermeable membrane allows solvent particles to pass, not solute particles.

The *osmotic pressure* (Π) is:

$$\Pi = MRT{\cdot}i$$

where Π is the osmotic pressure, M is the molarity (mol/L), R is the ideal gas constant (0.08206 L·atm/mol·K), and T is the temperature (K)

Osmotic pressure determines whether and in what direction osmosis occurs.

Osmosis is a solvent's movement across a semi-permeable membrane from a low solute concentration (i.e., high solvent concentration) to an area of high solute concentration (i.e., low solvent concentration).

The solvent moves from a low Π value to an area across the semipermeable membrane with a high Π value.

For example, what is the osmotic pressure of a solution prepared by adding 10.5 g of sucrose ($C_{12}H_{22}O_{11}$) to enough water to make 300.0 mL of solution at 25 °C? (Use the molecular mass of sucrose = 342.0 g/mol and the conversion of 300.0 mL = 0.30 L)

Calculate the number of moles of $C_{12}H_{22}O_{11}$ by dividing the mass of the sample by the molar mass of the sample.

moles = (mass) / (molar mass)

moles = (10.5 g $C_{12}H_{22}O_{11}$) / (342.0 g/mol $C_{12}H_{22}O_{11}$)

moles = 0.03 mol $C_{12}H_{22}O_{11}$

Calculate the molarity, *M*:

Molarity = (moles solute) / (liters of solution)

Molarity = (0.03 mol $C_{12}H_{22}O_{11}$) / (0.30 L solution)

Molarity = 0.10 mol/L

Convert temperature to Kelvin:

$T =$ °C + 273

$T = 25 + 273$

$T = 298$ K

Since sucrose does not dissociate, calculate osmotic pressure (Π) using $i = 1$.

$$\Pi = MRT{\cdot}i$$

$$\Pi = (0.10 \text{ mol/L}){\cdot}(0.08206 \text{ L}{\cdot}\text{atm/mol}{\cdot}\text{K}){\cdot}(298 \text{ K}){\cdot}(1)$$

$$\Pi = 2.45 \text{ atm}$$

The osmotic pressure for the sucrose solution is 2.45 atm.

Colloids

A *solution* is a homogeneous mixture that consists of one phase (i.e., the solution stays mixed).

A *colloid* is a solution with microscopic insoluble particles dispersed throughout the solution.

A colloid has a dispersed phase (i.e., suspended particles) and a continuous phase (i.e., a medium of suspension that holds the particles).

A colloid stays mixed (i.e., the particles will not settle) unless centrifuged under the influence of gravity.

A typical colloid le is homogenized milk, consisting of butterfat globules dispersed within a water-based solution.

For example, a colloid forms when water and oil are shaken together.

Henry's Law

In 1803, English chemist William Henry (1774-1836) showed that at a constant temperature, the solubility of a gas in a liquid is directly proportional to the gas's partial pressure above the liquid (when the gas is in equilibrium with the liquid).

Henry's Law is described by :

$$P_{\text{solute}} = k_{\text{H}}{\cdot}c$$

where P_{solute} is the partial pressure of the solute at the solution's surface, k_{H} is Henry's Law constant (unique for each solute-solvent pair), and c is the concentration of the dissolved gas

For example, how many grams of carbon dioxide (CO_2) gas is dissolved in a 0.5 L bottle of carbonated water if the manufacturer uses a pressure of 2.5 atm in the bottling process? (Use the K_{H} of CO_2 in water = 29.76 atm/(mol/L) and molar mass of CO_2 = 44 g/mol)).

Determine the concentration of CO_2 using Henry's Law:

$$P_{\text{solute}} = k_{\text{H}}{\cdot}c$$

$$2.7 \text{ atm} = [29.76 \text{ atm/(mol/L)}]{\cdot}c$$

$$c = 0.09 \text{ mol/L}$$

Calculate the moles of CO_2 in 0.5 L by dividing the mol/L by 2.

moles / grams = molar mass

0.09 / 2 = 0.045 mol

Convert moles to grams using the molar mass of CO_2, 44 g/mol (from the periodic table).

0.045 mol × (44 g/mol) = 1.98 g

There are 1.98 grams of CO_2 in a 0.5 L bottle of carbonated water.

Notes for active learning

Gas Phase

Gaseous state

Gas is a fundamental state of matter, distinguished by the vast separation of the individual gas particles.

Unlike solids and liquids, gases have an indefinite shape and volume because they expand to fill their containers.

As a result, gases are subject to pressure, volume, and temperature changes.

Some common elements and compounds exist in the gaseous state under normal conditions of pressure and temperature, such as oxygen (O_2), nitrogen (N_2), and carbon dioxide (CO_2).

A vapor is a gaseous form usually existing as a liquid or solid at ordinary pressures and temperatures (e.g., water vapor).

Four variables are needed to describe the gaseous state:

1. Quantity of gas, n (in moles)

2. The temperature of the gas, T (in Kelvin)

3. The volume of gas, V (in liters)

4. The pressure of gas, P (in kPa)

Absolute temperature

Temperature is an objective measure of particles' kinetic energy, and the volume of a gas (i.e., the space that it occupies) is proportional to its temperature.

In the 19th century, experiments revealed that gases expand as temperature rises and compress when temperature drops, which led to the determination of the expansion coefficient of gases per degree Celsius.

This posed an interesting question: what happens if the temperature is low enough to reach a point where the gas's calculated volume is zero?

This *absolute zero* temperature was calculated to be –273.15 °C.

Gases do not reach zero volume because gases liquefy or solidify before reaching this temperature.

Additionally, gas cannot have a zero volume because the gas molecules' atoms occupy a certain amount of space.

Kelvin scale

In 1848, British physicist Lord William Thomson Kelvin (1824-1907) proposed a temperature scale based on this absolute zero value.

The Kelvin (K) scale starts with absolute zero as 0 K, and the increments are identical to the Celsius scale.

To convert from Kelvin to Celsius, add 273.15; and for the opposite conversion, subtract 273.15.

Kelvin should be converted to Celsius before converting it into Fahrenheit.

Kelvin temperatures are not written with a degree symbol (273 K *vs.* 273 °C or 273 °F).

Common temperatures expressed in K, °C and °F

	K	°C	°F
Absolute zero	0	–273	–460
Freezing point of water / melting point of ice	273	0	32
Room temperature	298	25	77
Body temperature	310	37	99
Boiling point of water / condensation of steam	373	100	212

Notes for active learning

Gas Pressure

Pressure as force

The *pressure* is the average force applied to or experienced by a given area.

Gas molecules are in motion, so in a container, the gas molecules continuously move around and collide with the container's walls. Since the force is applied over a specific area, this force is the container's gas pressure.

Newton's Second Law, $F = ma$, can be used to determine the units of pressure.

The SI units of mass (m) and acceleration (a) are the kilogram (kg) and the meter per second squared (m/s^2).

The unit of force (F) is derived from the equation:

$$F = ma, \text{ is kg·m/s}^2$$

This derived unit is Newton (N).

The pressure is the force exerted over a specific area.

The SI unit of length is a meter (m), so the SI unit for area is m^2.

Therefore, the unit for pressure is N/m^2, known as the Pascal (Pa).

Simple mercury barometer

A *barometer* measures Earth's atmospheric pressure. 30 inHg is considered normal.

Dry air creates more pressure than wet air, so dry air is more massive than the same amount of wet air.

Consequently, *high barometric readings* (i.e., 30.70 inHg) indicate dry air and fair weather.

Low barometric readings (i.e., 29.80 inHg) indicate an increased chance of rain or showers.

Hurricane conditions are indicated by barometric readings of 27.30 inHg.

A *mercury barometer* (shown below) is made by inverting a column filled with mercury and placing it in a mercury dish without allowing air to enter the tube. Some mercury flows out when the tube is inverted, leaving a space in the tube above the mercury. This space is nearly a vacuum. Therefore, there is no pressure pushing down on the column.

The pressure exerted by the mercury column balances the pressure of the atmosphere.

The mercury column's height is about 760 mm above the surface of the mercury in the dish at sea level.

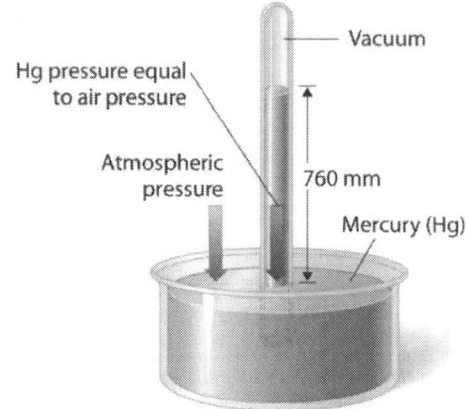

Expressing pressure

Any liquid may be used to measure atmospheric pressure; the height of the column depends on the density of the liquid. A comparison of mercury (density = 13.6 g/cm^3) and water (density = 1 g/cm^3) indicates that the column would be almost 34 feet high if a barometer were filled with water.

The pressure exerted by the mercury barometer is recorded in units of millimeters of mercury (mmHg), a unit sometimes called the torr in honor of the Italian physicist Evangelista Torricelli (1608-1647), who invented the mercury barometer in 1643.

The standard atmosphere (atm) is another unit of pressure, and its conversion factors are listed:

1 atm = 101,325 Pa

1 atm = 760 mm Hg = 760 torr = 76 cm Hg

1 atm = 14.7 lb/in^2 (psi, or pound per square inch)

A bar is a metric unit of pressure equal to 100,000 Pa; it is about equal to atmospheric pressure (101,325 Pa).

Closed-tube and open-tube manometers

A *manometer* is an instrument that measures the pressure of gases. It is simply a bent piece of tubing and, in principle, operates like a barometer.

There are two major types of manometers.

A *closed-tube manometer* (shown on the left below) measures pressures below atmospheric pressures, usually the container's gas pressure. Since the end is sealed, it contains a vacuum.

The *open-tube manometer* (shown on the right below) measures gas pressures near atmospheric pressures.

The *pressure* is the difference in the mercury levels' heights in the two arms. The mercury levels in the manometer's two arms relate the gas pressure to the atmospheric pressure.

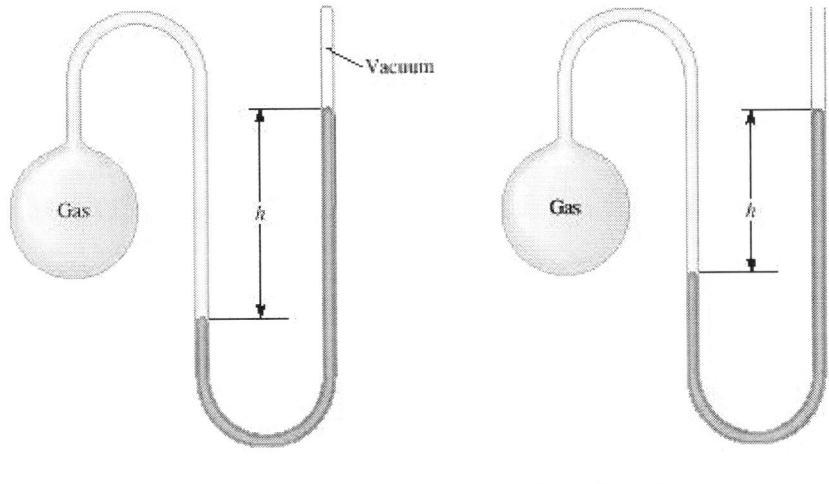

$$P_{gas} = Ph \qquad\qquad P_{gas} = P_h + P_{atm}$$

Left: gas pressure is less than atmospheric pressure (closed tube)

Right: gas pressure is higher than atmospheric pressure (open tube)

Molar volume at STP

In 1811, Italian scientist Amedeo Avogadro suggested that "*equal volumes of gases measured at the same temperature and pressure conditions contain the same number of molecules.*"

In chemistry, *molar volume* is usually measured at "standard temperature and pressure" (i.e., STP), which is a temperature of 273.15 K (0 °C, 32 °F) and absolute pressure of exactly 10^5 Pa (100 kPa, 1 bar).

At these standard conditions, a mole of gas occupies 22.7 L of space.

Before 1982, when the International Union of Pure and Applied Chemistry (IUPAC, the organization that governs chemistry standards) changed the definition of STP, the standard pressure was defined as 101,325 Pa (1 atm).

In these conditions, the volume of 1 mol of gas occupies 22.4 L.

For problems, use the volume of one mole of gas as 22.7 L unless indicated otherwise.

Notes for active learning

Kinetic Molecular Theory of Gases

Macroscopic properties of gas

The *kinetic molecular theory of gases* relates the microscopic values of atomic kinetic energy to quantities like temperature and pressure.

The *kinetic molecular theory* (KMT) uses atomic theory to describe gases' behavior. This theory developed over 100 years, culminating in an 1857 paper by German physicist Rudolf Clausius (1822-1888).

KMT explains the macroscopic properties of a gas (e.g., pressure and temperature) in terms of its microscopic components (e.g., molecules or atoms).

The KMT of gases originated in the ancient idea that matter consists of tiny invisible atoms in rapid motion.

The idea was developed between the 17th and 19th centuries to explain gases' properties.

The kinetic theory of gases requires *five critical assumptions*.

1. Gas molecules move in random molecular motion and continue straight until they collide with something.

2. The gas molecule is a "point mass," a particle so small that mass is nearly zero.

    An ideal gas particle has a *negligible volume*.

3. No attractive intermolecular molecular forces for gases participate.

4. Collisions between gas molecules are "perfectly elastic."

    Therefore, the gas molecules' total kinetic energy (KE) remains *constant before and after* the collision since intermolecular attractive and repulsive forces are nonexistent.

5. The average kinetic energy (*KE*) for gas particles is proportional to the absolute temperature, regardless of the chemical identity or atomic mass.

    At 0 K (absolute zero or –273 °C), the molecules (and orbiting electrons) are not moving and have no volume.

These rules display gases as ideal substances and only approximate their actual behavior.

However, their description is remarkably accurate.

From these assumptions, laws are derived to describe the behavior of gases.

Boltzmann's constant

The final assumption (#5) for KE is expressed by:

$$KE = \tfrac{1}{2}\, mv^2$$

$$\tfrac{1}{2}\, mv^2 = (3/2)\, k_B T$$

so,

$$KE = (3/2)\, k_B T$$

where *KE* is kinetic energy, *m* is mass, *v* is velocity, *T* is the temperature (Kelvin) and k_B is *Boltzmann's constant* ($k_B = 1.38 \times 10^{-23}$ J/K)

Boltzmann's constant relates the macroscopic and microscopic behavior of gases to their particles

The KE = (3/2) $k_B T$ equation is significant because it states that a gas particle's average kinetic energy is proportional to its absolute temperature.

Increasing the temperature of the gas particles increases the total speed and overall energy.

Gas atoms are infinitesimally small; nearly impossible to accurately measure a particle's speed.

The speed of gas particles is defined by the root-mean-square speed (u_{rms}).

The *root-mean-square speed* equation:

$$u_{rms} = \sqrt{\frac{3RT}{MM}}$$

where *R* is the ideal gas constant, *T* is the absolute temperature (K), and *MM* is the molar mass of the gas sample

Root-mean-square speed written with the *Boltzmann constant*, k_B:

$$u_{rms} = \sqrt{\frac{3kT}{m}}$$

where *k* is the Boltzmann constant, *T* is the absolute temperature (K), and *m* is the mass of one gas molecule

Boltzmann distribution plot

A plot of each gas particle's distribution of speeds at a given temperature shows a slightly asymmetric curve.

This speed-distribution curve is the *Maxwell-Boltzmann distribution curve.*

The *Boltzmann distribution plot* shows that the curve's peak corresponds to the most probable speed.

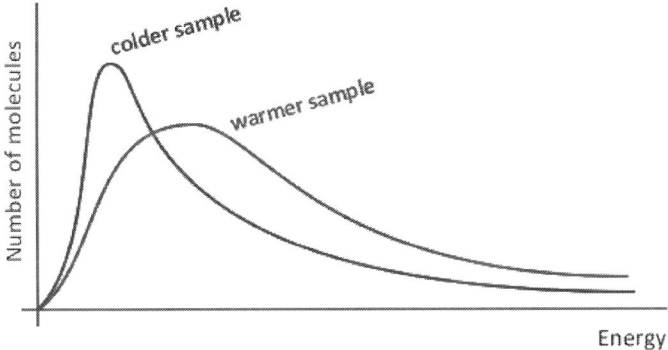

Maxwell-Boltzmann Distribution curves of molecular speeds at two temperatures

If the temperature (K) remains constant, the total kinetic energy remains unchanged.

However, the energy is distributed in several ways, and the gas particles travel at many speeds.

This distribution changes continually as the gas atoms collide with each other and with the container walls.

The curve flattens (i.e., net area under the curve is greater) and shifts to the right at higher temperatures, indicating that more gas molecules move at higher speeds and therefore possess more kinetic energy.

Effusion and diffusion

Effusion is the flow of gas particles *under pressure* from one compartment through a small opening, as shown.

Diffusion, the process whereby a substance (solute or particle) spreads from a region of high concentration to one of lower concentration.

Diffusion is explained using the kinetic theory of motion.

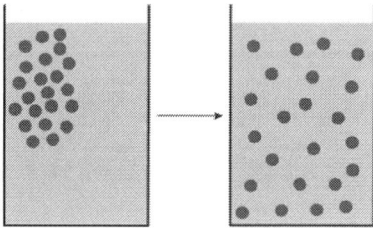

Diffusion of solutes in a solvent

According to the kinetic theory, heavier gases diffuse more slowly than lighter gases because of the difference in their travel speed.

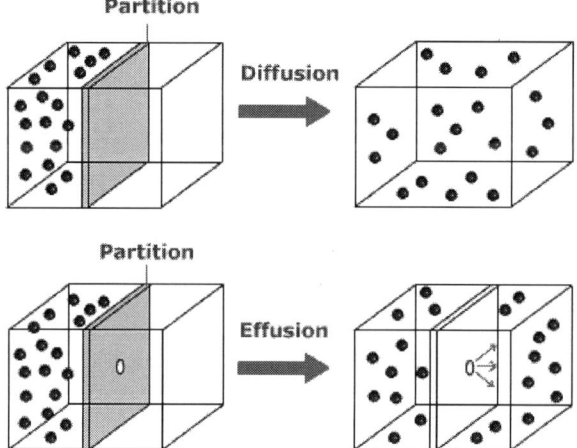

Diffusion and effusion of gas particles

Graham's Law

Scottish chemist Thomas Graham (1805-1869) developed *Graham's Law*, which states that *at constant temperature and pressure conditions, the rates at which two gases diffuse are inversely proportional to the square root of their molar masses.*

Graham's Law is expressed by:

$$\frac{r_1}{r_2} = \sqrt{\frac{MM_2}{MM_1}}$$

where r_1 and r_2 are the diffusion rates of gas 1 and gas 2, respectively, and MM_1 and MM_2 are the molar masses of the gases

Graham's Law applies to the effusion of gas particles.

The equation for effusion is the same as for diffusion.

Gas Laws

Ideal gas law

The *ideal gas law* was proposed in 1834 by French physicist Émile Clapeyron (1799-1864). It is a combination of the gas laws discussed, and it is the equation regarding the state of a hypothetical gas.

The *ideal gas law* explains the relationship between pressure (P), volume (V), and temperature (T).

The gas law is expressed as:

$$PV = nRT$$

where P is the pressure, V is the volume, T is the temperature, n is the number of moles, and R is the gas constant (use $R = 0.082$ L·atm/K·mol)

One mole $= 6.023 \times 10^{23}$ molecules; technically, the number of H atoms in one gram of H.

Because atoms are so small, it is more practical to count atoms in moles than to count them individually.

The ideal gas law applies to the pressure exerted by a gas on a cylinder with a moving wall.

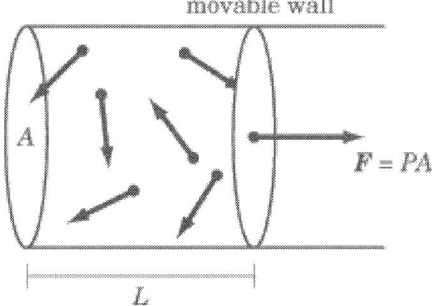

Since pressure is equal to $P = F/A$, the force that the gas exerts on the wall is equal to $F = PA$.

If this force moves the wall back a length of L, the volume of the cylinder increases by $\Delta V = LA$.

By solving for A and substituting it into the equation for force, the result is $F = P\Delta V L$, equal to $P\Delta V = FL$.

For example, suppose 2.5 moles of gas are at standard temperature and pressure. What is the volume occupied by the gas? (Use standard pressure $= 1.00$ atm and standard temperature $= 273.0$ K)

Use the ideal gas law:

$$PV = nRT$$

Solve for V

$$V = (nRT) / P$$

$$V = [(2.5 \text{ mol}) \cdot (0.082 \text{ L·atm/K·mol}) \cdot (273.0 \text{ K})] / (1.00 \text{ atm})$$

$$V = 56 \text{ L}$$

Several useful expressions are derived from the ideal gas law ($PV = nRT$) that relates the molar mass and density of gases to the pressure and temperature, substituting a known expression for one of the ideal gas law variables.

Work (w) is *force multiplied by the distance traveled*:

$$w = Fd$$

By pushing the wall, a distance of L with a force of F, the gas has done work equal to FL.

When gas does work, it symbolizes a change in energy.

If a change in PV is equal to a change in energy, then PV is the total energy of the gas.

For the ideal gas law, this means that nRT is the expression for the total kinetic energy of the gas molecules.

The ideal gas law ($PV = nRT$) is related to the number of molecules (N) and the Boltzmann's constant (k), which has a value of 1.381×10^{-23} J/K:

$$PV = NkT$$

The number of moles (n) and the gas constant (R) are constant.

Other gas laws derive from the ideal gas law by holding specific variables constant:

> Boyle's Law
>
> Charles' Law
>
> Combined gas law
>
> Closed container law

Ideal gases follow KMT (isothermal process)

The *kinetic molecular theory* (KMT) described previously is based on a theoretical ideal gas.

An *ideal gas* follows the *five* assumptions of KMT, whereby gas molecules:

> move in random motion
>
> are point masses with no volume
>
> have no intermolecular forces
>
> exhibit elastic collisions
>
> kinetic energy of the molecules is proportional to temperature

Compared to an ideal gas, the behavior of a real gas is *complicated.*

By conceptualizing ideal gas behavior, real gas behavior is easier to comprehend.

In the 17th and 18th centuries, scientists realized several relationships between gases' properties.

Boyle's, Charles', and Avogadro's Laws relate the properties of pressure, volume, and temperature developed empirically as individual cases of the ideal gas equation: PV = nRT.

Boyle's Law (isothermal process)

In 1661, Robert Boyle (1627-1691) studied the compressibility of gases. He observed that *the volume of a fixed amount of gas at a given temperature is inversely proportional to the pressure exerted on the gas.*

For *Boyle's Law*, the temperature of a gas is held constant.

It states that an *increase in pressure* causes a *decrease in volume.*

Likewise, a *decrease in pressure* causes an *increase in volume.*

Boyle's Law is expressed as:

$$P_1V_1 = P_2V_2$$

where subscripts 1 and 2 refer to the same gas sample under two sets of pressure and volume conditions

A plot of volume *vs.* pressure for a gas illustrates that Boyle's Law is a particular case of the ideal gas law, where *n* (moles of gas) and *T* (temperature) are constant (figure below).

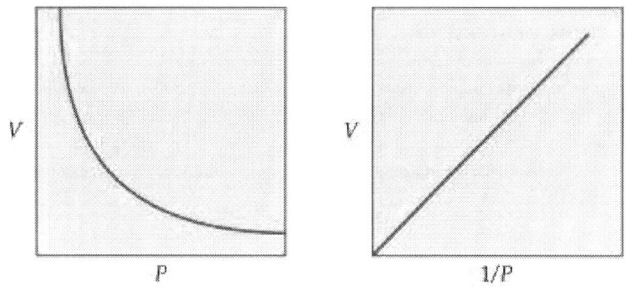

Boyle's Law: as pressure increases, volume decreases

For example, what is the new pressure in a container of gas in a 15.0 L container at a pressure of 5.00 atm when the volume decreases to 0.500 L?

Substitute the known quantities (P_1, V_1, and V_2) into Boyle's Law equation to solve P_2.

Note: the volume units need to be consistent. In this example, volumes are expressed in units of liters.

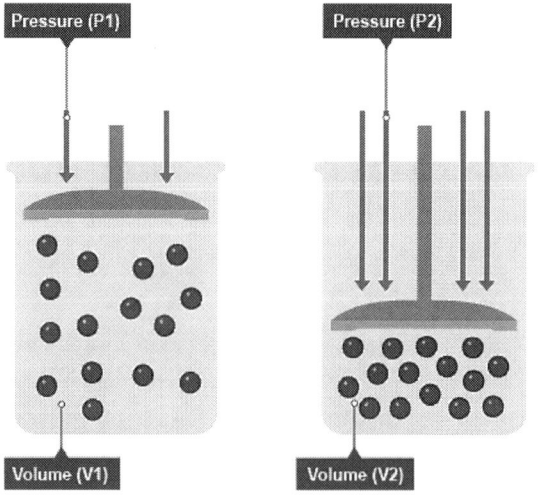

Boyle's Law:

$$P_1V_1 = P_2V_2$$

Solve for P_2:

$$P_1V_1 / V_2 = P_2$$

$$[(5.00 \text{ atm}) \cdot (15.0 \text{ L})] / 0.500 \text{ L} = P_2$$

$$P_2 = 150 \text{ atm}$$

Charles' Law (isobaric process)

French scientist Jacques Charles (1746-1823) developed the relationship between temperature and gas volume.

Charles' Law states that *at constant pressure, the volume of a gas is directly proportional to its absolute temperature (Kelvin)*. It describes the behavior of a gas under *constant pressure.*

The volume and temperature are directly proportional; when the temperature increases, the volume increases; when the temperature decreases, the volume decreases.

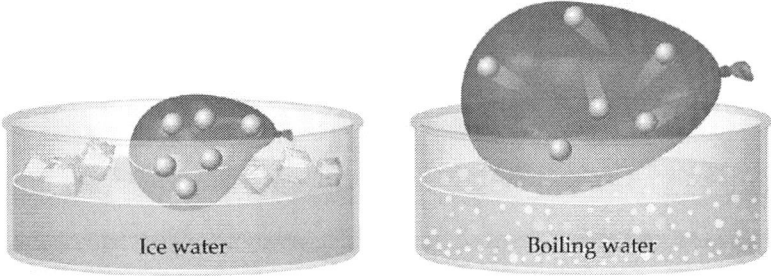

Charles' Law states that the volume of a gas is proportional to temperature

Charles' Law is expressed as:

$$\frac{V_1}{T_1} = \frac{V_2}{T_2}$$

where subscripts 1 and 2 refer to the same gas sample but under different temperature and volume conditions

A plot of *temperature vs. volume* for a gas illustrates that Charles' Law is another application of the ideal gas law, where n and P are constant, as shown.

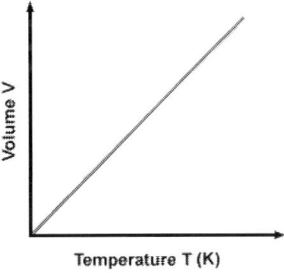

Charles' Law states that as temperature increases, volume increases

For example, if there is 25.0 L of gas at 0 °C, and the temperature is raised to 100 °C, what is the volume of the gas?

First, the temperature is converted to Kelvin.

$$T_1 = 0\ °C + 273 = 273\ K$$

$$T_2 = 100\ °C + 273 = 373\ K$$

Substitute the known quantities into the equation for Charles' Law.

$$V_1 / T_1 = V_2 / T_2$$

Rearrange and solve for V_2:

$$V_1\ T_2 / T_1 = V_2$$

$$(25.0\ L)·(373\ K) / 273\ K = V_2$$

$$V_2 = 34.2\ L$$

Gay-Lussac's Law

In 1802, French chemist Louis Gay-Lussac (1778-1850) proposed *Gay-Lussac's Law* that states that *the pressure of a given sample of gas is directly proportional at a constant volume to its absolute temperature (Kelvin)*.

Gay-Lussac's Law is expressed by:

$$\frac{P_1}{T_1} = \frac{P_2}{T_2}$$

where subscripts 1 and 2 refer to the same gas sample but under different temperature and pressure conditions.

A plot of *temperature vs. pressure* for a gas illustrates that Gay-Lussac's Law is another particular case of the ideal gas law, where *n* and V are constant.

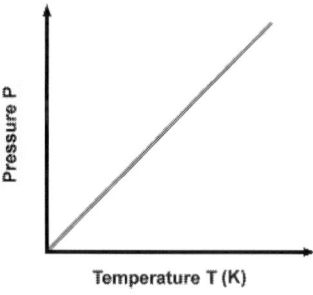

Gay-Lussac's Law states that as temperature increases, pressure increases

For example, suppose a gas is at 30.0 atm pressure and 100 °C, and the temperature changes to 400 °C. What is the new pressure of the gas?

Convert the temperature to units of Kelvin.

$T_1 = 100\ °C + 273 = 373\ K$

$T_2 = 400\ °C + 273 = 673\ K$

Gay-Lussac's Law solves P_2 by substituting the quantities into the equation.

$$\frac{P_1}{T_1} = \frac{P_2}{T_2}$$

$P_2 = [(30.0\ \text{atm}) \cdot (673\ K)] / 373\ K$

$P_2 = 54.1\ \text{atm}$

Closed container law (constant volume)

The *closed container law* describes the behavior of a gas under constant volume. In such cases, pressure and temperature are directly proportional:

$$\frac{P_i}{T_i} = \frac{P_f}{T_f}$$

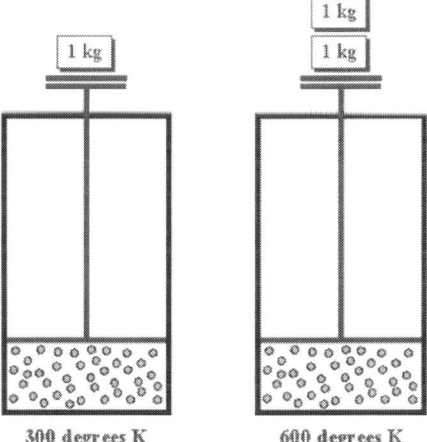

Avogadro's Law

The volume of a gas in a container is affected by pressure, temperature, and moles of gas.

The relationship between the quantity of gas and its volume was derived by Italian scientist Amadeo Avogadro (1776-1856).

Avogadro's Law states that *gases at a given temperature and pressure occupying a volume are directly proportional to the number of moles of gas present.*

Avogadro's Law is expressed as:

$$\frac{n_1}{V_1} = \frac{n_2}{V_2}$$

where n_1 and n_2 are moles of gas 1 and gas 2, and V_1 and V_2 are the volumes of gas 1 and gas 2

A plot of *volume vs. moles* of gas (shown below) illustrates that Avogadro's Law is a particular case of the ideal gas law where T and P are held constant.

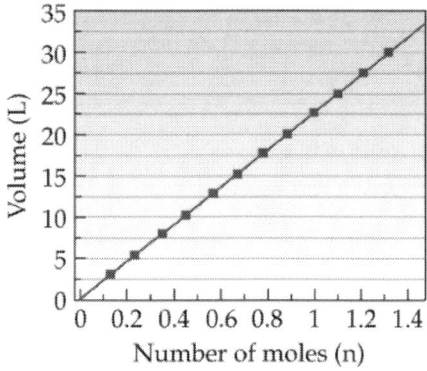

Avogadro's Law states that as moles of a gas increase, volume increases

For example, if 8.00 moles of a gas occupy a volume of 4.00 L at constant pressure and temperature, what volume of gas would 16.0 moles of this gas occupy at the same temperature and pressure?

Use *Avogadro's Law*

$$n_1 / V_1 = n_2 / V_2$$

where n_1 and n_2 are the moles of gas 1 and 2, and V_1 and V_2 are the volumes of gas 1 and gas 2

Solve for the V_2

$$V_2 = (V_1 \times n_2) / n_1$$

$$V_2 = [(4.00 \text{ L}) \cdot (16.0 \text{ mol})] / 8.00 \text{ mol}$$

$$V_2 = 8 \text{ L}$$

Combined gas law

The *combined gas law* is an amalgam of the other gas laws.

The combined gas law is *not* the ideal gas law (PV = nRT); it is expressed as an equation to relate changes in temperature, volume, and gas pressure.

The *combined gas law* is:

$$\frac{P_1 V_1}{T_1} = \frac{P_2 V_2}{T_2}$$

where subscripts 1 and 2 refer to the same gas sample but under different temperatures, pressure, and volume

For example, suppose a gas is at 15.0 atm pressure, with a volume of 25.0 L and a temperature of 300.0 K. What would the volume of the gas be at standard temperature and pressure? (Use standard pressure = 1.00 atm and standard temperature = 273.0 K)

Use the *combined gas law* at standard temperature (T_2) and pressure (P_2):

$$(P_1 V_1) / T_1 = (P_2 V_2) / T_2$$

Solve for V_2:

$$(P_1 \times V_1 \times T_2) / (T_1 \times P_2) = V_2$$

$$V_2 = [(15.0 \text{ atm}) \cdot (25.0 \text{ L}) \cdot (273.0 \text{ k})] / [(300.0 \text{ K}) \cdot (1.00 \text{ atm})]$$

$$V_2 = 341 \text{ L}$$

Molar mass of a gas

The moles of a gas n can be expressed as the mass of gas in grams over the molar mass of gas:

$$n = m / MM$$

where MM = molar mass, m = mass of the gas in grams, and n = moles of gas

Substitute expression (m/MM) into the ideal gas law for moles (n):

$$PV = nRT$$

$$PV = mRT / MM$$

Multiplying each side by the molar mass (MM):

$$(MM)PV = mRT$$

This derived equation determines the molar mass of gas from experimental data, where the gas's mass, pressure, volume, and temperature are measured.

Divide each side of the above expression by the volume (V):

$(MM)P = gRT / V$

Density:

$\rho = g/V$

Substitute density for g/V in the derived equation:

$(MM)P = \rho RT$

This derived equation relates the gas's pressure, density, and temperature, like the other empirical gas laws.

For example, what is the molar mass of a gas with the density (ρ) of 1.855 g/L at 0.950 atm and 297.0 K.? (Use $R = 0.0820$ L·atm/K·mol)

Substitute the quantities into the derived equation above.

$(MM)P = \rho RT$

$(MM) \cdot (0.950 \text{ atm}) = (1.855 \text{ g/L}) \cdot (0.0820 \text{ L·atm/K·mol}) \cdot (297.0 \text{ K})$

Solve for molar mass (MM) by dividing each side of the equation by P (0.95 atm).

$MM = [(1.855 \text{ g/L}) \cdot (0.0820 \text{ L·atm/K·mol}) \cdot (297.0 \text{ K})] / (0.950 \text{ atm})$

$MM = 47.6 \text{ g/mol}$

Deviation of real-gas behavior

The *kinetic molecular theory* and the *ideal gas law* are approximations of gas behavior.

These theories are based on the theoretical ideal gas, while real gases have deviations from this behavior (five critical assumptions of an ideal gas).

When molecules are far apart (i.e., low pressure and high temperature), a real gas behaves more like an ideal gas.

When molecules are brought close (under high pressure and low temperature), gas molecules experience an intermolecular attraction. Therefore, the gas may deviate significantly from the behavior of ideal gas.

The most substantial deviations from the ideal occur at high pressure and low temperature. The gas molecules are "squished" at these conditions and thus experience intermolecular interactions. The molecular volume becomes significant when the total volume is *squished*.

The intermolecular attractions cause collisions to be sticky and inelastic.

Furthermore, gases *condense* into liquids at extremely high pressures and low temperatures.

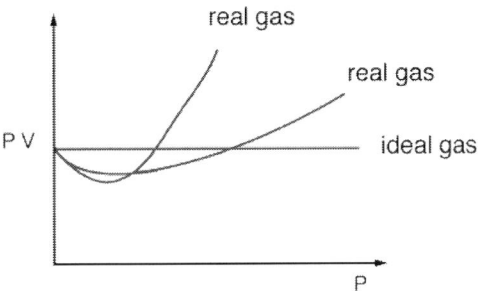

An ideal gas is a *point mass*, a particle so small that the particle's *volume is negligible*.

A real gas particle has a volume because real gases liquefy as they cool and cannot compress to zero volume; the molecules of a gas are not dimensionless points.

The *collisions* between gas particles are "elastic" with *no attractive or repulsive forces* for an ideal gas.

Thus, no energy is exchanged during the collisions.

For a real gas, collisions are *inelastic* (i.e., objects stick and lose kinetic energy).

Van der Waals equation describes deviations from ideal gas

In 1873, Dutch theoretical physicist Johannes Diderik van der Waals (1873-1923) derived a mathematical relationship describing real gases' *behavior* and *condensation to the liquid phase*.

The *ideal gas equation* for comparison:

$$PV = nRT$$

"Corrections" are made to the pressure (P) and volume (V) terms,

where P becomes:

$$P + (n^2 a) / V^2$$

and V becomes:

$$V - nb$$

The *van der Waals equation* is expressed by:

$$[P + (n^2 a) / V^2] \times (V - nb) = nRT$$

Since the collisions of real gases are inelastic, the term $n^2 a / V^2$ corrects the interactions between gas particles.

Since real gas particles occupy space (volume), the *nb* term corrects the volume neglected by the ideal gas law.

The values of *a* and *b* are constant and are determined *experimentally* for each gas. The constants *a* and *b* would be given in a problem, and pressure or temperature can be easily solved.

Solving for the volume is nontrivial and involves solving a cubic polynomial equation.

For pressure, the van der Waals equation is rearranged by dividing by the volume term ($V - nb$):

$$P + (n^2 a) / (V^2) = (nRT) / (V - nb)$$

Subtract the intermolecular interaction term ($n^2 a / V^2$) from both sides of the equation:

$$P = [(nRT) / (V - nb)] - [(n^2 a) / (V^2)]$$

Repulsion   Attraction

For example, use the real gas law to find the pressure of 2.00 moles of carbon dioxide (CO_2) gas at 298 K in a 5.00 L container. (Use van der Waals constants for CO_2 whereby $a = 3.592$ $L^2 \cdot atm/mol^2$ and $b = 0.04267$ L/mol).

Substitute the variables into the equation and solve for the pressure (*P*):

$$P = [(nRT) / (V - nb)] - [(n^2 a) / (V^2)]$$

$$P = [(2\ mol) \cdot (0.0821\ L \cdot atm) \cdot (298\ K)] / [(5\ L) - (2.00\ mol)mol \cdot K \cdot (0.04267\ L)]$$
$$- [(2\ mol)^2 \cdot (3.592\ L^2 \cdot atm) / (5.00\ L)^2 \cdot mol^2]$$

$$P = 9.38\ atm$$

Compare this value (9.38 atm) to the calculated pressure using the ideal gas law:

$$PV = nRT$$

$$P = nRT / V$$

$$P = [(2.00\ mol) \cdot (0.0821\ L \cdot atm) \cdot (298\ K)] / (5.00\ L)mol \cdot K$$

$$P = 9.77\ atm$$

Although the results from the two equations are similar, they are not identical, illustrating the difference in the behavior of a real gas (not obeying the five stated assumptions) from an ideal gas.

Partial Pressure

Mole fraction

There is so much distance between the gas molecules that other gas molecules share the space.

Different gases combine to form homogeneous mixtures.

Each gas behaves *independently* and makes its unique contribution to the total pressure.

The *gas's partial pressure* is as if it were the *only gas* in the container.

The *mole fraction* (X_A) indicates the ratio between moles of a specific gas component and the total number of moles in the homogeneous mixture.

The symbol X_A expresses the component gas's mole fraction in the mixture.

X_A = (moles of gas A) / (total moles of the gas mixture)

Dalton's Law of partial pressures

When two or more nonreactive gases are present in the same container, they behave relatively independently.

Dalton's Law of partial pressures states that *the total pressure of a mixture of gases equals the sum of the individual gas pressures.*

It states that a mixture of gases' total pressure equals the sum of the individual gas components' partial pressures.

Dalton's Law of partial pressures is expressed by:

$$P_T = P_A + P_B + P_C \ldots$$

where P_T is the total pressure in the container, and P_A, P_B and P_C equal the partial pressures of gases *A, B,* and *C*

The partial pressure of a gas is related to its mole fraction by:

$$P_A = X_A P_T$$

where X_A = (moles of gas A) / (total moles of gas)

For example, suppose there is 1.0 L oxygen at 1.0 atm pressure in a container, 1.0 L nitrogen at 0.5 atm pressure in another container, and 1.0 L hydrogen at 3.0 atm pressure in a third container. What is the total pressure if the gases combine in a single 1.0 L container?

 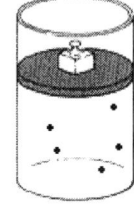 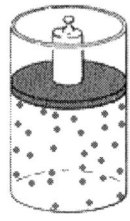

$$P_{O_2} = 1.0 \text{ atm} \qquad P_{N_2} = 0.5 \text{ atm} \qquad P_{H_2} = 3.0 \text{ atm}$$

For a single 1.0 L container, the total pressure is the sum of the partial pressures (X_X) of each gas component:

$$X_X = 1.0 \text{ atm} + 0.5 \text{ atm} + 3.0 \text{ atm} = 4.5 \text{ atm}$$

$$P_{Total} = P_{O_2} + P_{N_2} + P_{H_2}$$

$$P_{Total} = 4.5 \text{ atm}$$

Measuring vapor pressure

A critical application of Dalton's Law of partial pressure is to determine *vapor pressure* (i.e., the pressure exerted by a vapor in thermodynamic equilibrium with its liquid or solid phase).

The vapor pressure determines the amount of a water-insoluble gaseous reaction product or a slightly soluble gas such as hydrogen or oxygen.

A typical experimental method (illustrated below) to determine the amount of gas present is by collecting it over water and measuring the height of displaced water; this is accomplished by placing a tube into an inverted bottle, the opening of which is immersed in a larger container of water. The gas bubbles into the test tube and displace the water until the test tube is full.

The collected gas is not the only gas in the test tube since liquid water is in equilibrium with vapor, so the collected gas is a mixture of two gases: the gas collected and water vapor.

The partial pressure of water is the *vapor pressure of water* and is dependent on temperature.

The *volume* of gas collected consists of a mixture of the *gas collected* and *water vapor*.

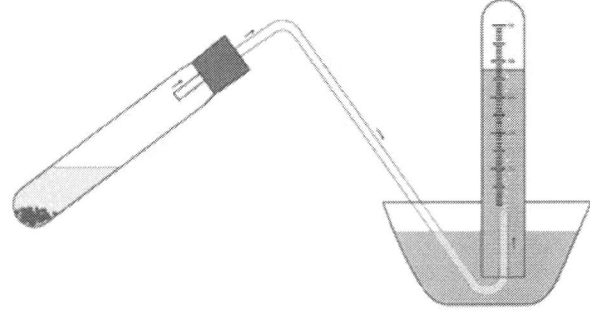

An experimental method to determine the amount of gas present

The total pressure is the sum of the two contributing partial pressures.

The *pressure of dry gas* collected is calculated from a table of water vapor pressure values at various temperatures. Partial pressure of the dry gas is calculated by subtracting the reference vapor pressure of H_2O.

The equalizing of the atmospheric pressure yields the partial pressure of the gaseous product collected. The amount of gas product can be determined with known volume and temperature.

Vapor pressure of water at specific temperatures

Temperature (°C)	Vapor Pressure (kPa)	Temperature (°C)	Vapor Pressure (kPa)
0	0.61	26	3.36
5	0.87	27	3.57
10	1.23	28	3.78
15	1.71	29	4.00
16	1.82	30	4.24
17	1.94	35	5.62
18	2.06	40	7.38
19	2.20	45	9.58
20	2.34	50	12.33
21	2.49	60	19.92
22	2.64	70	31.16
23	2.81	80	47.34
24	2.98	90	70.10
25	3.17	100	101.30

Determining partial pressure

For example, a small piece of zinc (Zn) reacts with dilute hydrochloric acid (HCl) to form hydrogen (H_2) gas, which is collected over water at 16.0 °C. The total pressure is adjusted to a barometric pressure of 100.24 kPa, and the volume of hydrogen gas is measured as 1,495 cm^3. What are the partial pressure and mass of the hydrogen gas? (Use $R = 8.314$ L·kPa·K^{-1}·mol^{-1})

The partial pressure of hydrogen gas:

$P_{HCl} = 100.24$ kPa – 1.82 kPa

$P_{HCl} = 98.42$ kPa

To calculate the mass of the hydrogen gas, use the ideal gas equation (PV = nRT) and solve the number of moles.

For hydrogen (H_2) gas, multiply the mass of one hydrogen atom (1.008 g/mol) by two to obtain the molecular mass of 2.016 g/mol.

$H_2 = 2 \times 1.008$ g/mol

$H_2 = 2.016$ g/mol

Temperature and volume were given.

The partial pressure of hydrogen was calculated earlier; the ideal gas equation can now be used.

Determine the number of moles of H_2 gas:

$PV = nRT$

$n = PV / RT$

$n = [(98.42$ kPa)·(1.495 L)] / [(8.314 L·kPa·K^{-1}·mol^{-1})·(289 K)]

$n = 0.061$ mol of H_2 gas

Calculate the mass of the H_2 gas formed:

mass = (moles of H_2 gas) × (molecular mass of H_2 gas)

mass = 0.061 mol H_2 gas × 2.016 g/mol H_2 gas

mass = 0.123 g of H_2 gas

Intermolecular Forces

Forces between molecules

Intermolecular forces are *attractive or repulsive* forces between molecules (e.g., hydrogen bonding, London dispersion).

However, some other intermolecular interactions are not as strong as covalent or ionic bonds but have a significant role in determining substances' properties.

Dipole interactions

Molecules with dipole moments (from differences in electronegativity of bonded atoms) are attracted to each other. The more polar the molecule is, the stronger the dipole attraction.

Four types of dipole attractions:

- *Ion-dipole attractions* form between an ion and a polar molecule. Since a polar molecule has a slight charge (i.e., dipole) and an ion is a charged atom, these particles are attracted to each other.

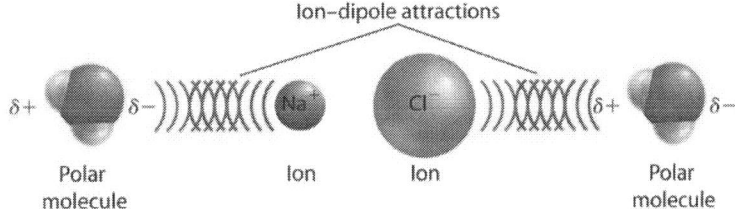

- *Dipole-dipole attraction* forms between two polar molecules. The partially positive side of one polar molecule attracts the partially negative side of another polar molecule.

- *Dipole-induced dipole attraction* forms between partially polar and nonpolar molecules.

  For example, the polar H_2O molecules induce the nonpolar O_2 molecules, shifting electron density and inducing temporary partial positive (δ^+) and negative (δ^-) regions in the nonpolar O_2 molecules.

  positive pole ⊢⟶ negative pole

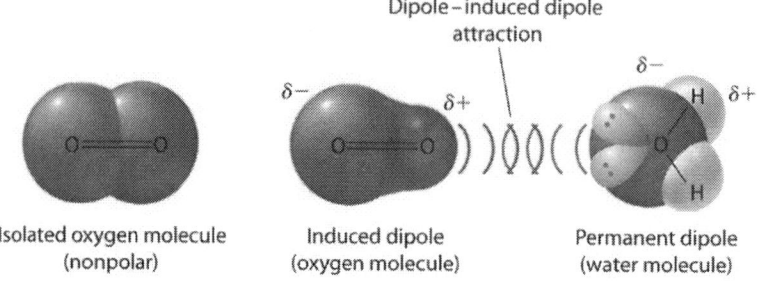

Dipole–induced dipole attraction

Isolated oxygen molecule (nonpolar)

Induced dipole (oxygen molecule)

Permanent dipole (water molecule)

- *Induced dipole-induced dipole* attraction (or *London dispersion force*) forms between two (or more) nonpolar molecules. These nonpolar molecules become temporarily polar (i.e., momentary flux) due to random, instantaneous induction by another polar molecule that results in the probability of finding the negative electron in a specific region within its orbital.

Longer molecules have more *surface contact* and experience higher London dispersion forces.

For example, octane (C_8H_{18}) has a stronger London dispersion force than methane (CH_4).

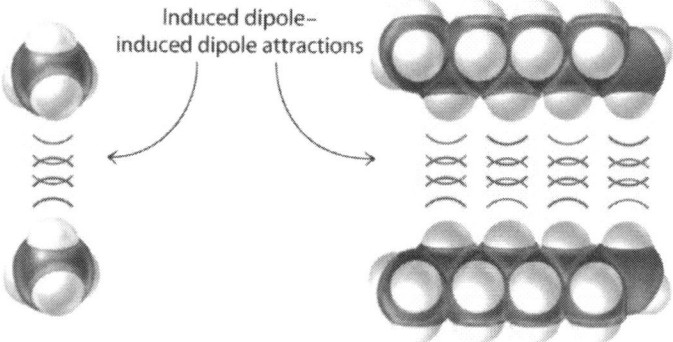

Induced dipole–induced dipole attractions

Molecular attractions involving dipoles

Attraction	Relative Strength
Ion–dipole	Strongest
Dipole–dipole	
Dipole–induced dipole	
Induced dipole–induced dipole	Weakest

Hydrogen bonding

Hydrogen bonding is a specific type of bond that, based on its strength, is categorized as a *dipole-dipole* attraction.

Hydrogen (hydrogen bond donor with one valence electron) bonding occurs when hydrogen is covalently bonded directly to the electronegative atom of N, O, or F (hydrogen bond donors).

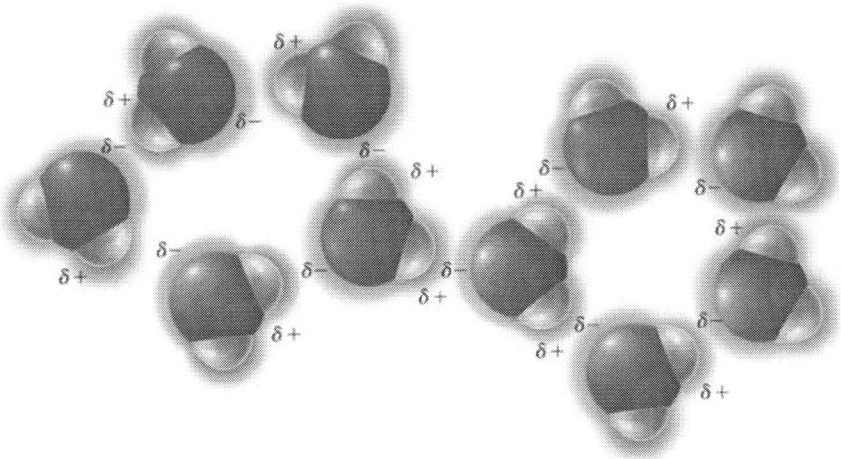

Hydrogen bonding produces differences in physical properties

The bonding electrons are closer to the more electronegative N, O, or F atom.

Due to these factors, the hydrogen nucleus is exposed.

The highly positive hydrogen nucleus attracts the electronegative atom from a neighboring molecule. The neighboring molecule needs a lone electron pair to form a hydrogen bond. This creates a relatively powerful intermolecular force.

The more polar a bond is, the stronger the hydrogen bond.

The H−F bond is most polar, then H−O bonds, and then H−N bonds.

Intermolecular forces in polar molecules

Hydrogen bonding significantly increases the boiling point of compounds.

For example, methanol (CH_3OH) and methane (CH_4) have similar molecular structures.

However, methanol is liquid at room temperature, while methane is a gas.

Methanol is liquid at room temperature because of hydrogen bonding between methanol molecules, which is *much stronger* than the London dispersion (weak intermolecular) forces between the nonpolar methane molecules.

Therefore, methane is a gas at room temperature.

Hydrogen bond donors and acceptors

The molecule bonding (by lone pair electrons) to the hydrogen (bonded to F, O, N) is the hydrogen bond *acceptor*.

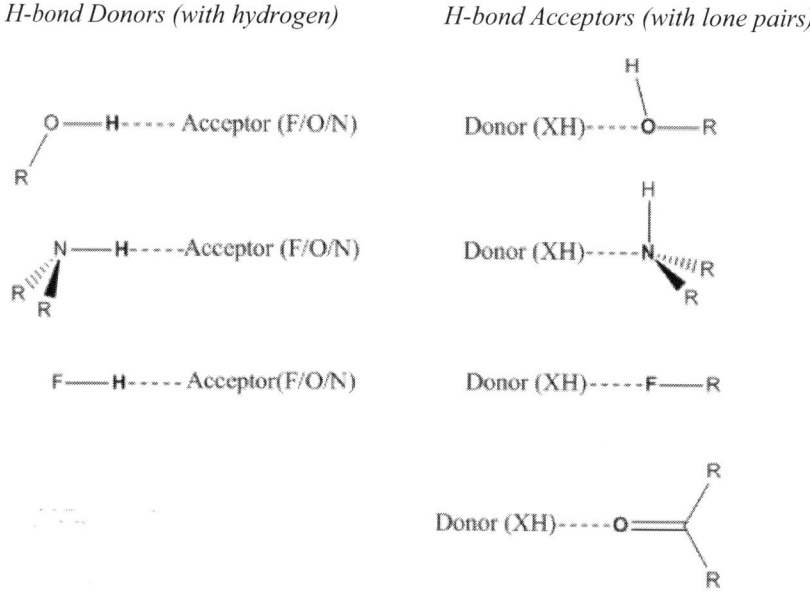

The hydrogen bond donor has a δ^+ H bonded to the electronegative F, O, N.

The hydrogen bond *acceptor* has a *lone pair of electrons* to bond to the δ^+ hydrogen of the donor.

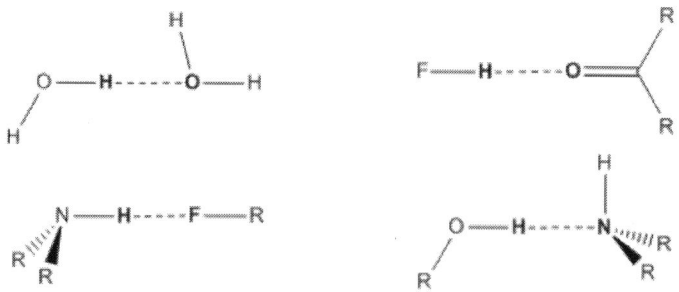

Examples of hydrogen-bonded molecules

Van der Waals' forces

Atoms and molecules exhibit weak intermolecular attractions, known as *van der Waals forces*. This results from a molecule's nuclei's weak attraction to another molecule's valence electrons.

The van der Waals equation (described previously) accounts for these *attractive intermolecular forces*.

The van der Waals force affects the *boiling point* because the boiling point reflects the kinetic energy needed to release a molecule from the liquid's cooperative attractions.

A larger *molecular mass* reduces the kinetic energy ($KE = \frac{1}{2}mv^2$) for the same energy input.

London dispersion forces

Nonpolar molecules have fluctuating dipoles that align with other molecules from one instant to the next.

The *momentary flux of electron density* is attracted to nearby nuclei's positive charge, and this attraction decreases dramatically as the distance between the molecules increases.

London dispersion forces are a subset of Van der Waals' attractive forces resulting from temporarily fluctuating molecules ("induced dipole or induced dipole attraction").

London dispersion forces exist for many molecules but are a significant source of intermolecular forces for nonpolar molecules that do not have dipole-dipole, dipole-induced dipole, or hydrogen bonding.

London dispersion forces are weak compared to the forces generated between polar molecules.

For polar molecules, dipole forces are strong, influential, and predominant.

In general, larger nonpolar molecules have higher boiling points than smaller nonpolar molecules due to increased London dispersion forces.

Branched molecules (i.e., hydrocarbons) have weaker London dispersion forces.

Notes for active learning

Practice Questions

1. Consider the phase diagram for H_2O. The termination of the gas-liquid transition at which distinct or liquid phases do NOT exist is the:

A. critical point

B. endpoint

C. triple point

D. condensation point

E. inflection point

2. When liquids and gases are compared, liquids have [ ] compressibility compared to gases and a [ ] density.

A. lower… lower

B. higher … higher

C. higher… lower

D. lower … higher

E. same … same

3. How does a real gas deviate from an ideal gas?

I. Molecules occupy a significant amount of space

II. Intermolecular forces may exist

III. Pressure is created from molecular collisions with the walls of the container

A. I only

B. II only

C. I and II only

D. II and III only

E. I, II and III

4. Under which conditions does a real gas behave most like an ideal gas?

A. High temperature and high pressure

B. High temperature and low pressure

C. Low temperature and low pressure

D. Low temperature and high pressure

E. None of the above

5. Which of the following compounds has the highest boiling point?

A. CH_3OH

B. $CH_3CH_2CH_2CH_2CH_2OH$

C. $CH_3OCH_2CH_2CH_2CH_3$

D. $CH_3CH_2OCH_2CH_2CH_3$

E. $CH_3CH_2CH_2C(OH)HOH$

6. 15.0 liters of O_2 gas is at a temperature of 23 °C. If the temperature of the gas is raised to 45 °C at constant pressure, the new volume is:

A. 16.1 liters

B. 11.4 liters

C. 8.20 liters

D. 22.8 liters

E. 6.80 liters

7. The boiling point of a liquid is the temperature:

A. where sublimation occurs

B. where the vapor pressure of the liquid equals the atmospheric pressure over the liquid

C. equal to or greater than 100 °C

D. where the rate of sublimation equals evaporation

E. where the vapor pressure of the liquid equals the atmospheric pressure over the liquid

8. Which of the following statements about gases is correct?

A. Formation of homogeneous mixtures, regardless of the nature of non-reacting gas components

B. Relatively long distances between molecules

C. High compressibility

D. No attractive forces between molecules

E. All of the above

9. What is the proportionality relationship between a gas's pressure and volume?

A. directly

B. inversely

C. pressure is raised to the 2nd power

D. pressure raised to the $\sqrt{2}$ power

E. none of the above

10. How does the volume of a fixed sample of gas change if the pressure is doubled?

A. Decreases by a factor of 2

B. Increases by a factor of 4

C. Doubles

D. Remains the same

E. Requires more information

11. What is the ratio of the diffusion rate of O_2 molecules to the diffusion rate of H_2 molecules if six moles of O_2 gas and six moles of H_2 gas are placed in a large vessel, and the gases and vessels are at the same temperature?

    A. 4:1

    B. 1:4

    C. 12:1

    D. 1:1

    E. 2:1

12. Under ideal conditions, which of the following gases is least likely to behave as an ideal gas?

    A. CF_4

    B. CH_3OH

    C. N_2

    D. O_3

    E. CO_2

Notes for active learning

Detailed Explanations

1. A is correct.

At a pressure and temperature corresponding to the *triple point* (point D on the graph) of a substance, the three states (gas, liquid and solid) exist in equilibrium.

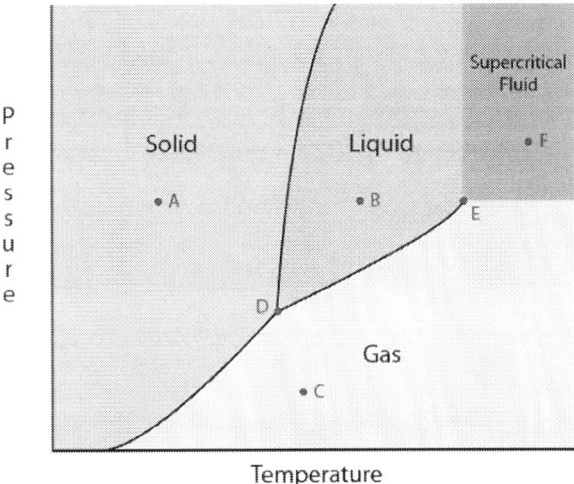

Phase diagram of pressure vs. temperature

The *critical point* (point E on the graph) is the phase equilibrium curve's endpoint, where the liquid and its vapor become indistinguishable.

2. D is correct.

Density = mass / volume

Gas molecules have a large amount of space between them; therefore, they can be pushed together, and thus, gases are very compressible.

There is a large amount of space between molecules in a gas, the extent to which gas molecules can be pushed together is much greater than the extent to which liquid molecules can be forced together.

Therefore, gases have greater *compressibility* than liquids.

Gas molecules are further apart than liquid molecules, so gases have a smaller density.

3. C is correct.

The molecules of an ideal gas *do not occupy a significant amount of space* and exert *no intermolecular forces*.

In contrast, real gas molecules occupy space and exert (weak attractive) intermolecular forces.

An ideal and real gas have pressure created by molecular collisions with the container walls.

4. B is correct.

The molecules of an ideal gas exert *no attractive forces*. Under these conditions, the molecules are far apart and exert little or no attractive forces on each other. Therefore, a real gas behaves nearly like an ideal gas at high temperature and low pressure.

5. E is correct.

Hydroxyl (~OH) groups significantly increase the boiling point because they form hydrogen bonds with ~OH groups of neighboring molecules.

Hydrocarbons are nonpolar molecules, which means that the dominant intermolecular force is London dispersion. This force gets stronger as the number of atoms in each molecule increases, and the stronger force increases the boiling point.

Branching of the hydrocarbon affects the boiling point, and straight molecules have slightly higher boiling points than branched molecules with the same number of atoms. The reason is that straight molecules can align parallel against each other.

All atoms in the molecules participate in London dispersion forces.

Another factor is *heteroatoms* (i.e., atoms other than carbon and hydrogen).

For example, the electronegative oxygen atom between carbon groups or in an ether (C–O–C) slightly increases the boiling point.

6. A is correct.

Charles' Law (or *law of volumes*) explains how, at constant pressure, gases behave when heated.

$$V \, \alpha \, T$$

or

$$V \, / \, T = \text{constant}$$

Convert the initial temperature from Celsius to Kelvin:

$$23 \, °C = 273 + 23 = 296 \, K$$

Convert the initial temperature from Celsius to Kelvin:

45 °C = 273 + 45 = 318 K

Charles' Law:

(V_1 / T_1) = (V_2 / T_2)

Solve for the final volume:

V_2 = (V_1 / T_1) / T_2

V_2 = [(15.0 L) / (296 K)] × (318 K)

V_2 = 16.1 L of O_2

7. B is correct.

Boiling occurs when the vapor pressure of a liquid equals atmospheric pressure.

Vapor pressure is the pressure exerted by a vapor in equilibrium with its condensed phases (i.e., solid or liquid) in a closed system at a given temperature.

Atmospheric pressure is the pressure exerted by the weight of air in the atmosphere.

Vapor pressure is inversely correlated with the strength of the intermolecular force.

The molecules are more likely to stick together in liquid form from stronger intermolecular forces. Fewer molecules participate in the liquid-vapor equilibrium; therefore, the molecule boils at a higher temperature.

8. E is correct.

Gases form *homogeneous* mixtures, regardless of the identities or relative proportions of the component gases.

There is a relatively *large distance* between gas molecules (as opposed to solids or liquids where the molecules are much closer).

When pressure is applied to gas, its volume readily decreases, and thus, gases are highly compressible.

There are no attractive forces between gas molecules, which is why molecules of a gas can move about freely.

9. B is correct.

Boyle's Law (i.e., pressure-volume law) states that *pressure and volume are inversely proportional*:

$$(P_1V_1) = (P_2V_2)$$

or

$$P \times V = \text{constant}$$

If the volume of a gas increases, its pressure decreases proportionally.

10. A is correct.

Boyle's Law (i.e., pressure-volume law) states that *pressure and volume are inversely proportional*:

$$(P_1V_1) = (P_2V_2)$$

or

$$P \times V = \text{constant}$$

If the pressure of a gas increases, its volume decreases proportionally.

Doubling the pressure reduces the volume by half.

11. B is correct.

Graham's Law of Effusion states that the rate of effusion (i.e., escaping through a small hole) of a gas is inversely proportional to the square root of the molar mass of its particles.

$$\text{Rate 1 / Rate 2} = \sqrt{(\text{molar mass gas 1 / molar mass gas 2})}$$

The diffusion rate is the inverse root of the molecular weights of the gases.

Therefore, the rate of effusion is:

$$O_2 / H_2 = \sqrt{(2 / 32)}$$

rate of diffusion = 1 : 4

12. B is correct.

Methanol (CH_3OH) is alcohol that participates in hydrogen bonding.

Therefore, this gas experiences the strongest intermolecular forces.

Notes for active learning

Notes for active learning

CHAPTER 4

Stoichiometry

- Chemical Formulae and Molecular Mass

- Naming Molecular Compounds

- Common Metric Units

- Mole Concept

- Density and Concentration

- Oxidation States

- Chemical Equations

- Redox Equations

- Limiting Reactant

- Practice Questions & Detailed Explanations

Chemical Formulae and Molecular Mass

Chemical formula

Chemical formulas are commonly expressed as molecular and empirical formulas

> The *molecular formula* describes the atomic composition of a molecule in its naturally occurring form; it specifies the number of each type of atom present.

> The *empirical formula* describes the *simplest integer ratio* of atoms.

For example,

> The *molecular formula* of hydrogen peroxide is H_2O_2.

> The *empirical formula* of hydrogen peroxide is HO.

The structure, molecular, and empirical formula of glucose are below.

Molecular Structure	Molecular Formula	Empirical Formula
CHO | H——C——OH | H——C——OH | H——C——OH | H——C——OH | CH_2OH	$C_6H_{12}O_6$	CH_2O

Molecular compounds

A *molecular compound* is an electrically neutral particle consisting of two or more nonmetals covalently bonded.

A covalent bond shares electrons between two atoms during chemical bonding.

Groups of atoms within a molecule behave like single particles or discrete units.

Based on the composition and type of atoms, molecules have a specific molecular weight (MW).

For example, a water molecule consists of two hydrogen atoms bonded to one oxygen atom.

Particle diagrams

A particle diagram has elements and compounds represented as distinctive shapes (e.g., balls shaded or unshaded). Particle diagrams can represent elements and compounds and their molecular composition by the types of shapes and how they are connected.

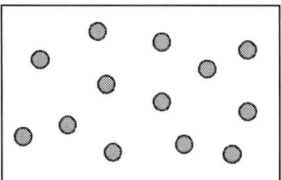 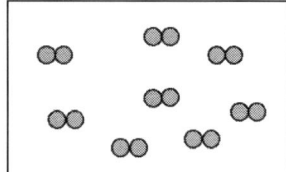

Elements

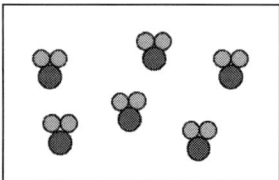 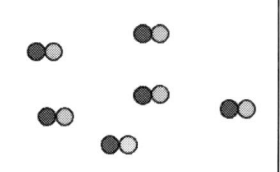

Compounds

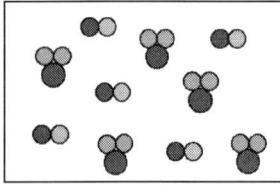 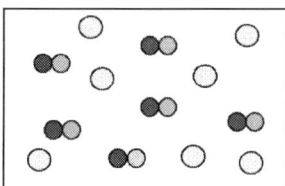 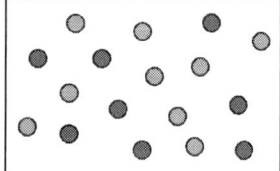

Mixtures

Avogadro's number

Molecular weight (MW) is the weight of 1 mole of molecules, where 1 mole equals 6.02×10^{23} particles.

The 6.02×10^{23} particles value is Avogadro's number (N_A).

The molecular weight of a substance is the sum of the atomic masses of its atoms. Typically, molecular weight is expressed in grams per mole (g/mol).

Molecular weight can be expressed in *atomic mass units* (amu or u), where 1 u = 1 g/mol.

$\qquad$ 1 amu = 1 Dalton (Da).

For example, ^{12}C weighs 12 amu = 12 g/mol.

Atomic mass

The *atomic mass* is the mass of a single atom.

The *molecular weight* is the sum of the weights of the atoms in the molecule.

The *atomic mass* is listed on the periodic table; it is directly below the symbol for the element, as indicated by the arrow for lithium (Li).

Molar mass

The *molar mass* is the mass of a given substance divided by the amount of a substance and is expressed in g/mol.

Atomic mass gives the mass of atoms, while molecular weight gives the mass of molecules.

Formula weight is the sum of weights of the atoms in an empirical formula of a molecule.

For example, the molar mass of methane (molecular formula = CH_4)

CH_4 = 1 carbon (12.0108 g/mol) + 4 hydrogen (1.0079 g/mol)

CH_4 = 12.0108 g/mol + 4(1.0079 g/mol)

CH_4 = 16.0424 g/mol

Molecular mass

The *molecular mass* (m) is the mass of a given molecule, measured in Daltons (Da or u).

The molecular mass is the *ratio of the mass of a molecule to the atomic mass unit*. The molecular mass is more commonly used when referring to the mass of a single molecule.

The molecular mass is often used interchangeably with molecular weight (MW).

Molecules of the same compound have different molecular masses with different isotopes of an element.

Molecular weight

Molecular weight (MW) is the collection of weights of the atoms in a molecular formula.

Molecular weight commonly refers to a weighted average of a sample.

For example, determine the molecular weight of water (H_2O).

Water, from its chemical formula, has two hydrogens and one oxygen.

From the periodic table, the hydrogen (H) atom's mass is 1.008 amu, and oxygen (O) is 15.999 amu.

For example, the equation to determine the MW of water (H_2O):

The chemical formula for water:

2 (H) + 1 (O)

Molecular weight for water:

MW= 2 H (1.008 g/mol) + 1 O (15.999 g/mol)

MW = 18.015 g/mol

Another example of atomic mass is seen in the graphic below.

A single carbon atom is approximately 12 amu (12 g/mol).

Oxygen exists as a diatomic molecule.

O_2 (2 × 15.999 g/mol = 32 g/mol)

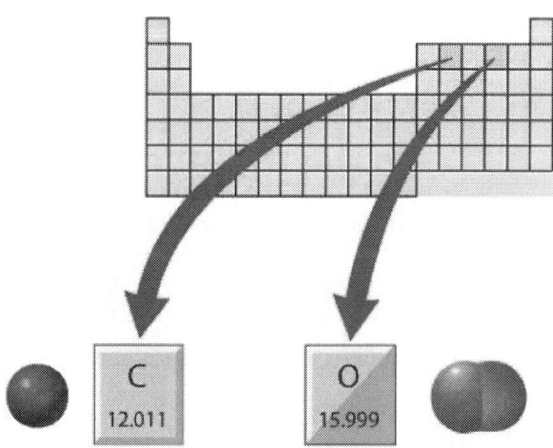

The mass of one carbon atom is approximately 12 amu.

The mass of an O_2 molecule is approximately (16 + 16) 32 amu.

A carbon (C) atom is [(12 g/mol) / (32 g/mol)], or 3/8 as massive as an O_2 molecule.

Naming Molecular Compounds

Naming binary molecular compounds

Binary molecular compounds are covalently bonded compounds composed of two elements, generally nonmetals.

Writing the molecular formula of a binary molecular compound is straightforward using three rules:

1. The first element in the formula is first, using the element's full name.

2. The second element is named as if it were an anion.

3. The number of each element in the compound is indicated by a prefix (table below).

Numerical prefixes for naming compounds

Prefix	Meaning
mono-	1
di-	2
tri-	3
tetra-	4
penta-	5
hexa-	6
hepta-	7
octa-	8

For example, the formula for carbon dioxide is CO_2.

The name indicates a single carbon (no prefix because the *mono-* prefix is omitted when the first element exists as a single atom).

The name indicates two oxygens because of *di-* (2).

Disulfur tetrafluoride has the formula S_2F_4 :

      The *di-* prefix indicates two sulfur atoms.

      The *tetra-* prefix indicates four fluorine atoms.

Naming ionic compounds

An *ionic compound* consists of a metal atom and a nonmetal atom that transfers valence electron(s) and is bonded by electrostatic forces between oppositely charged particles.

Ionic compounds are named differently. Identify the ions (e.g., cations, monatomic, polyatomic anions) for writing a chemical formula from an ionic compound. The inability to recognize these ions is a leading cause of difficulty in writing chemical formulae of inorganic compounds; memorization of the ions is essential.

The cation is listed first and the anion second in an ionic compound (or chemical formula).

For example, sodium chloride (NaCl) forms from a sodium cation (Na^+) and chlorine anion (Cl^-).

Some cations form one ion (Group 1 (IA), 2 (IIA), and 3 (IIIA) elements, except Thallium). Cations forming more than one ion have a stock number, a Roman numeral in parentheses (e.g., Cu (I), Fe (II), or Fe (III)).

For monatomic anions, the charge equals the Group number minus 8.

For example, oxygen has the monatomic anion O^{2-}. Oxygen's group number is 6, so the charge is $6 - 8 = -2$.

Common polyatomic ions

$C_2H_3O_2^-$	acetate	OH^-	hydroxide
NH_4+	ammonium	ClO^-	hypochlorite
CO_3^{2-}	carbonate	NO_3^-	nitrate
ClO_3^-	chlorate	NO_2^-	nitrite
ClO_2^-	chlorite	$C_2O_4^{2-}$	oxalate
CrO_4^{2-}	chromate	ClO_4^-	perchlorate
CN^-	cyanide	MnO_4^-	permanganate
$Cr_2O_7^{2-}$	dichromate	PO_4^{3-}	phosphate
HCO_3^-	bicarbonate	SO_4^{2-}	sulfate
HSO_4^-	bisulfate	SO_3^{2-}	sulfite
HSO_3^-	bisulfite		

Learn the polyatomic anions in the table above

Common Metric Units

Seven base units of SI measurements

The metric system is a decimal system of measurement known as the *International System of Units* (commonly abbreviated as SI).

The SI has seven base units:

> length (meters)
>
> mass (kilograms)
>
> time (seconds)
>
> electric current (amperes)
>
> temperature (Kelvin)
>
> amount of substance (moles)
>
> luminous intensity (candelas)

Other SI units of measurements, such as volume (liters), density (kg/m^3), and pressure (atmospheres), are derived from the base units.

The SI units are based on 10 (or multiples of 10); conversion between the units is uniform and straightforward. The SI system uses prefixes for the magnitude of a measured quantity; the prefix gives the conversion factor.

Common prefixes, symbols, and powers

Learn the common prefixes shown:

Prefix	Symbol	Power	Prefix	Symbol	Power
mega-	M	10^6	centi-	c	10^{-2}
kilo-	k	10^3	milli-	m	10^{-3}
hecto-	h	10^2	micro-	μ	10^{-6}
deca-	D	10^1	nano-	n	10^{-9}
deci-	d	10^{-1}	pico-	p	10^{-12}

The SI units do not allow for double prefixes (e.g., $1,000 \text{ g} \neq 1$ hectodecagram); instead, one prefix is used for the quantity of base units (e.g., $1,000 \text{ g} = 1$ kilogram).

Common conversions

Memorize a few typical metric-imperial conversions:

Length: 2.54 cm = 1 inch

Mass: 454 g = 1 pound

Volume: 0.946 L = 1 quart

Temperature: °C = (°F − 32) / 1.8

Significant figures

Significant figures indicate the number of digits known for a measured (or calculated) quantity within a *degree of uncertainty*.

For example, if a thermometer indicates a boiling point of 36.2 °C and has an uncertainty of ± 0.2 °C, three figures in 36.2 are significant (including the 0.2, which is uncertain).

The notation 36.2 ± 0.2 °C indicates the known quantity and the uncertainty digits (± 0.2 °C).

Converting between units

Conversions between metric units involve adding or subtracting zeros.

For example, suppose a 250 mg aspirin tablet's mass needs to be converted to grams. Start by using the units to set up the problem. If a unit is to be converted (in this case, mg), it is placed in the numerator. That unit must be in the denominator of the conversion factor for it to cancel:

$(250 \text{ mg}/1) \cdot (1 \times 10^{-3} \text{ g}/1 \text{ mg})$

$(250 \text{ mg}/1) \cdot (1 \times 10^{-3} \text{ g}/1 \text{ mg}) = 0.250 \text{ g}$

Units cancel to give grams.

The conversion factor is shown as a numerator of 1×10^{-3} as it must be entered on most calculators, not 10^{-3}.

The mg is a value of 1, and the prefix "*milli-*" applies to the gram unit. 1 mg = 1×10^{-3} g.

Conversions between English/imperial and metric units work similarly.

The difference is the conversion factors; they are not powers of ten and are different for each unit, making conversions between English/imperial and metric units more complex.

Therefore, the imperial system is not ideal for scientific calculations.

Converting with multiple units

For example, convert the mass of a 23 lbs. object to kilograms?

Pound units convert to grams, and grams are converted to kilograms.

Use the proper units to set up the problem:

23 lbs. = (23 lbs. / 1) × (454 g/ 1 lbs.) × (1 kg/1 × 10^3 g)

23 lbs. = (23 ~~lbs.~~ / 1) × (454 ~~g~~/ 1 ~~lbs.~~) × (1 kg/1 × 10^3 ~~g~~)

23 lbs. = 10 kg

A conversion problem may include multiple units.

For example, convert the pressure 14 lb/in^2 to g/cm^2.

For problems involving multiple conversions, work with one unit at a time.

Convert the pounds to gram units:

14 lb/in^2 × 454 g/lb

Convert the in^2 unit to cm^2 units.

Set up the conversion without the exponent first, using the conversion factor:

1 in = 2.54 cm

Since in^2 and cm^2 are needed, raise each value to the second power:

= 14 lb/in^2 × 454 g/lb × 1^2 in^2/2.54 cm^2

= 9.9 × 10^2 g/cm^2

When the units are squared, the numbers associated with them must be squared.

For 1 in^2, it does not make a difference because of $1^2 = 1$.

Square the quantities (numbers) when the units are squared.

Check the units to verify that the problem has been set up correctly.

Percent mass

Percent mass (% mass) is concentration *comparing the mass of one part of a substance to the mass of the whole.*

The percent mass unit is used for solutions, especially concentrated acid, or base solution, where the concentration is expressed as percent by mass on the bottle.

Percent mass is:

% mass = (mass of species of interest / total mass) × 100%

For example, find the % mass of sodium (Na) in sodium bicarbonate ($NaHCO_3$).

Use the periodic table to find the atomic mass of the elements.

Na = 22.99 g/mol

H = 1.01 g/mol

C = 12.01 g/mol

O = 16.00 g/mol

Use the molecular mass of the atoms to calculate the molecular weight of $NaHCO_3$:

MW $NaHCO_3$ = 1 Na (22.99 g/mol) + 1 H (1.01 g/mol) + 1 C (12.01 g/mol) + 3 O (3 × 16.00 g/mol)

MW $NaHCO_3$ = 84.01 g/mol

Find the percent mass of Na:

% mass Na = (mass of species of interest / total mass) × 100%

% mass Na = (22.99 g/mol / 84.01 g/mol) × 100%

% mass Na = 27.4%

Concentration can be expressed as % volume, calculated like % mass.

Mole Concept

Stoichiometry

Stoichiometry uses quantitative relationships between products and reactants.

The number of molecules is expressed in the mole unit.

Avogadro's number (N_A) is the number of particles in 1 mole of a substance and equals 6.02×10^{23} particles.

The substance's mass and moles are determined for reactants and products for a reaction.

$\underline{2\ H_2}$	$+$	$\underline{1\ O_2}$	$\rightarrow$	$\underline{2\ H_2O}$
2 moles = 4 g		1 mole = 32 g		2 moles = 36 g
12.04×10^{23} molecules		6.02×10^{23} molecules		12.04×10^{23} molecules

Most stoichiometry problems follow a set strategy using moles:

$$\text{Quantity A} \rightarrow \text{Moles A} \rightarrow \text{Moles B} \rightarrow \text{Quantity B}$$

Many stoichiometry problems are solved using this strategy.

Each step is examined and combined to solve complicated problems.

Moles to moles conversion

For example, how many moles of $CaCO_3$ are in a 25.0 g sample?

Calculate the molar mass of $CaCO_3$ using atomic mass information from the periodic table.

Calculate the molar mass with the same significant figures as the quantity converted:

1 Ca = 40.08 g/mol × 1 = 40.08 g/mol

1 C = 12.01 g/mol × 1 = 12.01 g/mol

3 O = 16.00 g/mol × 3 = 48.00 g/mol

$CaCO_3$ = 40.08 g/mol + 12.01 g/mol + 48.00 g/mol

$CaCO_3$ = 100.09 = 100.1 g/mol

Use the molar mass to convert the 25.0 g mass of $CaCO_3$ to moles $CaCO_3$.

$CaCO_3$ = 25.0 g × (1 mol $CaCO_3$ / 100.09 g $CaCO_3$)

$CaCO_3$ = 0.250 mol

Moles to grams conversion

For example, what is the number of grams of $CaCO_3$ in 0.750 mol $CaCO_3$? (Use molecular mass of $CaCO_3$ = 100.09)

$CaCO_3$ = 0.750 mol × (100.09 g $CaCO_3$ / 1 mol $CaCO_3$)

$CaCO_3$ = 75.1 g $CaCO_3$

Moles A to moles B conversion

For example, how many moles of sodium ion (Na^+) does 0.100 mol of sodium carbonate (Na_2CO_3) have?

One mole of sodium carbonate Na_2CO_3 contains 2 moles of Na, 1 mole of C, and 3 moles of O. Na_2CO_3 completely dissociates (i.e., electrolyte) into ions in a solution so the sodium atoms are ions. Compare the moles of Na_2CO_3 and moles of Na^+:

= 0.100 mol Na_2CO_3 × (2 mol Na^+ / 1 mol Na_2CO_3)

= 0.200 mol Na^+

Moles of reactants to moles of products conversion

A reaction may be involved in other examples, and a comparison of two compounds is necessary.

The stoichiometric coefficient is the mole ratio between species for a balanced reaction.

For example, a typical dissociation reaction:

$2 KClO_3 \rightarrow 2 KCl + 3 O_2$

In this reaction, 2 moles of potassium chlorate ($KClO_3$) decompose into 2 moles of potassium chloride (KCl) and 3 moles of oxygen (O_2).

How many O_2 molecules are produced during the dissociation reaction above?

If there are 0.400 mol of $KClO_3$ and the number of moles of O_2 is needed, use stoichiometric coefficients to arrange a mole ratio, so moles of $KClO_3$ cancel, and moles of O_2 remain:

= (0.400 mol $KClO_3$) × (3 mol O_2 / 2 mol $KClO_3$)

= 0.600 mol O_2

Avogadro's number converts the number of moles to the number of particles.

= 0.600 mol O_2 × (6.02 × 10^{23} molecules O_2 / 1 mol O_2)

= 3.61 × 10^{23} O_2 molecules

The mole units of O_2 successfully cancel in the answer.

Density and Concentration

Density and specific gravity

Density (ρ) is the ratio of mass over the volume of a substance.

The SI unit of density is kg/m^3.

Specific gravity is a unitless ratio between the density of a sample and a *reference substance* (often water).

Some units of density:

- Density of water = 1 g/mL = 1 g/cm^3

- Specific gravity of water = 1 g/cm^3 / 1 g/cm^3 = 1

- Density of lead = 11 g/cm^3

- Specific gravity of lead = 11 g/cm^3 / 1 g/cm^3 = 11

The densest known element is osmium at 22.59 g/cm^3.

The least dense known element is hydrogen at 8.99 × 10^{-5} g/cm^3.

Hydrogen is about 250,000 times less dense than osmium.

Molarity and molality

Molarity (*M*) is the *moles of solute per liter of solution*.

$$\text{Molarity} = \text{moles of solute / L of solution}$$

Molarity is the most common concentration unit used for solutions.

The molality (*m*) is the moles of solute per kilogram of solvent (Phase Equilibria chapter).

$$\text{Molality} = \text{moles of solute / kg of solvent}$$

Concentration of dilute solutions

Very dilute solutions use the unit of *parts per million* (ppm).

$$\text{ppm} = \frac{\text{grams of solute}}{\text{grams of solution}} \times 10^6$$

The amount of solute relative to the amount of solvent is typically minimal.

Therefore, the density of the solution is, to a first approximation, the same as the density of the solvent.

Parts per million may be expressed as:

$$ppm = \frac{mg\ solute}{kg\ solution}$$

If the solvent is water (density of 1.00 kg/L), the expression for ppm:

$$ppm = \frac{mg\ solute}{L\ solution}$$

Another measurement unit to express the concentration of more dilute solutions is parts per billion (ppb).

The expression for *parts per billion*:

$$ppb = \frac{grams\ of\ solute}{grams\ of\ solution} \times 10^9$$

The density of a dilute solution approximates the solvent's density.

Thus, parts per billion may be expressed as:

$$ppb = \frac{\mu g\ solute}{kg\ solution}$$

$$ppb = \frac{\mu g\ solute}{L\ solution}$$

Mole fraction and mole percent

The *mole fraction* (χ) is the ratio between the moles of a specific molecule over the total moles of components in the mixture.

The *mole percent* (%χ) is the mole ratio multiplied by 100.

$$\chi_{solute} = \frac{mol\ solute}{total\ moles\ of\ all\ components}$$

$$\chi_{solute}\ \% = \frac{mol\ solute}{total\ moles\ of\ all\ components} \times 100$$

When converting between units, start by choosing an arbitrary amount of solution in the denominator of the concentration to be converted.

Percent mass to molarity

For example, if converting percent mass to molarity, assume 100 grams of solution.

If converting molarity to percent mass, assume one liter of solution.

For example, what are the molarity, molality, and mole fraction of HCl for a concentrated HCl solution known to be 37.0% HCl by mass, and its density is 1.19 g/ml?

Begin with the valid assumption that the HCl solution is 100 g.

Since the HCl percent mass is 37%, 37.0 g of the solution is HCl (grams of solute), the remaining 63.0 g is water (grams of solvent).

To find molarity, determine the moles of HCl (solute) per liter of solution.

First, convert the mass of HCl to moles:

$$\text{mol HCl} = 37.0 \text{ g HCl} \times 1 \text{ mol HCl} / 36.5 \text{ g HCl}$$

$$\text{mol HCl} = 1.01 \text{ mol HCl}$$

Using the solution's density, convert the known mass of the solution (100 g) to liters of solution.

$$\text{L solution} = 100 \text{ g solution} \times \frac{1 \text{ mL solution}}{1.19 \text{ g solution}} \times \frac{1 \text{ L solution}}{1000 \text{ mL solution}}$$

$$\text{L solution} = 100 \, \cancel{\text{g solution}} \times \frac{1 \, \cancel{\text{mL solution}}}{1.19 \, \cancel{\text{g solution}}} \times \frac{1 \text{ L solution}}{1000 \, \cancel{\text{mL solution}}}$$

$$\text{L solution} = 0.0840 \text{ L solution}$$

Using the moles of solute (HCl) and volume of solution in liters, calculate molarity (M) of the solution as moles of solute per liter of solution:

$$M = \frac{1.01 \text{ mol HCl}}{0.0849 \text{ L solution}}$$

$$M = 12.0 \; M \text{ HCl/L solution}$$

Molality to mole fraction

From above, find the molality of the HCl solution.

The moles of solute are known (1.01 mol HCl).

Determine the mass of the solvent (H_2O) in kilograms:

$$63.0 \text{ g } H_2O \times (1 \text{ kg } H_2O \text{ / } 1,000 \text{ g } H_2O) = 0.0630 \text{ kg } H_2O$$

Using the moles and the mass of solvent, calculate molality (m or b):

$$m = 1.01 \text{ mol HCl / } 0.0630 \text{ kg } H_2O$$

$$m = 16.0 \text{ mol HCl/kg } H_2O$$

$$m = 16.0 \text{ m } H_2O$$

Finally, determine the mole fraction of HCl.

From before, there are 1.01 moles of HCl. Calculate moles of H_2O.

$$\text{mol } H_2O = 63.0 \text{ g } H_2O \times (1 \text{ mol } H_2O \text{ / } 18.0 \text{ g } H_2O)$$

$$\text{mol } H_2O = 63.0 \; \cancel{\text{g } H_2O} \times (1 \text{ mol } H_2O \text{ / } 18.0 \; \cancel{\text{g } H_2O})$$

$$\text{mol } H_2O = 63.0 \times (1 \text{ mol } H_2O \text{ / } 18.0)$$

$$\text{mol } H_2O = 3.50 \text{ mol}$$

The moles of the molecules in the problem are known.

Calculate the mole fraction of HCl:

$$\chi_{solute} = \frac{\text{mol solute}}{\text{total moles of all components}}$$

$$\chi_{HCl} = 1.01 \text{ mol HCl / } [(1.01 \text{ mol HCl} + 3.50 \text{ mol } H_2O)]$$

$$\chi_{HCl} = 0.244$$

Oxidation States

Oxidation number

The *oxidation number* (or *oxidation state*) is the charge an atom has or appears to have when a set of rules counts the compound's electrons.

Oxidation numbers are typically used in ionic compounds for oxidation-reduction reactions, which involve the transfer of electrons.

For nonionic compounds, oxidation numbers determine if the compound's chemical formula is written correctly.

Assigning oxidation numbers

1. The oxidation state of an element is zero.

2. For main group metals (Groups 1-2, 13-18), the ion's charge equals the valence electrons or group number.

   Transition metals do not obey this rule and may have multiple oxidation states.

   Oxidation states are sometimes not by group number

3. The oxygen ion in a compound is typically -2 except the peroxide ion (O_2^{2-}), which is -1.

4. The sum of oxidation states equals the polyatomic ion's (or neutral molecule's) overall charge for polyatomic ions (or compounds).

Determining oxidation states

For example, determine the oxidation states of the elements in potassium permanganate ($KMnO_4$).

Solution: apply rule two.

Since potassium is a Group IA alkali metal, its oxidation state is $+1$.

Assign x to Mn for now since manganese may exist in several oxidation states.

There are four oxygen atoms in the permanganate ion, with oxidation states of -2 per O atom.

The overall charge of the neutral compound equals zero:

K   Mn O_4

+1 x 4(–2)

The algebraic expression is:

$1 + x - 8 = 0$

Solving for x gives the oxidation state of manganese:

$x - 7 = 0$

$x = +7$

K   Mn O_4

+1   +7   4(–2)

For example, what is the oxidation state of chromium in dichromate ion $Cr_2O_7^{2-}$?

Start by assigning –2 as the oxidation state for oxygen.

Since the oxidation state for chromium is not known, and two chromium atoms are present, assign the algebraic value of $2x$ for chromium:

Cr_2 $O_7{}^{2-}$

$2x$ 7(– 2)

Use the algebraic equation to solve for x.

Since the overall charge of the ion is –2, the expression is set equal to –2, rather than 0:

$2x + 7(-2) = -2$

Solve for x:

$2x - 14 = -2$

$2x = 12$

$x = +6$

Each chromium in the ion has an oxidation state of +6.

What are the oxidation states of the elements in the polyatomic compound $Fe_2(CO_3)_3$?

Two elements (iron and carbon) have more than one oxidation state.

When considering molecules formed from cations (positive ions) and anions (negative ions), the molecule is split into constituent ions. Determine the charge on each ion.

For example, iron (Fe) has more than one oxidation state, but carbonate ion has an oxidation state of -2 (CO_3^{2-}).

With this information, Fe's oxidation state can be determined:

$$Fe_2 \quad (CO_3)^3$$

$$2x \quad 3(-2)$$

$$2x - 6 = 0$$

$$2x = 6$$

$$x = 3$$

Each iron ion in the compound has an oxidation state of $+3$.

Next, consider the carbonate ion independent of the iron (III) ion:

$$CO_3^{2-}$$

$$x \quad 3(-2)$$

$$x - 6 = -2$$

$$x = +4$$

The oxidation state of carbon is $+4$, and each oxygen is -2.

Oxidation-reduction reactions

An *oxidation-reduction reaction* occurs when a transfer of electrons occurs between two chemicals.

For an oxidation-reduction (or redox) reaction to proceed, one substance in a reaction is oxidized while another substance is reduced; reduction or oxidation processes cannot occur separately.

Oxidation is a loss of electrons, which is typically associated with metals and may involve the addition of oxygen or the removal of hydrogen.

Reduction is a gain of electrons, which is typically observed in nonmetals and may involve the addition of hydrogen or removal of oxygen.

In biological systems, cells oxidize and reduce metals. Cytochrome c protein plays a vital role in ATP production. The cytochrome c protein contains a Fe^{2+} cation that undergoes oxidation to form Fe^{3+}, then reduced into Fe^{2+}.

In organic reactions, look for the movement of oxygen and hydrogen.

Combustion (i.e., burning) is a reaction with oxygen as an example of a redox reaction.

Oxidizing and reducing agents

An *oxidizing agent* (or *oxidant*) is a substance that oxidizes another substance by removing electrons.

The oxidizing agent gains the removed electrons, thus reducing itself.

Thus, the oxidation number of the oxidizing agent becomes *less positive*.

A *reducing agent* is a substance that reduces another by donating its electrons to the substance being reduced.

The reducing agent loses these electrons and is oxidized.

Thus, the reducing agent's oxidation number becomes *more positive* (or less negative).

Identifying reduced and oxidized species

The number line below helps identify the reduced or oxidized substances and whether they are oxidizing or reducing agents, respectively.

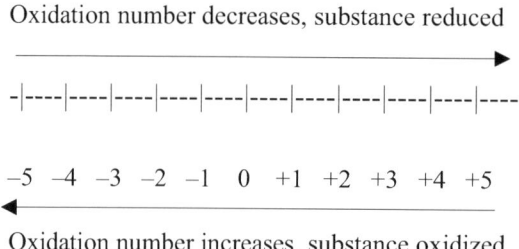

The image below summarizes the relationship between reducing agents and oxidizing agents.

Essentially, atoms that *lose electrons* during a chemical reaction undergo oxidation, and atoms that *gain electrons* during a chemical reaction undergo reduction.

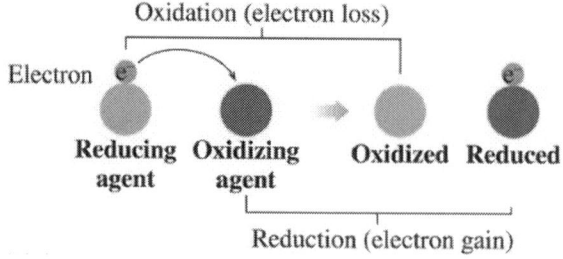

The oxidized *species* is the reducing *agent*, and the reduced *species* is the oxidizing *agent*.

Loss and gain of electrons

The OIL RIG mnemonic helps remember what is happening to the electrons in a redox reaction.

OIL RIG = **O**xidation **I**s **L**oss; **R**eduction **I**s **G**ain

Remember, this mnemonic device refers to the *loss and gain of electrons*.

For example. identify the substance being oxidized, the substance being reduced, the oxidizing agent, and the reducing agent in the following reaction:

$$Cu\ (s) + 4\ HNO_3\ (aq) \rightarrow Cu(NO_3)_2\ (aq) + 2\ NO_2\ (g) + 2\ H_2O\ (l)$$

Determine the oxidation state for the reactants and products.

Elemental copper, Cu (*s*), has an oxidation state of zero (0).

For nitric acid (HNO_3) set N equal to *x*:

H   N O_3

+1   *x*   3(–2)

$1 + x - 6 = 0$

$x = +5$

so,

H = +1, N = +5, O = –2

For copper (II) nitrate, $Cu(NO_3)_2$:

$Cu(NO_3)_2$

Nitrates are –1, while Cu might be +1 or +2.

$Cu(NO_3)_2$

Cu $(NO_3)_2$

$x + 2(-1) = 0$

$x = 2$

So,

Cu's oxidation state is +2, Oxygen is –2

For nitrate ion (NO_3^-) set N equal to *x*:

N O_3^-

x   3(–2)

$$x - 6 = -1$$

$$x = +5$$

So, $Cu = +2$, $N = +5$, $O = -2$

For nitrogen dioxide (NO_2) set N equal to x:

N O_2

x $2(-2)$

$x - 4 = 0$

$x = +4$

So, $N = +4$, $O = -2$

For water (H_2O):

H_2 O

$2(+1)$ -2

So, $H = +1$, $O = -2$

The reaction is summarized in the following table.

$Cu\ (s) + 4\ HNO_3\ (aq) \rightarrow Cu(NO_3)_2\ (aq) + 2\ NO_2\ (g) + 2\ H_2O\ (l)$

Element	Oxidation State: reactants	Oxidation State: products
Cu	0	+2
H	+1	+1
N	+5	+5 (in NO_3^-), +4 (in NO_2)
O	−2	−2

Oxidation states of reactants and products for the example reaction above

Identify the molecules that changed their oxidation state.

Copper's oxidation number increased from 0 to +2, so copper has been oxidized (loses electrons), and therefore, is the reducing agent.

Copper reduced nitric acid by donating electrons to the nitrogen in nitric acid.

Nitrogen in nitric acid changed from +5 to +4 in nitrogen dioxide, which means that copper has reduced nitric acid. Since nitric acid is reduced, it is the oxidizing agent. Oxidation cannot occur without a reduction.

Common oxidizing and reducing agents

Common oxidizing agents	Common reducing agents
Oxygen O_2	Hydrogen H_2
Ozone O_3	metals (such as K)
Permanganates MnO_4^-	Zn/HCl
Chromates $CrO4_2^-$	Sn/HCl
Dichromates $Cr_2O_7^{2-}$	LAH (lithium aluminum hydride)
peroxides H_2O_2	$NaBH_4$ (sodium borohydride)
Lewis acids	Lewis bases
compounds with many oxygens	compounds with many hydrogens

Common oxidizing and reducing oxidation agents

Disproportionation reactions

In *disproportionation reactions*, an atom undergoes oxidation and reduction to form two atoms with different oxidation states.

For example, consider:

$$2\ Cu^+ \rightarrow Cu + Cu^{2+}$$

The Cu^+ acts as oxidizing and reducing agents and simultaneously reduces and oxidizes itself.

The oxidized Cu^+ becomes Cu^{2+}

The reduced Cu^+ becomes Cu

For example, which of the following equations is a disproportionation reaction?

$$2\ H_2O \rightarrow 2\ H_2 + O_2$$

$$H_2SO_3 \rightarrow H_2O + SO_2$$

$$HNO_2 \rightarrow NO + HNO_3$$

$$Mg + H_2SO_4 \rightarrow MgSO_4 + H_2$$

Disproportionation is a redox reaction in which a species is simultaneously reduced and oxidized to form two different products.

Consider: $\qquad\qquad$ $HNO_2 \rightarrow NO + HNO_3$

Unbalanced reaction: $\qquad$ $HNO_2 \rightarrow NO + HNO_3$

Balanced reaction: $\qquad$ $3\ HNO_2 \rightarrow 2\ NO + HNO_3 + H_2O$

The other reactions are classified as:

$\qquad$ $2\ H_2O \rightarrow 2\ H_2 + O_2$: decomposition

$\qquad$ $H_2SO_3 \rightarrow H_2O + SO_2$: decomposition

$\qquad$ $Mg + H_2SO_4 \rightarrow MgSO_4 + H_2$: single replacement

Redox titration

Titration (or *volumetric analysis*) is a standard laboratory procedure to determine an *unknown concentration*.

The *titrant* is the analytical reagent of known concentration carefully added to the sample from a buret until the *equivalence point* (i.e., the point of neutralization) is reached.

If the titrant's volume and concentration are known, its number of moles can be calculated.

The *endpoint* is marked by an indicator's color change when a solution transitions from acidic to slightly basic.

An *indicator* is added to the reaction to determine the endpoint of a redox titration (i.e., when sample molecules have been depleted).

A *redox indicator* undergoes a color change at the neutralization reaction's equivalence point.

For example, ascorbic acid (vitamin C) titrations use iodine undergoing a redox reaction.

Ascorbic acid is *oxidized* (i.e., loses electrons) to dehydroascorbic acid, while iodine is *reduced* (i.e., gains electrons) to an iodide ion.

Starch is an *indicator* because excess iodine reacts with starch and turns the solution from clear to dark blue.

The ascorbic acid concentration is calculated using the initial volume of ascorbic acid (analyte) and iodine's measured volume (titrant).

Chemical Equations

Classifying chemical reactions

Predicting the products formed in each reaction is no easy task.

Since many chemical compounds exist, memorizing every reaction is not necessary.

Most chemical reactions are classified into several groups.

Identifying the reaction type is a significant step towards predicting a chemical reaction's products.

The important classes of reactions encountered in general chemistry are shown:

Reaction Type	General Reaction
Synthesis	$A + B \rightarrow AB$
Decomposition	$AB \rightarrow A + B$
Replacement	$AB + C \rightarrow AC + B$ (single) $AB + CD \rightarrow AD + CB$ (double)

1. ***Combination reactions*** (or *synthesis reactions*): two (or more) substances form a single product.

    The general form is $A + B \rightarrow AB$, in which two substances combine to form one compound.

2. ***Decomposition reactions***: one reactant forms two or more products, the reverse of a combination reaction.

    The general form is $AB \rightarrow A + B$, compound decomposes to form constituent elements.

3. ***Replacement reactions*** (*substitution reactions* or *displacement reactions*): one atom (or group) replaces another species in a compound.

    o   In a ***single replacement reaction***, an element replaces the corresponding element.

    The general form is $AB + C \rightarrow AC + B$, which replaces element B.

o In a ***double replacement reaction*** (or *metathesis reaction*), two compounds react and switch partners; cations and anions in one compound exchange with counterparts in the other compound.

The general form is $AB + CD \rightarrow AD + CB$

Note: In each type of replacement reaction, the number of substances on the reactant side of the equation is the *same* as on the product side.

Specific reaction types

- **Oxidation-reduction reactions** (or *redox*) *transfer electrons* between the two reactants.

  A *loss of electrons* (or increase in oxidation number) is oxidation.

  A *gain of electrons* (or decrease in oxidation number) is reduction.

  o Many combination and decomposition reactions use oxidation/reduction and single displacement reactions.

- ***Combustion*** involves a carbon-containing molecule (e.g., a hydrocarbon) reacting with oxygen to produce carbon dioxide (CO_2) and water (H_2O).

  o Combustion reactions tend to be highly exothermic and are considered irreversible.

  o Combustion is decomposition and oxidation-reduction reactions.

  o Combustion uses O_2 to produce energy, dissipating as heat or performing work.

  o General formula for combustion of a hydrocarbon:

  $$C_xH_y + O_2 \rightarrow CO_2 + H_2O$$

- ***Condensation*** involves water produced from a double replacement reaction.

  Condensation reactions occur among functional groups that contain ~H and ~OH that break from their compounds and form H_2O.

Adenosine triphosphate (ATP)

Adenosine diphosphate (ADP) + Energy

Hydrolysis (addition of water) for ATP conversion to ADP + energy (inorganic phosphate)

- *Hydrolysis* (hydro- means "water," lysis means "break") is the reverse process of condensation, where water is a reactant, while the other reactant molecule is split into two smaller product molecules.

The general form for *condensation and hydrolysis* is shown below.

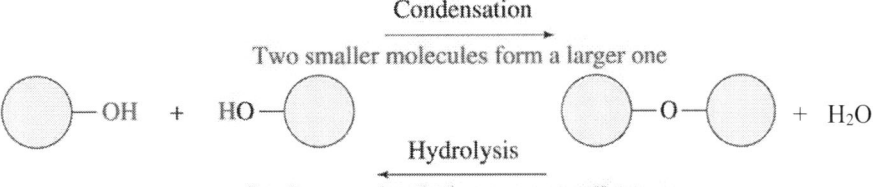

Condensation
Two smaller molecules form a larger one

Hydrolysis
One large molecule forms two smaller ones

- *Carboxylation reactions* add a carboxyl (C=O) or carboxylic acid (~COOH) group.

  For example, as carbon dioxide (CO_2) moves through the cell, it is added to biomolecules by carboxylase and removed by decarboxylase (*~ase* for enzymes).

Carboxylation is shown below.

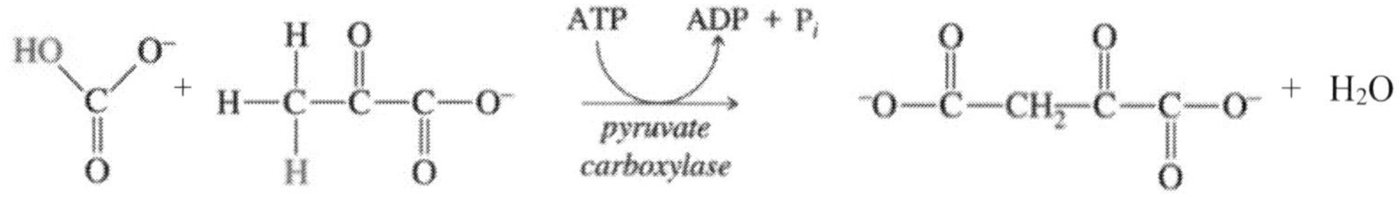

bicarbonate   pyruvate   oxaloacetate   water

Conventions for writing chemical equations

Chemical equations have several important parts:

$$H_2SO_4\ (aq) + 2\ NaOH\ (aq) \rightarrow 2\ Na^+\ (aq) + SO_4^{2-}\ (aq) + 2\ H_2O$$

phase   *coefficient*   *direction*   *charge*

The *phase* is indicated with a subscript, solid (*s*), liquid (*l*), gas (*g*), or aqueous (*aq*).

The *coefficient* indicates the relative number of moles of reactants or products.

The *direction* is represented as a single-headed arrow, denoting the forward direction.

Reversible reactions are at a state of *chemical equilibrium* and represented with a double-headed arrow.

At equilibrium, the *rates* of the forward and reverse reactions are equal and constant, and no change in the net number of reactants or products.

A double-sided arrow with one side *larger* denotes a *nonequilibrium condition*, which spontaneously favors the larger arrow's direction.

The *charge* is indicated with a numerical superscript and +/– sign.

It is common not to indicate the charge on a neutral compound or substance.

Balancing chemical equations

The *law of mass conservation* (refer to the chapter on thermodynamics) states that atoms are neither created nor destroyed in a chemical reaction—they are simply rearranged.

The reactants and products coefficients must balance; there are *equivalent amounts of atoms* on each side of the reaction.

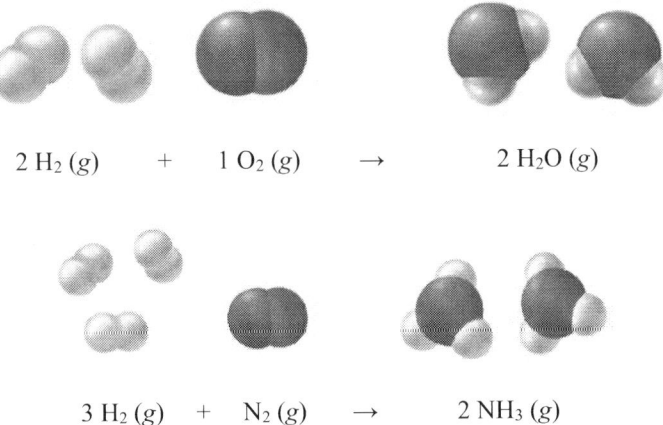

$$2 \, H_2 \, (g) \quad + \quad 1 \, O_2 \, (g) \quad \rightarrow \quad 2 \, H_2O \, (g)$$

$$3 \, H_2 \, (g) \quad + \quad N_2 \, (g) \quad \rightarrow \quad 2 \, NH_3 \, (g)$$

Balance a chemical equation by assigning coefficients to each species until atoms balance on both sides.

For example, balance the combustion of propanol (C_3H_8O):

$$C_3H_8O + O_2 \rightarrow CO_2 + H_2O$$

Select the atom (or ion) present in one species on each side of the equation.

For the combustion of propanol, start with carbon.

Add the coefficient 3 to CO_2 on the right side since there are three carbons on the left side, indicated by C_3.

$$C_3H_8O + O_2 \rightarrow \mathbf{3} \, CO_2 + H_2O$$

Hydrogen is the other species present in one molecule on each side so balance the hydrogens.

Add the coefficient 4 to H_2O on the right because $4 \times 2 = 8$ equals the number of H on the left, indicated by H_8.

$$C_3H_8O + O_2 \rightarrow 3 \, CO_2 + \mathbf{4} \, H_2O$$

Starting with hydrogen before balancing carbon yields the same result.

First, balance elements present in one species on each side to avoid changing coefficients previously balanced species.

Oxygen is present every term, so it would be more difficult if O were balanced first.

Because of the adjustments to balance other atoms, one oxygen-containing species that has not yet been balanced.

Count the oxygen atoms:

one from C_3H_8O

six from $3\ CO_2$ (because $3 \times 2 = 6$)

four from $4\ H_2O$

Set equation:

$1 + 2x = (3 \times 2) + 4$, where x is the coefficient of the last term, O_2.

Solve for x, which equals $^9/_2$.

$C_3H_8O + x\ O_2 \rightarrow 3\ CO_2 + 4\ H_2O$

$C_3H_8O + {}^9/_2\ O_2 \rightarrow 3\ CO_2 + 4\ H_2O$

Remove fractions, so multiply every term by 2.

$2\ C_3H_8O + 9\ O_2 \rightarrow 6\ CO_2 + 8\ H_2O$

Remove non-integer coefficients, and the equation is balanced.

Redox Equations

Balancing redox reactions

When solving oxidation-reduction equations, a different approach is required.

Balancing redox reactions is more complicated, splitting the redox reaction into half-reactions.

Two methods for balancing half-reactions:

> The ***ion-electron method*** balances the elements, then the charge by adding electrons.

> The ***oxidation-state method*** treats the species of interest as a single element (i.e., changes oxidation number) and balances it.

Half-reactions balance oxidation-reduction reactions

Use the following procedure when balancing oxidation-reduction reactions:

1. Separate the equation into two *half-reactions*. A half-reaction contains only the species of interest (i.e., those containing the atom that changes oxidation state).

Each half-reaction corresponds to the oxidation half-reaction or reduction half-reaction.

Covalently attached moieties are not part of the species of interest.

A species that does not change its oxidation state is a *spectator ion* (i.e., an ion present in solution but not involved in the reaction of interest).

2. Balance each half-reaction for the *charge* and *number* of atoms.

Under *acidic conditions*: add H_2O to the side needing the oxygen atom, then add H^+ to the other.

Under *basic conditions*: add 2 OH^- to the side needing oxygen atoms, then add H_2O to the other.

3. After the half-reactions are balanced, recombine the half-reactions:

Multiply each half-reaction by a factor, so the electrons cancel.

Like solving a simultaneous equation, the electron term must be eliminated.

4. Lastly, perform these additional steps:

Combine identical species on the same side of the equation.

Cancel identical species on opposite sides of the equation.

Add back in the spectator ions.

For the oxidation-state method, balance the oxygen and hydrogen elements.

Check that each side of the equation has an equal number of atoms and a neutral net charge.

For example, how many electrons are needed to balance the following half-reaction in a basic solution?

$$C_8H_{10} \rightarrow C_8H_4O_4^{2-}$$

Rules for balancing a half-reaction in basic conditions:

The first few steps are identical to balancing reactions in acidic conditions.

Step 1: Balance atoms except for H and O

$$C_8H_{10} \rightarrow C_8H_4O_4^{2-} \qquad \text{C is balanced}$$

Step 2: To balance oxygen, add H_2O to the side with fewer oxygen atoms

$$C_8H_{10} + 4\ H_2O \rightarrow C_8H_4O_4^{2-}$$

Step 3: To balance hydrogen, add H^+ to the opposing side of H_2O added in the previous step

$$C_8H_{10} + 4\ H_2O \rightarrow C_8H_4O_4^{2-} + 14\ H^+$$

This next step is the unique additional step for basic conditions.

Step 4: Add equal amounts of OH^- on both sides. The number of OH^- should match the number of H^+ ions. Combine H^+ and OH^- on the same side to form H_2O. If there are H_2O molecules on each side, subtract accordingly to end up with H_2O on one side only.

There are 14 H^+ ions on the right, so add 14 OH^- ions on both sides:

$$C_8H_{10} + 4\ H_2O + 14\ OH^- \rightarrow C_8H_4O_4^{2-} + 14\ H^+ + 14\ OH^-$$

Combine H^+ and OH^- ions to form H_2O:

$$C_8H_{10} + 4\ H_2O + 14\ OH^- \rightarrow C_8H_4O_4^{2-} + 14\ H_2O$$

H_2O molecules are on both sides, which cancel, and some H_2O remain on one side:

$$C_8H_{10} + 14\ OH^- \rightarrow C_8H_4O_4^{2-} + 10\ H_2O$$

Step 5: Balance charges by adding electrons to the side with a greater/more positive total charge

Total charge on left side: $14(-1) = -14$

Total charge on the right side: -2

Add 12 electrons to the right side:

$$C_8H_{10} + 14\ OH^- \rightarrow C_8H_4O_4^{2-} + 10\ H_2O + 12\ e^-$$

For example, how many electrons are needed to balance the charge for the following half-reaction in an acidic solution?

$$C_2H_6O \rightarrow HC_2H_3O_2$$

Balancing half-reaction in acidic conditions:

Step 1: Balance atoms except for H and O

$$C_2H_6O \rightarrow HC_2H_3O_2 \text{ (C is balanced)}$$

Step 2: To balance oxygen, add H₂O to the side with fewer oxygen atoms

$$C_2H_6O + H_2O \rightarrow HC_2H_3O_2$$

Step 3: To balance hydrogen, add H⁺ to the opposing side of H₂O added in the previous step

$$C_2H_6O + H_2O \rightarrow HC_2H_3O_2 + 4\ H^+$$

Step 4: Balance charges by adding electrons to the side with a greater/more positive total charge

Total charge on the left side: 0

Total charge on right side: $4(+1) = +4$

Add 4 electrons to the right side:

$$C_2H_6O + H_2O \rightarrow HC_2H_3O_2 + 4\ H^+ + 4\ e^-$$

The reaction requires adding 4 electrons to the right side.

Ion-electron method for balancing redox reactions

For example, balance the following redox reaction using the ion-electron method:

$$K_2Cr_2O_7\ (aq) + HCl\ (aq) \rightarrow KCl\ (aq) + CrCl_3\ (aq) + H_2O\ (l) + Cl_2\ (g)$$

Step 1 – Separate into half-reactions:

Reduction: $Cr_2O_7^{2-} \rightarrow Cr^{3+}$

Oxidation: $Cl^- \rightarrow Cl_2$

The species of interest for the oxidation reaction is Cl^- (not HCl) because the H^+ is not covalently attached to Cl^- and the ions separate in an aqueous solution.

$Cr_2O_7^{2-}$ is used, not $K_2Cr_2O_7$, because K^+ and H^+ are spectator ions.

Step 2 – Balance each of half-reactions:

For the ion-electron method, balance the elements, then balance the charge.

Balance elements for the reduction half-reaction (ion-electron method):

$$Cr_2O_7^{2-} \rightarrow Cr^{3+}$$

$$Cr_2O_7^{2-} \rightarrow 2Cr^{3+}$$

$$Cr_2O_7^{2-} + 14\ H^+ \rightarrow 2\ Cr^{3+} + 7\ H_2O$$

Balance charge for the reduction half-reaction (ion-electron method):

$$Cr_2O_7^{2-} + 14\ H^+ + 6\ e^- \rightarrow 2\ Cr^{3+} + 7\ H_2O$$

Balance charge for the oxidation half-reaction (ion-electron method):

$$Cl^- \rightarrow Cl_2$$

$$2\ Cl^- \rightarrow Cl_2$$

$$2\ Cl^- \rightarrow Cl_2 + 2\ e^-$$

Step 3 – Recombine the half-reactions:

$$Cr_2O_7^{2-} + 14\ H^+ + 6\ e^- \rightarrow 2\ Cr^{3+} + 7\ H_2O$$

$$2\ Cl^- \rightarrow Cl_2 + 2\ e^-$$

Multiply each species in the second equation by 3:

$$6\ Cl^- \rightarrow 3\ Cl_2 + 6\ e^-$$

Add the two equations:

$$Cr_2O_7^{2-} + 14\ H^+ + 6\ e^- + 6\ Cl^- \rightarrow 2\ Cr_3+ + 7\ H_2O + 3\ Cl_2 + 6\ e^-$$

Step 4 – Complete the process:

Except for electrons, there are no identical species to combine or cancel.

$$Cr_2O_7^{2-} + 14\ H^+ + 6\ Cl^- \rightarrow 2\ Cr^{3+} + 7\ H_2O + 3\ Cl_2$$

Step 5 – For the ion-electron method, the equation is balanced. The spectator ions need to be added to the equation.

When adding the spectator ions, add equal numbers of ions to each side of the reaction.

To the left side: the dichromate ion was paired with K^+, so add 2 K^+ for the dichromate.

To the right side: match the left side by adding 2 K^+.

$$K_2Cr_2O_7 + 14\ H^+ + 6\ Cl^- \rightarrow 2\ Cr^{3+} + 7\ H_2O + 3\ Cl_2 + \textbf{2\ K}^+$$

Step 6 – There are 14 H^+ on the left side and 14 on the right, so they are balanced. Referring to the original equation, the H and Cl elements on the left originated from the HCl.

Add 8 Cl elements to the product side.

$$K_2Cr_2O_7 + 14\ HCl \rightarrow 2\ Cr^{3+} + 7\ H_2O + 3\ Cl_2 + 2\ K^+ + \textbf{8\ Cl}^-$$

Step 7 – The right side shows two Cl^- to be combined with the 2 K^+, and the remaining 6 Cl^- goes with the Cr.

The final balanced redox equation is:

$$K_2Cr_2O_7\ (aq) + 14\ HCl\ (aq) \rightarrow \textbf{2}\ \textbf{CrCl}_3\ (aq) + 7\ H_2O\ (l) + 3\ Cl_2\ (g) + \textbf{2\ KCl}\ (aq)$$

Oxidation-state method for balancing redox reactions

For example, balance the redox reaction (from above) using the oxidation-state method:

$$K_2Cr_2O_7\ (aq) + HCl\ (aq) \rightarrow KCl\ (aq) + CrCl_3\ (aq) + H_2O\ (l) + Cl_2\ (g)$$

Step 1 - Separate into half-reactions (same as the ion-electron method):

      Reduction: $Cr_2O_7{}^{2-} \rightarrow Cr^{3+}$

      Oxidation: $Cl^- \rightarrow Cl_2$

Step 2 - Balance each half-reaction:

Balance the elements of interest first when using the oxidation-state method.

Balance the elements for the reduction half-reaction (oxidation-state method):

1. $Cr_2O_7{}^{2-} \rightarrow Cr^{3+}$

2. $Cr_2O_7{}^{2-} \rightarrow 2\ Cr^{3+}$

3.   Each oxygen is 2^- so the 2 Cr on the left must be 6^+

4. $2\ Cr^{6+} \rightarrow 2\ Cr^{3+}$

Balance charge for the reduction half-reaction (oxidation-state method):

  1. $2\,Cr^{6+} + 6\,e^- \rightarrow 2\,Cr^{3+}$

Balance charge for the oxidation half-reaction (oxidation-state method):

  1. $Cl^- \rightarrow Cl_2$

  2. $2\,Cl^- \rightarrow Cl_2$

  3. $2\,Cl^- \rightarrow 2\,Cl^\circ$

  4. $2\,Cl^- \rightarrow 2\,Cl^\circ + 2\,e^-$

Step 3 - Recombine the half-reactions:

  $2\,Cr^{6+} + 6\,e^- \rightarrow 2\,Cr^{3+}$

  $2\,Cl^- \rightarrow 2\,Cl^\circ + 2\,e^-$

To cancel the electrons, multiply each term in the second equation by 3:

  $2\,Cr^{6+} + 6\,e^- \rightarrow 2\,Cr^{3+}$

  $6\,Cl^- \rightarrow 6\,Cl^\circ + 6\,e^-$

Add the two equations:

  $2\,Cr^{6+} + 6\,e^- + 6\,Cl^- \rightarrow 2\,Cr^{3+} + 6\,Cl^\circ + 6\,e^-$

Step 4 - Except for the electrons, no like terms combine or cancel.

  $2\,Cr^{6+} + 6\,Cl^- \rightarrow 2\,Cr^{3+} + 6\,Cl^\circ$

Convert the elements into species by referring to the original equation.

  $K_2Cr_2O_7 + 6\,HCl \rightarrow 2\,CrCl_3 + 3\,Cl_2$

Unlike the ion-electron method, where the equation is balanced, and spectator ions are added, the oxidation-state method requires balancing the equation again. After the elements are combined to recreate the molecules, the equation is no longer balanced.

Step 5 - Oxygen: there are seven O atoms on the left, so add 7 H_2O molecules to the right.

(Remember that this method applies to acidic reactions – see explanation for basic reactions).

  $K_2Cr_2O_7 + 6\,HCl \rightarrow 2\,CrCl_3 + 3\,Cl_2 + \mathbf{7\,H_2O}$

Step 6 - Hydrogen: there are 6 H atoms on the left but 14 H atoms on the right.

Eight H atoms should be added to the left for 14 H atoms.

14 H atoms on the left should be HCl (refer to the original equation).

$$K_2Cr_2O_7 + \textbf{14 HCl} \rightarrow 2\ CrCl_3 + 3\ Cl_2 + 7\ H_2O$$

Important: HCl is the species of interest *and* spectator ion.

Some HCl contributes to the $Cl^- \rightarrow Cl_2$ oxidation, but other HCl does not undergo redox; it provides the H^+ ions for water and Cl^- ions for the KCl and $CrCl_3$.

Step 7 - Chlorine: 14 Cl atoms on the left and 12 Cl atoms on the right.

Add 2 Cl atoms to the right. From the original equation, the right-sided Cl atoms come in KCl; do not modify the Cl_2 since it has already been balanced by the oxidation-state method.

When balancing equations at this stage, only manipulate the water and spectator species.

$$K_2Cr_2O_7 + 14\ HCl \rightarrow 2\ CrCl_3 + 3\ Cl_2 + 7\ H_2O + \textbf{2 KCl}$$

The balanced redox equation is:

$$K_2Cr_2O_7\ (aq) + 14\ HCl\ (aq) \rightarrow 2\ CrCl_3\ (aq) + 3\ Cl_2\ (g) + 7\ H_2O\ (l) + 2\ KCl\ (aq)$$

Ten steps for balancing redox half-reactions

1. Split the equation into two half-reactions.

2. Balance atoms other than O or H.

3. Balance O using H_2O. Add water as needed to balance oxygens on the side deficient in O.

4. If in an acidic solution, balance hydrogens using H^+.

    Add H^+ ions as needed to the side deficient in H.

5. In basic solution, add OH^- ions equal to the number of H^+ ions to each side of the chemical equation.

6. The mass is balanced. To balance charge, determine the charge on each side of the reaction and add needed electrons to the positive side, so the charge is the same.

7. Repeat steps 1 through 4 for each half-reaction.

8. If electrons lost do not equal the number gained, multiply each half-reaction by the necessary factor.

9. Sum half-reactions to obtain the balanced net ionic reaction. Inspect H^+ ions and H_2O molecules to cancel.

10. Check the final equation for mass and charge balance.

Important: If expressing H^+ as H_3O^+, change the number of H^+ into the same number of H_3O^+.

Add that same number of H_2O to the opposite side of the equation.

Limiting Reactant

Identifying limiting reactants

The *limiting reactant* is a component in the lowest stoichiometric quantity, limiting product formed.

The limiting reactant is the reagent depleted first and terminates the reaction.

For example, identify the limiting reactant when a 50.6 g sample of magnesium hydroxide $Mg(OH)_2$ reacts with 45.0 g of hydrogen chloride HCl, according to the reaction below:

$$Mg(OH)_2 + 2\ HCl \rightarrow MgCl_2 + 2\ H_2O$$

Notice that quantities of each reactant are known.

There are three common methods of solving limiting reagent problems.

Moles to determine the limiting reactant

To identify the limiting reagent, *convert grams of reactants to moles* using molar mass:

$$50.6\ g\ Mg(OH)_2 \times (1\ mol\ Mg(OH)_2\ /\ 58.3\ g\ Mg(OH)_2) = 0.868\ mol\ Mg(OH)_2$$

$$45.0\ g\ HCl \times (1\ mol\ HCl\ /\ 36.5\ g\ HCl) = 1.23\ mol\ HCl$$

Select one reactant and calculate the moles of the other reactant needed to deplete the reactant.

For example, begin with magnesium hydroxide ($Mg(OH)_2$):

$$0.868\ mol\ Mg(OH)_2 \times (2\ mol\ HCl\ needed/1\ mol\ Mg(OH)_2) = 1.74\ mol\ HCl\ needed$$

Compare the moles of HCl needed to the actual moles of HCl available.

In this example, 1.74 moles of HCl are needed, and 1.23 moles of HCl are present.

Even though HCl has more moles than $Mg(OH)_2$, HCl is the limiting reagent.

The HCl is consumed before the $Mg(OH)_2$, limiting the product formed.

Theoretical to actual ratio to determine the limiting reactant

The *theoretical yield* is the amount of product expected from a reaction.

Compare the *theoretical ratio* to the *actual ratio* of reactants.

Determine the moles of each reactant using molar mass:

$$50.6\ g\ Mg(OH)_2 \times (1\ mol\ Mg(OH)_2\ /\ 58.3\ g\ Mg(OH)_2) = 0.868\ mol\ Mg(OH)_2\ available$$

$$45.0\ g\ HCl \times (1\ mol\ HCl\ /\ 36.5\ g\ HCl) = 1.23\ mol\ HCl\ available$$

For example, consider the balanced reaction:

$$Mg(OH)_2 + 2\ HCl \rightarrow MgCl_2 + 2\ H_2O$$

From the balanced equation, the theoretical mole ratio is:

2 moles of HCl needed / 1 mol $Mg(OH)_2$

The actual mole ratio, based on the amounts of reactants present:

1.23 mol HCl / 0.868 $Mg(OH)_2$ = 1.42 mol HCl present / 1 mol $Mg(OH)_2$

There is not enough HCl from these ratios, so HCl is the limiting reagent.

2 mol HCl needed / 1 mol $Mg(OH)_2$ *vs.* **1.42 mol HCl present** / 1 mol $Mg(OH)_2$

Comparing theoretical yield to determine the limiting reactant

Calculate each *reactants' theoretical yield* and choose the lesser as the limiting reagent.

$$\textit{theoretical yield} = 50.6\ g\ Mg(OH)_2 \times \frac{1\ mol\ Mg(OH)_2}{58.3\ g\ Mg(OH)_2} \times \frac{1\ mol\ MgCl_2}{1\ mol\ Mg(OH)_2} \times \frac{95.3\ g\ MgCl_2}{1\ mol\ MgCl_2}$$

theoretical yield = 82.7 g $MgCl_2$

$$\textit{theoretical yield} = 45.0\ g\ HCl \times \frac{1\ mol\ HCl}{36.5\ g\ HCl} \times \frac{1\ mol\ MgCl_2}{2\ mol\ HCl} \times \frac{95.3\ g\ MgCl_2}{1\ mol\ MgCl_2}$$

theoretical yield = 58.6 g $MgCl_2$

HCl produced less product (i.e., limiting reagent); 58.6 g $MgCl_2$ as theoretical yield.

Theoretical yield and percent yield

Reactions often yield less product than predicted for several reasons (e.g., equipment inefficiency or human error).

The *percent yield* is the experimental (actual) ratio to theoretical yields.

percent yield = (experimental yield / theoretical yield) × 100%

For example, if the reaction of 30.0 grams of calcium carbonate ($CaCO_3$) produces 15.0 grams of calcium oxide (CaO), what is the percent yield for the following reaction?

$$CaCO_3 \rightarrow CaO + CO_2$$

The *experimental yield* was 15.0 g, so the theoretical yield must be determined.

For every mole of $CaCO_3$ reacted, one mole of CaO was produced.

From the periodic table, the molar mass of $CaCO_3$ is 100.0896, and the molar mass of CaO is 56.0774.

Use these conversion factors to perform the stoichiometric calculations and determine the theoretical yield.

theoretical yield = 30.0 g CaCo₃ × (1 mol $CaCO_3$ / 100.0869 g) × (1 mol CaO / 1 mol $CaCO_3$)

× (56.0774 g CaO / 1 mol CaO)

theoretical yield = 16.8 g CaO

Use the formula above to calculate the *percent yield*.

% yield = (actual yield / theoretical yield) × 100%

% yield = (15.0 g CaO / 16.8 g CaO) × 100%

% yield = 89.3%

Notes for active learning

Practice Questions

1. Which substance listed is the strongest reducing agent, given the following spontaneous redox reaction?

$$\text{Mg } (s) + \text{Sn}^{2+} (aq) \rightarrow \text{Mg}^{2+} (aq) + \text{Sn } (s)$$

A. Sn

B. Mg^{2+}

C. Sn^{2+}

D. Mg

E. None of the above

2. Which of the following is a guideline for balancing redox equations by the oxidation number method?

A. Verify that the total number of atoms and the ionic charge is the same for reactants and products

B. In front of the substance reduced, place a coefficient corresponding to the number of electrons lost by the substance oxidized

C. In front of the substance oxidized, place a coefficient corresponding to the number of electrons gained by the substance reduced

D. Determine the electrons lost by the substance oxidized and gained by the substance reduced

E. All of the above

3. Which of the following represents the oxidation of Co^{2+}?

A. $\text{Co} \rightarrow \text{Co}^{2+} + 2 \text{ e}^-$

B. $\text{Co}^{3+} + \text{e}^- \rightarrow \text{Co}^{2+}$

C. $\text{Co}^{2+} + 2 \text{ e}^- \rightarrow \text{Co}$

D. $\text{Co}^{2+} \rightarrow \text{Co}^{3+} + \text{e}^-$

E. $\text{Co}^{3+} + 2 \text{ e}^- \rightarrow \text{Co}^+$

4. What is the molecular formula of a compound with an empirical formula of CHCl with a molar mass of 194 g/mol?

A. $\text{C}_4\text{H}_4\text{Cl}_4$

B. $\text{C}_2\text{H}_4\text{Cl}_3$

C. $\text{C}_3\text{H}_5\text{Cl}_3$

D. CHCl

E. $\text{C}_4\text{H}_6\text{Cl}_4$

5. What is the term for the amount of substance that contains 6.02×10^{23} particles?

A. molar mass

B. mole

C. Avogadro's number

D. formula mass

E. none of the above

6. How many grams are in 0.7 moles of $CaCO_3$?

A. 25 g   **C.** 70 g
B. 40 g   **D.** 48 g
                                                     E. 57 g

7. What is the coefficient for CO_2 in the balanced reaction?

$$\_\_C_5H_{12} + \_\_O_2 \rightarrow \_\_CO_2 + \_\_H_2O$$

A. 5   **C.** 8
B. 7   **D.** 10
                                                     E. 12

8. What are the products for this double-replacement reaction?

$$BaCl_2\,(aq) + K_2SO_4\,(aq) \rightarrow$$

A. $BaSO_3$ and $KClO_4$ **C.** BaS and $KClO_4$
B. $BaSO_4$ and 2 KCl   **D.** $BaSO_3$ and KCl
                                                     E. $BaSO_4$ and $KClO_4$

9. Which coefficients balance the following equation: $\_\_P_4\,(s) + \_\_H_2\,(g) \rightarrow \_\_PH_3\,(g)$?

A. 2, 10, 8   **C.** 1, 6, 4
B. 1, 4, 4   **D.** 4, 2, 3
                                                     E. 1, 3, 4

10. After balancing the following redox reaction in acidic solution, what is the coefficient of H^+?

$$Mg\,(s) + NO_3^-\,(aq) \rightarrow Mg^{2+}\,(aq) + NO_2\,(aq)$$

A. 1   **C.** 4
B. 2   **D.** 6
                                                     E. None of the above

11. How many atoms are in a sample of phosphorus trifluoride (PF_3) that contains 1.40 moles?

A. 3.37×10^{24} **C.** 3.46
B. 5.38   **D.** 2.218×10^{24}
                                                     E. 8.98×10^{23}

12. Which of the following is the percent mass composition of acetic acid (CH_3COOH)?

    A. 48% carbon, 8% hydrogen, and 44% oxygen

    B. 52% carbon, 12% hydrogen, and 36% oxygen

    C. 32% carbon, 6% hydrogen, and 62% oxygen

    D. 40% carbon, 7% hydrogen, and 53% oxygen

    E. 34% carbon, 4% hydrogen and 62% oxygen

13. If one mole of Ag is produced in the following reaction, how many grams of O_2 gas is produced?

$$2\ Ag_2O \rightarrow 4\ Ag + O_2$$

A. 6 g	**C.** 2 g
B. 8 g	**D.** 12 g
	E. 5 g

14. Which metal in the free state has an oxidation number of zero?

A. Mg	**C.** Al
B. Na	**D.** Ag
	E. All of the above

15. What is the oxidation number of Cr in $K_2Cr_2O_7$?

A. +6	**C.** +4
B. +5	**D.** +2
	E. +1

Notes for active learning

Detailed Explanations

1. D is correct.

Use the mnemonic OIL RIG: <u>O</u>xidation <u>I</u>s <u>L</u>oss, <u>R</u>eduction <u>I</u>s <u>G</u>ain (of electrons).

Oxidation is the loss of electrons, while reduction is the gain of electrons.

An oxidizing agent undergoes reduction, while a reducing agent undergoes oxidation.

Because the reaction is spontaneous, the reducing agent reactant is the strongest reducing agent.

Mg is the reducing agent being oxidized because it went from 0 as a reactant to +2 as a product.

2. E is correct.

All statements are correct for balancing redox equations by the oxidation number method.

3. D is correct.

Use the mnemonic OIL RIG: <u>O</u>xidation <u>I</u>s <u>L</u>oss, <u>R</u>eduction <u>I</u>s <u>G</u>ain (of electrons).

Oxidation is the loss of electrons, while reduction is the gain of electrons.

An *oxidizing agent* undergoes reduction, while a *reducing agent* undergoes oxidation.

Co^{2+} is the starting reactant because it must lose electrons and produce an ion with a higher oxidation number.

4. A is correct.

Formula mass is a synonym for molecular mass/molecular weight (MW).

Start by calculating the mass of the formula unit (CHCl):

$$CHCl = (12.01 \text{ g/mol} + 1.01 \text{ g/mol} + 35.45 \text{ g/mol})$$

$$CHCl = 48.47 \text{ g/mol}$$

Divide the molar mass by the formula unit mass:

$$194 \text{ g/mol} / 48.47 \text{ g/mol} = 4.0025$$

Round it to the closest whole number: 4

Multiply the formula unit by 4:

$$(CHCl)_4 = C_4H_4Cl_4$$

5. B is correct.

A *mole* is a unit of measurement used to express amounts of a chemical substance.

Avogadro's number of molecules in a mole is 6.02×1023.

However, *Avogadro's number* relates to the number of molecules, not the amount of substance.

Molar mass refers to the mass per mole of a substance.

Formula mass may be used for molecular mass or molecular weight; it refers to the mass of a specific molecule.

6. C is correct.

MW of $CaCO_3$:

$$(Ca = 40.08 \text{ g/mol}) + (C = 12 \text{ g/mol}) + (O = 3 \times 16 \text{ g/mol}) = 100 \text{ g/mol}$$

Mass of 0.7 moles of $CaCO_3$:

$$0.7 \text{ mole } CaCO_3 \times 100 \text{ g/mole} = 70 \text{ g } CaCO_3$$

7. A is correct.

The number of oxygens on each side of the reaction equation must be equal.

There are 5 CO_2 molecules; since each CO_2 molecule contains 2 oxygens, multiply $5 \times 2 = 10$.

On the right side are 6 H_2O molecules; $6 \times 1 = 6$.

Add 10 and 6 to get the total oxygens on the right side: $10 + 6 = 16$.

Since there are 16 oxygens on the right side, there should be 16 oxygens on the left side.

Each O_2 molecule contains 2 oxygens; $16 / 2 = 8$.

Therefore, coefficient 8 is needed to balance the equation.

Balanced equation (combustion):

$$C_5H_{12} + 8 O_2 \rightarrow 5 CO_2 + 6 H_2O$$

8. B is correct.

Balanced reaction:

$$BaCl_2 (aq) + K_2SO_4 (aq) \rightarrow BaSO_4 \text{ and } 2 KCl$$

A *double replacement reaction* indicates an exchange of cations and anions between the reactants.

Separate the reactants into ions and then exchange the cation and anion pairings.

9. C is correct.

Balanced equation (synthesis):

$$P_4 \, (s) + 6 \, H_2 \, (g) \rightarrow 4 \, PH_3 \, (g)$$

10. C is correct.

Balancing Redox Equations

From the balanced equation, the coefficient for the proton can be determined.

Balancing a redox equation requires balancing the atoms in the equation, but the charges must be balanced.

The equations must be separated into two different half-reactions. One half-reaction addresses the *oxidizing component*, and the other addresses the *reducing component*.

Magnesium is oxidized; therefore, the unbalanced oxidation half-reaction is:

$$Mg \, (s) \rightarrow Mg^{2+} \, (aq)$$

Furthermore, the nitrogen is reduced; therefore, the unbalanced reduction half-reaction is:

$$NO_3^- \, (aq) \rightarrow NO_2 \, (aq)$$

Each half-reaction must be balanced for each atom, and the net electric charge on each side of the equations must be balanced.

Order of operations for balancing half-reactions:

1) Balance atoms except for oxygen and hydrogen.

2) Balance the oxygen atoms by adding water.

3) Balance the hydrogen atoms by adding protons.

   a) If in basic solution, add equal amounts of hydroxide to each side to cancel the protons.

4) Balance the electric charge by adding electrons.

5) If necessary, multiply the coefficients of one half-reaction equation by a factor that cancels the electron count when the equations are combined.

6) Cancel ions or molecules that appear on each side of the overall equation.

After determining the balanced overall redox reaction, the stoichiometry indicates the moles of protons.

For magnesium:

$$Mg \, (s) \rightarrow Mg^{2+} \, (aq)$$

continued...

The magnesium is already balanced with a coefficient of 1.

There are no hydrogen or oxygen atoms present in the equation.

The magnesium cation has a +2 charge, so 2 moles of electrons are added to the right to balance the charge.

The balanced half-reaction for oxidation:

$$Mg\ (s) \rightarrow Mg^{2+}\ (aq) + 2\ e^-$$

For nitrogen:

$$NO_3^-\ (aq) \rightarrow NO_2\ (aq)$$

Equation is balanced for nitrogen because one nitrogen atom appears on each side of the reaction.

The nitrate reactant has three oxygen atoms, while the nitrite product has two oxygen atoms.

To balance the oxygen, one mole of water should be added to the right side:

$$NO_3^-\ (aq) \rightarrow NO_2\ (aq) + H_2O$$

Adding water to the right side of the equation introduces hydrogen atoms to that side.

Therefore, the hydrogen atom count needs to be balanced.

Water possesses two hydrogen atoms; therefore, two protons need to be added to the left side of the reaction:

$$NO_3^-\ (aq) + 2\ H^+ \rightarrow NO_2\ (aq) + H_2O$$

The reaction occurs in *acidic conditions*.

If the reaction were basic, OH^- would need to be added to each side to cancel the protons.

The net charge needs to be balanced.

The left side has a net charge of +1 (+2 from the protons and −1 from the electron), while the right is neutral.

Therefore, one electron should be added to the left side:

$$NO_3^-\ (aq) + 2\ H^+ + e^- \rightarrow NO_2\ (aq) + H_2O$$

When half-reactions are recombined, the electrons in the overall reaction must cancel.

The *reduction half-reaction* contributes *one electron* to the left side of the overall equation.

The *oxidation half-reaction* contributes *two electrons* to the product side.

continued...

Therefore, the *coefficients* of the reduction half-reaction should be doubled:

$$2 \times [NO_3^- \, (aq) + 2 \, H^+ + e^- \rightarrow NO_2 \, (aq) + H_2O]$$
$$= 2 \, NO_3^- \, (aq) + \mathbf{4 \, H^+} + 2 \, e^- \rightarrow 2 \, NO_2 \, (aq) + 2 \, H_2O$$

The answer is 4 at this step.

Combining half reactions:

$$Mg \, (s) + 2 \, NO_3^- \, (aq) + 4 \, H^+ + 2 \, e^- \rightarrow Mg^{2+} \, (aq) + 2 \, e^- + 2 \, NO_2 \, (aq) + 2 \, H_2O$$

Cancel electrons on each side of the reaction in the following balanced net equation:

$$Mg \, (s) + 2 \, NO_3^- \, (aq) + 4 \, H^+ \rightarrow Mg^{2+} \, (aq) + 2 \, NO_2 \, (aq) + 2 \, H_2O$$

This equation is now fully balanced for mass, oxygen, hydrogen and electric charge.

11. A is correct.

Number of molecules = moles × Avogadro's constant

Number of molecules = $1.40 \, \text{mol} \times 6.02 \times 10^{23}$ molecules/mol

Number of molecules = 8.428×10^{23} molecules

4 atoms in each PF_3 molecule:

$$4 \times 8.428 \times 10^{23} = 3.37 \times 10^{24} \text{ atoms}$$

12. D is correct.

CH_3COOH:

mass of O atoms = 2×16 g/mol = 32 g/mol

mass of H atoms = 4×1 g/mol = 4 g/mol

Therefore, the mass % of O is 8 times (32 / 4) greater than the mass % of H.

Calculate the molecular mass (MW) of acetic acid:

MW of CH_3COOH: = (2 × atomic mass of C) + (4 × atomic mass of H) + (2 × atomic mass of O)

MW of CH_3COOH = $(2 \times 12.01$ g/mole$) + (4 \times 1.01$ g/mole$) + (2 \times 16.00$ g/mole$)$

MW of CH_3COOH = 60.05 g/mole

continued...

To obtain the *percent mass composition* of each element, calculate the total mass of each element, divide it by the molecular mass, and multiply by 100%.

% mass composition of carbon = [(2 × 12.01 g/mole) / 60.05 g/mole] × 100%

% mass composition of carbon = 40.00%

% mass composition of hydrogen = [(4 × 1.01 g/mole) / 60.05 g/mole] × 100%

% mass composition of hydrogen = 6.71%

% mass composition of oxygen = [(2 × 16.00 g/mole) / 60.05 g/mole] ×100%

% mass composition of oxygen = 53%

13. B is correct.

From the balanced equation:

4 moles of Ag are produced for each mole of O_2.

Therefore, if 1 mole of Ag is produced, ¼ mole of O_2 is produced.

The mass of ¼ mole of O_2 is:

(¼ mol)·(2 × 16 g/mol) = 8 g

14. E is correct.

All metals (and elements) have an oxidation number of zero in their elemental state.

15. A is correct.

Assign *oxidation number* to each species:

K_2	Cr_2	O_7
2(+1)	2x	7(−2)

The sum of charges in a neutral molecule is zero:

2 + (2x) + (7 × −2) = 0

2 + (2x) + (−14) = 0

2x = −2 + 14

2x = +12

Oxidation number of Cr =+6

Notes for active learning

Notes for active learning

CHAPTER 5

Solution Chemistry

- Solutions

- Ions in Solution

- Solubility Concepts

- Solution Equilibrium

- Common and Complex Ions

- Practice Questions & Detailed Explanations

Solutions

Solutions as homogenous mixtures

A *solution* is a homogeneous mixture of one or more solutes and a solvent with indistinguishable components.

The formation of a solution involves disrupting the crystalline lattice structure of the solute and mixing solute particles with solvent molecules.

The *solvent* is the substance present in the most substantial quantity.

Solute particles diffuse into the solution and become uniformly dispersed.

Aqueous solutions have water as the solvent.

Enthalpy considerations for solutions

There must be sufficiently strong interactions between solute particles and solvent molecules for solute particles to diffuse into the solution.

The interactions between solute and solvent molecules must overcome the attractive intermolecular forces between solute particles in the liquid or solid.

Solution processes may be divided into three main stages:

1. pure solvent $\to$ separated solvent molecules (endothermic);
   $\Delta H_1 > 0$

2. pure solute $\to$ separated solute particles (endothermic);
   $\Delta H_2 > 0$

3. separated solvent and solute molecules $\to$ solution (exothermic);
   $\Delta H_3 < 0$

For example:

Solute(s) + Solvent $\to$ Solution

$\Delta H_{soln} = \Delta H_1 + \Delta H_2 + \Delta H_3$

Depending on the magnitude of ΔH_1, ΔH_2 and ΔH_3, the solution process is:

Exothermic: if $| \Delta H_3 | > | \Delta H_1 + \Delta H_2 |$

Endothermic: if $| \Delta H_3 | < | \Delta H_1 + \Delta H_2 |$

For solutions, remember the phrase "like dissolves like."

Polar solutes dissolve in polar solvents, while nonpolar solutes dissolve in nonpolar solvents.

Notes for active learning

Ions in Solution

Hydration energy

Many ionic compounds dissolve in polar water with negative enthalpies (release energy) as *hydration energy*.

> *Solvation* is the attraction of a solvent with solute molecules or ions.

> *Hydration* is the process of attraction of water with solute molecules or ions.

The *hydration energy* produced from ion-dipole interactions provides the energy to overcome the lattice energy and disrupt ions in the crystalline solids (e.g., NaCl).

Water molecules are polar and interact strongly with ionic cations and anions through ion-dipole interactions.

The overall solute–solution formation process may be exothermic or endothermic, depending on if the solute–solute or solute–solution enthalpy is larger.

Ionic compounds dissolve in water and dissociate (i.e., ionize) to produce free ions (cations and anions).

For example, NaCl dissolves and produces Na^+ and Cl^- ions, which become hydrated with water molecules.

Sodium chloride and many other ionic compounds dissolve in water because of the strong ion-dipole interactions between the charged solute particles and the polar solvent (e.g., water) molecules.

For example, for NaCl, the *lattice energy* (i.e., ionic solid to gaseous ions) is slightly higher than the sum of *hydration energy* for Na^+ and Cl^- ions. The solution process for NaCl is slightly endothermic.

Electrolytes

Strong electrolytes (solutions of ionic compounds) completely dissociate, and the dissolved ions are good conductors of electric current.

Weak electrolytes (solutions of polar covalent compounds) partially dissociate in solution.

Nonelectrolytes do not dissociate into ions (e.g., sugar as a nonelectrolyte) and dissolve in the solution as the original molecules rather than separate into cations and anions.

Anions and cations

Anions are *negatively charged* species, with a single negative sign indicating a charge of -1.

Cations are *positively charged* species, with a single positive sign indicating a charge of $+1$.

The charge is written as a superscript.

Example ions include:

ammonium (NH^{4+})

phosphate (PO_4^{3-})

sulfate (SO_4^{2-})

Common anions and cations	Formula
Anions	
Hydroxide	OH^-
Chloride	Cl^-
Hypochlorite	ClO^-
Chlorite	ClO_2^-
Chlorate	ClO_3^-
Perchlorate	ClO_4^-
Halide, hypohalite, etc.	X^-, XO^-, etc.
Carbonate	CO_3^{2-}
Hydrogen Carbonate (Bicarbonate)	HCO_3^-
Sulfate	SO_4^{2-}
Hydrogen Sulfate (Bisulfate)	HSO_4^-
Sulfite	SO_3^{2-}
Thiosulfate	$S_2O_3^{2-}$
Nitrate	NO_3^-
Nitrite	NO_2^-
Phosphate	PO_4^{3-}

Hydrogen Phosphate	HPO_4^{2-}
Dihydrogen Phosphate	$H_2PO_4^-$
Phosphite	PO_3^{3-}
Cyanide	CN^-
Thiocyanate	SCN^-
Peroxide	O_2^{2-}
Oxalate	$C_2O_4^{2-}$
Acetate	$C_2H_3O_2^-$
Chromate	CrO_4^{2-}
Dichromate	$Cr_2O_7^{2-}$
Permanganate	MnO_4^-
Cations	
Hydronium	H_3O^+
Ammonium	NH_4^+
Metal	M^{n+}

Common ions, associated formulas, and charges shown

Identifying ions experimentally

Qualitative analysis is a laboratory method for separating and identifying ions in a mixture.

The technique uses solubility differences for ionic compounds in an aqueous solution and a particular cation's ability to form *complex ions* with ligands.

The approach in qualitative analysis of cations separates them sequentially into *ion groups*.

The complex ions of many transition metals are often colored and used in visual identity.

Ion group 1: Insoluble chlorides

Treating the mixture with 6 M HCl precipitates Ag^+, Hg_2^{2+}, and Pb^{2+} as chlorides, with other cations in solution.

The formation of a *white precipitate* indicates at least one of these cations in the mixture.

Ion group 2: Acid-insoluble sulfides

The supernatant from the above treatment with HCl is adjusted to pH $\approx$ 0.5 and treated with aqueous H_2S.

The high $[H_3O^+]$ in solution keeps $[HS^-]$ low, which precipitates cation groups: Cu^{2+}, Cd^{2+}, Hg^{2+}, Sn^{2+}, and Bi^{3+}.

Centrifuging and decanting give the next solution.

Ion group 3: Base-insoluble sulfides

The supernatant from acidic sulfide treatment is treated with an NH_3/NH_4^+ buffer to make the solution slightly basic (pH $\approx$ 8).

The excess OH^- in solution increases $[HS^-]$, which causes precipitation of more soluble sulfides and hydroxides.

The cations that precipitate under this condition of base-insoluble sulfides are Zn^{2+}, Mn^{2+}, Ni^{2+}, Fe^{2+}, and Co^{2+}, as sulfides, and Al^{3+}, Cr^{3+}, and Fe^{3+} as hydroxides.

The precipitate is centrifuged, and the supernatant decanted to give the next solution.

Ion group 4: Insoluble phosphates

The slightly basic supernatant separated from the group 3 ions are treated with $(NH_4)_2HPO_4$, which precipitates $Mg_3(PO_4)_2$, $Ca_3(PO_4)_2$ and $Ba_3(PO_4)_2$.

Ion group 5: Alkali metal and ammonium ions

The final solution contains any of the following ions: Na^+, K^+, and NH_4^+.

Solvation and hydration

Solvation is the attraction of a solvent with solute molecules or ions.

Hydration is the attraction of water with solute molecules or ions, and water forms a shell around the ions in a solution.

For example, the oxygen atom on the water is partially negative, so it surrounds cations.

Conversely, the hydrogen atoms on the water are partially positive, surrounding anions.

Hydronium ions

H^+ does not exist as a bare proton in water; it exists as the *hydronium ion* (H_3O^+).

The high charge density of a proton attracts it to a nearby molecule with a partial or full negative charge.

Water has two lone pairs of electrons on the oxygen atom; the positive proton interacts with one of the lone pairs, forming a hydronium ion.

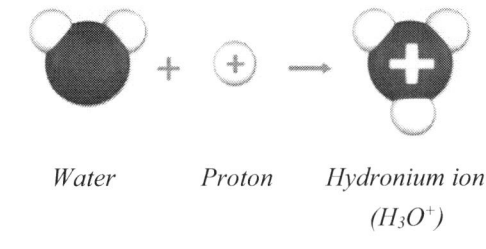

Water   *Proton*   *Hydronium ion*
                              (H_3O^+)

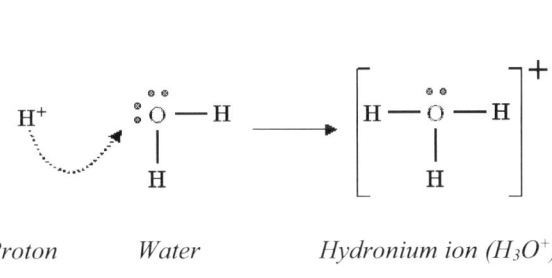

Proton   *Water*   *Hydronium ion (H_3O^+)*

Notes for active learning

Solubility Concepts

Solubility

A substance's solubility is the amount of solute (e.g., grams) that dissolves in a solvent to give a saturated solution at a specific temperature.

Solubility is *temperature-dependent*.

A *saturated* solution contains the maximum dissolved solute possible at a given temperature, with a state of dynamic equilibrium existing between dissolution and crystallization.

For example, the solubility of KNO_3 is ~30 g per 100 g water at 20 °C and 63 g at 40 °C.

Solubility rules for ionic compounds

Use the following rules to *predict the solubility of ionic compounds* in water. These rules generally relate to the compound's lattice and hydration energies of the individual ions.

One of the factors determining the *lattice energy* is the magnitudes of charges on the ions (i.e., the greater the magnitude of the charges, the greater the lattice energy). Simplified, the greater the lattice energy, the lower the solubility.

The rules focus on the magnitudes of the charges and ignore other influences on the lattice energies, such as hydration energy. In addition, the rules disregard the effect of temperature.

The solubility rules are listed in order of *decreasing importance*. For example, if the first rule applies, disregard the subsequent rules. Similarly, if the second rule is applicable, disregard the third rule. Thus, only use the third rule when rules one and two are not applicable.

The following ions are usually soluble: $C_2H_3O_2^-$, ClO_3^-, ClO_4^-. K^+, Na^+, NH_4^+, and NO_3^-.

General principle: Hydrides (H^-) decompose in water to yield H_2 and ^-OH. Many metal oxides react with water to produce hydroxides (^-OH).

Rule 1: compounds containing a +1 or –1 ion are typically soluble

Examples include: $AlCl_3$, $FeCl_2$, K_2SO_4, $NaBr$, Rb_3PO_4

Exceptions (Insoluble): OH^- (other than with cations that produce strong bases)

Hg_2^{2+} (other than with NO_3^-, $C_2H_3O_2^-$, ClO_3^-, and ClO_4^-)

Pb^{2+}, Hg^{2+} with most –1 ions

IB metals (column 11)

Rule 2: compounds containing a +3 or –3 or higher ion are typically NOT soluble

Examples include: $AlPO_4$, $Ca_3(PO_4)_2$, TiO2

Exceptions (Soluble): cations combined with sulfate or dichromate

Rule 3: compounds containing –2 ion are typically NOT soluble

Examples include: $CaCO_3$, FeS, ZnC_2O_4

Exceptions (Soluble):

Dichromates and sulfates (insoluble if combined with Ba, Ca, Hg, Pb, and Sr)

There are exceptions to the solubility rules besides those listed. However, the ions commonly encountered follow these rules.

Solubility examples for ionic compounds

AgBr	Insoluble	Rule 1	Ag^+
AgCl	Insoluble	Rule 1	$Ag^+ + Cl^-$
$AgIO_3$	Insoluble	Rule 1	$Ag^+ + IO_3^-$
AgOH	Insoluble	Rule 1	weak base
Ag_2S	Insoluble	Rule 1	Ag^+
Hg_2Cl_2	Insoluble	Rule 1	Hg_2^{2+}
$PbCl_2$	Insoluble	Rule 1	Pb^{2+}
$AgNO_3$	Soluble	Rule 1	$NO_3^- + Ag^+$
$Ba(NO_3)_2$	Soluble	Rule 1	NO_3^-
$Ba(OH)_2$	Soluble	Rule 1	strong base
NaCl	Soluble	Rule 1	$Na^+ + Cl^-$
$NaClO_3$	Soluble	Rule 1	$Na^+ + ClO_3^-$
NaOH	Soluble	Rule 1	strong base
Na_3PO_4	Soluble	Rule 1	Na^+
$(NH_4)_2CO_3$	Soluble	Rule 1	NH_4^+
NH_4IO_3	Soluble	Rule 1	$NH_4^+ + IO_3^-$
$FeAsO_4$	Insoluble	Rule 2	$Fe^{3+} + AsO_4^{3-}$
$Fe_2(SO_4)_3$	Soluble	Rule 2	SO_4^{2-}
$BaCO_3$	Insoluble	Rule 3	CO_3^{2-}
$BaSO_4$	Insoluble	Rule 3	$Ba^{2-} + SO_4^{2-}$

$MgSO_4$	Insoluble	Rule 3	SO_4^{2-}
$PbCrO_4$	Insoluble	Rule 3	CrO_4^{2-}
ZnS	Insoluble	Rule 3	S^{2-}
$Ba(IO_3)_2$	Insoluble	exception	

Solution concentration

When a solution that is almost saturated at a higher temperature is cooled slowly to a lower temperature where the solubility is less, the excess solute typically precipitates to give a saturated solution at the *lower* temperature.

However, if the solution is cooled rapidly, precipitates do not form, and the resulting solution contains more dissolved solute than it would in a typical saturated solution. This process produces a *supersaturated* solution.

> The *supersaturated solution* is unstable, and crystallization occurs by seeding (i.e., introducing particles providing nuclei for precipitation).

> A *concentrated solution* contains a relatively large amount of dissolved solute; contains 40 g or more of dissolved solute in 100 mL of water.

> A *dilute solution* contains few dissolved solutes in a relatively large amount of solvent; contains less than 10 g of dissolved solute per 100 mL of water.

A *concentrated* solution does *not* necessarily imply a *saturated* solution.

A *saturated* solution does *not* necessarily imply a *concentrated* solution.

Temperature and pressure affect solubility

The solubility of most liquids and solids in water *increases* with temperature.

Gas solubility *decreases* with temperature.

The pressure of a gas strongly affects its solubility, as described by *Henry's Law*, which states that solubility increases as pressure increases (i.e., directly proportional).

Factors affecting solubility include the solute's surface area (e.g., granulated sugar dissolves faster than a sugar cube) and heating or agitating a solution (e.g., stirring accelerates dissolving).

The terms *miscible* and *immiscible* describe liquids that dissolve or do not dissolve in another liquid, respectively.

For example, ethanol and water are *miscible* because they dissolve in each other freely when mixed.

Oil and water are *immiscible* since they remain separate.

Units of concentration

The solution's concentration is expressed for the solute amount dissolved and expressed as:

$$\text{Mass Percent, \% (w/w)} = \frac{\text{mass of solute}}{\text{mass of solution}} \times 100\%$$

$$\text{Volume Percent \% (v/v)} = \frac{\text{volume of solute}}{\text{volume of solution}} \times 100\%$$

$$\text{Molarity (M)} = \frac{\text{number of moles of solute}}{\text{liters of solution}}$$

$$\text{Molality (m)} = \frac{\text{number of moles of solute}}{\text{kg of solvent}}$$

$$\text{Mole fraction, } (X_i) = \frac{\text{moles of a component}}{\text{total moles of components in the solution}}$$

$$\text{Normality } (N) = \frac{\text{number of equivalents of solute}}{\text{liters of solution}}$$

where

$$\text{Number of equivalents of solute} = \frac{\text{grams of solute}}{\text{equivalent weight of solute}}$$

Normality and equivalent weight

Normality is based on the chemical mass unit of equivalent weight.

Equivalent weight is the amount of solute needed to equal one mole of hydrogen ions.

The equivalent weight is thus dependent on the *valence of the solute*.

For solutes with a valence of 1 (e.g., HCl), the molecular and equivalent weights are the same.

For solutes with a valence greater than 1 (e.g., H_3PO_4, valence equals 3), the equivalent weight equals the *molecular weight divided by the valence*.

Therefore, molarity is related to normality, as in the examples:

$$1 \text{ M HCl} = 1 \text{ N HCl}$$

$$1 \text{ M } H_2SO_4 = 2 \text{ N } H_2SO_4$$

$$1 \text{ M } H_3PO_4 = 3 \text{ N } H_3PO_4$$

Since ionic compounds dissociate into ions, the total number of particles in a solution of an ionic compound is greater than in a solution of a nonionic (molecular) compound.

The *total concentration* of ions in the solution is the sum of the concentrations of cations and anions, depending on the compound's formula.

For example, a 1.0 M aluminum nitrate solution, $Al(NO_3)_3$, contains 1.0 M of Al^{3+} and 3.0 M NO_3^- ions, with the total concentration of ions in the solution of 4.0 M.

Before dissolving		*After dissolving*	
$Al(NO_3)_3\,(aq)$	$\rightarrow$	$Al^{3+}\,(aq)\ +\ 3\,NO_3^-\,(aq)$	
(1.0 M)	$\rightarrow$	(1.0 M)	(3 × 1.0 M)

For example, a solution of 1.0 M $MgCl_2$ contains 1.0 M of Mg^{2+} and 2.0 M of Cl^- ions, and the total ion concentration of 3.0 M.

Notes for active learning

Solution Equilibrium

Solubility product constant (K_{sp})

Solubility product constants (K_{sp}) describe saturated solutions of relatively *low solubility* ionic compounds.

For example, when a slightly soluble salt such as silver chloride (AgCl) dissolves in water, a saturated solution is obtained rapidly.

A minimal number of solid dissolves, while most AgCl salt remains undissolved.

The equilibrium between AgCl (*s*) and Ag^+ (*aq*) + Cl^- (*aq*) ions occurs in solution:

$$AgCl\ (s) \rightleftarrows Ag^+ (aq) + Cl^- (aq)$$

$$K_{sp} = [Ag^+] \cdot [Cl^-]$$

A *general expression* of the solubility product constant (K_{sp}) for solubility equilibrium is:

$$M_aX_b\ (s) \rightleftarrows aM^{b+} (aq) + bX^{a-} (aq)$$

$$K_{sp} = [M^{b+}]^a \cdot [X^{a-}]^b$$

In some problems, the K_{sp} is given, and the solubility needs to be calculated, or vice versa.

How to proceed depends on the type of equilibria. Sample solubility problems are solved below.

Ionic equilibrium

A. Ionic equilibria of the type:

$$MX\ (s) \rightleftarrows M^{n+} (aq) + X^{n-} (aq)$$

$$K_{sp} = [M^{n+}] \cdot [X^{n-}]$$

If the solubility (*S*) of compounds is *S* mol/L:

$$K_{sp} = S^2$$

$$S = \sqrt{(K_{sp})}$$

For example, the solubility equilibrium (K_{sp}) for $BaSO_4$ is:

$$BaSO_4\ (s) \rightleftarrows Ba^{2+} (aq) + SO_4^{2-} (aq)$$

$$K_{sp} = [Ba^{2+}] \cdot [SO_4^{2-}]$$

$$K_{sp} = 1.5 \times 10^{-9}$$

If the solubility (S) of $BaSO_4$ is S mol/L, a saturated solution of $BaSO_4$ has:

$[Ba^{2+}] = [SO_4^{2-}] = S$ mol/L

$K_{sp} = S^2$

$S = \sqrt{(K_{sp})}$

$S = \sqrt{(1.5 \times 10^{-9})}$

$S = 3.9 \times 10^{-5}$ mol/L

B. Ionic equilibria of the type:

$MX_2(s) \rightleftarrows M^{2+}(aq) + 2\,X^-(aq)$

$K_{sp} = [M^{2+}]\cdot[X^-]^2$

For the type:

$M_2X(s) \rightleftarrows 2M^+(aq) + X^{2-}(aq)$

$K_{sp} = [M^+]^2\cdot[X^{2-}]$

For both types if the solubility is S mol/L:

$K_{sp} = 4S^3$

$S = (K_{sp}/4)^{1/3}$

For example:

$CaF_2(s) \rightleftarrows Ca^{2+}(aq) + 2\,F^-(aq)$

$K_{sp} = [Ca^{2+}]\cdot[F^-]^2$

$K_{sp} = 4.0 \times 10^{-11}$

The solubility (S) of calcium fluoride (CaF_2) is:

$S = (K_{sp}/4)^{1/3}$

$S = (4.0 \times 10^{-11}/4)^{1/3}$

$S = 2.2 \times 10^{-4}$ mol/L

C. Solubility equilibria of the type:

$MX_3(s) \rightleftarrows M^{3+}(aq) + 3\,X^-(aq)$

$K_{sp} = [M^{3+}]\cdot[X^-]^3$

Or of the type:

$$M_3X\,(s) \rightleftarrows 3\,M^+\,(aq) + X^{3-}\,(aq)$$

$$K_{sp} = [M^+]^3 \cdot [X^{3-}]$$

If the solubility (S) of the compound (MX_3 or M_3X) is S mol/L:

$$K_{sp} = 27S^4$$

$$S = \sqrt[4]{(K_{sp}/27)}$$

For example:

$$Ag_3PO_4\,(s) \rightleftarrows 3Ag^+\,(aq) + PO_4^{3-}\,(aq)$$

$$K_{sp} = [Ag^+]^3 \cdot [PO_4^{3-}]$$

$$K_{sp} = 1.8 \times 10^{-18}$$

The solubility (S) of silver phosphate (Ag_3PO_4) is:

$$S = \sqrt[4]{(K_{sp}/27)}$$

$$S = \sqrt[4]{(1.8 \times 10^{-18}/27)}$$

$$S = 1.6 \times 10^{-5}\ \text{mol/L}$$

D. Solubility equilibria of the type:

$$M_2X_3\,(s) \rightleftarrows 2\,M^{3+}\,(aq) + 3\,X^{2-}\,(aq)$$

$$K_{sp} = [M^{3+}]^2 \cdot [X^{2-}]^3$$

Or of the type:

$$M_3X_2\,(s) \rightleftarrows 3\,M^{2+}\,(aq) + 2\,X^{3-}\,(aq)$$

$$K_{sp} = [M^{2+}]^3 \cdot [X^{3-}]^2$$

If the solubility (S) of the compound M_3X_2 is S mol/L:

$$[M^{2+}] = 3S \text{ and } [X^{3-}] = 2S$$

$$K_{sp} = (3S)^3 \cdot (2S)^2$$

$$K_{sp} = 108S^5$$

$$S = \sqrt[5]{(K_{sp}/108)}$$

For example:

$$Ca_3(PO_4)_2 \; (s) \rightleftarrows 3 \; Ca^{2+} \; (aq) + 2 \; PO_4{}^{3-} \; (aq)$$

$$K_{sp} = [Ca^{2+}]^3 \cdot [PO_4{}^{3-}]^2$$

$$K_{sp} = 1.3 \times 10^{-32}$$

The solubility (S) of calcium phosphate ($Ca_3(PO_4)_2$) is:

$$S = 5\sqrt{(1.3 \times 10^{-32}) / 108}$$

$$S = 1.6 \times 10^{-7} \text{ mol/L}$$

Calculating K_{sp} from solubility

For example, the solubility (S) of $PbSO_4$ in water is 4.3×10^{-3} g/100 mL solution at 25 °C. What is the K_{sp} of $PbSO_4$ at 25 °C?

$$\text{Solubility } (S) \text{ of } PbSO_4 \text{ in mol/L} = \frac{4.3 \times 10^{-3} \text{ g}}{100 \text{ mL}} \times \frac{1000 \text{ mL/L}}{303.26 \text{ g/mol}}$$

$$S = 1.40 \times 10^{-4} \text{ mol/L}$$

A saturated solution of $PbSO_4$, contains:

$$[Pb^{2+}] = [SO_4{}^{2-}] = 1.4 \times 10^{-4} \text{ mol/L}$$

For the equilibrium:

$$PbSO_4 \; (s) \rightleftarrows Pb^{2+} \; (aq) + SO_4{}^{2-} \; (aq)$$

$$K_{sp} = [Pb^{2+}] \cdot [SO_4{}^{2-}]$$

$$S^2 = (1.4 \times 10^{-4} \text{ mol/L})^2$$

$$S = 2.0 \times 10^{-8}$$

Calculating solubility from K_{sp}

For example, if the K_{sp} of $Mg(OH)_2$ is 6.3×10^{-10} at 25 °C, what is its solubility in mol/L at 25 °C?

Determine the solubility equilibrium for magnesium hydroxide ($Mg(OH)_2$):

$$Mg(OH)_2 \; (s) \rightleftarrows Mg^{2+} \; (aq) + 2 \; OH^- \; (aq)$$

$$K_{sp} = [Mg^{2+}] \cdot [OH^-]^2$$

$$4S^3 = 6.3 \times 10^{-10}$$

Solubility of $Mg(OH)_2$:

$$S = \sqrt[3]{(K_{sp}/4)}$$

$$S = \sqrt[3]{(6.3 \times 10^{-10}/4)}$$

$$S = 5.4 \times 10^{-4} \text{ mol/L}$$

If the K_{sp} of $Mg(OH)_2$ is 6.3×10^{-10} at 25 °C, what is its solubility in a g/100 mL solution at 25 °C?

Solubility in g/100 mL solution:

$$S = (5.4 \times 10^{-4} \text{ mol/L}) \cdot (58.32 \text{ g/mol}) \cdot (0.1 \text{ L/100 mL})$$

$$S = 3.1 \times 10^{-3} \text{ g/100 mL solution}$$

Notes for active learning

Common and Complex Ions

Common ions

Common ions are produced by more than one solute in the solution.

According to *Le Châtelier's Principle*, the following equilibrium exists for acetic acid:

$$HC_2H_3O_2 \, (aq) + H_2O \, (l) \leftrightarrows H_3O^+ \, (aq) + C_2H_3O_2^- \, (aq)$$

The reaction shifts to the left if $C_2H_3O_2^-$ is introduced from another source. The acetate ion ($C_2H_3O_2^-$) is the common ion in sodium acetate and acetic acid solution.

This effect reduces the degree of dissociation of the acid, decreases $[H_3O^+]$, and increases the pH of the solution.

For example, the addition of ammonium chloride, NH_4Cl, causes the reaction to shift left in the equilibrium of ammonia in an aqueous solution:

$$NH_3 \, (aq) + H_2O \, (l) \leftrightarrows NH_4^+ \, (aq) + OH^- \, (aq)$$

NH_4Cl dissociates into NH_4^+ and Cl^-, where NH_4^+ is a common ion in ammonia equilibrium.

The equilibrium shift caused by NH_4^+ ion reduces the extent of ionization of ammonia, which decreases $[OH^-]$ in the system and lowers the pH of the solution.

Common-ion effect

The *common-ion effect* is Le Châtelier's Principle applied to K_{sp} reactions. It states that a common ion *decreases* a slightly soluble ionic compound's solubility.

A common ion is an ion in a solution common to the ionic compound.

For example, in the following equilibrium:

$$PbCl_2 \, (s) \rightleftharpoons Pb^{2+} \, (aq) + 2 \, Cl^- \, (aq)$$

If NaCl is added to a saturated solution of $PbCl_2$, the $[Cl^-]$ increases, and according to Le Châtelier's Principle, the equilibrium shifts in the direction that reduces $[Cl^-]$.

In this example, the equilibrium shifts left to form more $PbCl_2$ solid, decreasing $PbCl_2$ that dissolves into solution. More $PbCl_2$ dissolves in pure water than in water with Cl^- ions.

The common-ion effect precipitates one component in a mixture selectively in laboratory separations.

For example, add NaCl to separate AgCl from a mixture of AgCl and Ag_2SO_4. The addition of NaCl selectively displaces AgCl by the common-ion effect (Cl^- being the common ion).

Complex ion formation

A *complex ion* consists of a central metal ion covalently bonded to two or more *ligands*, which can be anions such as ^-OH, Cl^-, F^- and ^-CN, or neutral molecules such as H_2O, CO, and NH_3.

Metal ions form complex ions with water molecules as ligands in aqueous solutions.

Ligand exchanges occur when another ligand is introduced into the solution, establishing equilibrium.

For example, in the complex ion $[Cu(NH_3)_4]^{2+}$, Cu^{2+} is the central metal ion, with four NH_3 molecules covalently bonded to it.

Complex ions are Lewis adducts (i.e., the addition of a Lewis acid and a Lewis base).

The Lewis base can be charged or uncharged.

Lewis acids and bases

The metal ions act as *Lewis acids* (electron-pair acceptors).

The ligands are *Lewis bases* (electron-pair donors).

$$\text{Metal}^+ + \text{Lewis base} \rightarrow \text{Complex ion}$$

$$M^+ + L \rightarrow M - L_n^+$$

The K_{eq} for this reaction is K_f, or the *formation constant*.

Formation constant K_f

For example, when NH_3 is added to an aqueous solution containing Cu^{2+} ion, the following equilibrium occurs:

$$Cu(H_2O)_6^{2+} (aq) + 4\, NH_3 (aq) \rightleftarrows [Cu(NH_3)_4]^{2+} (aq) + 6\, H_2O$$

$$K_f = [Cu(NH_3)_4^{2+}] / [Cu(H_2O)_6^{2+}] \cdot [NH_3]^4$$

The ligand exchange process occurs stepwise; each water molecule is sequentially replaced with an NH_3 molecule to yield a series of intermediate species, each with its formation constant (K_f).

For convenience, the water molecule is omitted.

1. $Cu^{2+} (aq) + NH_3 (aq) \rightleftarrows Cu(NH_3)^{2+} (aq)$

    $$K_{f1} = [Cu(NH_3)^{2+}] / [Cu^{2+}] \cdot [NH_3]$$

2. $Cu(NH_3)^{2+} (aq) + NH_3 (aq) \rightleftarrows Cu(NH_3)_2^{2+} (aq)$

    $$K_{f2} = [Cu(NH_3)_2^{2+}] / [Cu(NH_3)^{2+}] \cdot [NH_3]$$

3. $Cu(NH_3)_2^{2+} (aq) + NH_3 (aq) \rightleftarrows Cu(NH_3)_3^{2+} (aq)$

$$K_{f3} = [Cu(NH_3)_3^{2+}] / [Cu(NH_3)_2^{2+}] \cdot [NH_3]$$

4. $Cu(NH_3)_3^{2+} (aq) + NH_3 (aq) \rightleftarrows Cu(NH_3)_4^{2+} (aq)$

$$K_{f4} = [Cu(NH_3)_4^{2+}] / [Cu(NH_3)_3^{2+}] \cdot [NH_3]$$

The *formation constant* (K_f) is the product of the intermediate formation constants:

$$K_f = K_{f1} \times K_{f2} \times K_{f3} \times K_{f4}$$

$$K_f = [Cu(NH_3)_4^{2+}] / [Cu^{2+}] \cdot [NH_3]^4$$

Solubility of complex ions

The *complex-ion effect* is the opposite of the *common* ion effect.

A ligand *increases* the solubility of slightly soluble ionic compounds if complex ions form with the metal ions.

For example, silver chloride (AgCl) is *more soluble* in ammonia (NH_3) solution because silver ions form complex ions with NH_3:

$$AgCl (s) \rightleftarrows Ag^+ (aq) + Cl^- (aq)$$

$$K_{sp} = 1.6 \times 10^{-10}$$

$$Ag^+ (aq) + 2\,NH_3 (aq) \rightleftarrows Ag(NH_3)_2^+ (aq)$$

$$K_f = 1.7 \times 10^7$$

$$AgCl (s) + 2\,NH_3 (aq) \rightleftarrows Ag(NH_3)_2^+ (aq) + Cl^- (aq)$$

$$K_{net} = K_{sp} \times K_f$$

$$K_{net} = (1.6 \times 10^{-10}) \cdot (1.7 \times 10^7)$$

$$K_{net} = 2.7 \times 10^{-3}$$

The Cl^- ion is reduced when a complex ion forms, so more AgCl dissolves.

Alternatively:

$$AgCl (s) \leftrightarrow Ag^+ (aq) + Cl^- (aq)$$

$$NH_3 + Ag^+ \leftrightarrow Ag\text{-}(NH_3)_n \text{ complex ion}$$

The complex ion formation reduces Ag^+, more AgCl dissolves.

pH and solubility

The pH affects the solubility of slightly soluble compounds containing anions as *conjugate bases of weak acids* (e.g., F^-, NO_2^-, OH^-, SO_3^{2-} and PO_4^{3-}).

The pH *does not* affect the solubility of slightly soluble compounds containing anions as *conjugate bases of strong acids* (SO_4^{2-}, Cl^- and Br^-).

In a saturated solution of calcium fluoride (CaF_2) equilibrium exists:

$$CaF_2 \,(s) \rightleftarrows Ca^{2+} \,(aq) + 2\,F^- \,(aq)$$

If a strong acid is added to the saturated solution, the following reaction occurs:

$$H^+ \,(aq) + F^- \,(aq) \rightarrow HF \,(aq)$$

The reaction has the net effect of *reducing* F^- ion concentration, causing the equilibrium to shift to the *right*, and more CaF_2 dissolves.

Acids are more soluble in bases.

$$HA \rightarrow H^+ + A^-$$

Placing the above reactions in a base reduces the H^+.

Thus, more HA dissolves, according to Le Châtelier's Principle.

Bases are more soluble in acids.

$$B + H^+ \rightarrow BH^+$$

Putting the above reactions in an acid adds H^+, and thus, more B dissolves, according to Le Châtelier's Principle.

Practice Questions

1. Which of the following statements describing solutions is NOT true?

    A. Solutions are colorless

    B. The particles in a solution are atomic or molecular

    C. Making a solution involves a physical change in size

    D. Solutions are homogeneous

    E. Solutions are transparent

2. Which of the following describes a saturated solution?

    A. When the ratio of solute to solvent is small

    B. When it contains less solute than it can hold at 25 °C

    C. When it contains as much solute as it can hold at a given temperature

    D. When it contains 1 g of solute in 100 mL of water

    E. When it is equivalent to a supersaturated solution

3. An ionic compound that strongly attracts atmospheric water is said to be:

    A. immiscible

    B. miscible

    C. diluted

    D. hygroscopic

    E. None of the above

4. What happens when molecule-to-molecule attractions in solutes are less than in the solvent?

    A. The material has only limited solubility in the solvent

    B. The solution will become saturated

    C. The solute can have infinite solubility in the solvent

    D. The solute does not dissolve in the solvent

    E. None of the above

5. What does a negative heat of solution indicate about solute-solvent bonds compared to solute-solute bonds and solvent-solvent bonds?

    A. Solute-solute and solute-solvent bond strengths are greater than solvent-solvent bond strength

    B. Solute-solute and solvent-solvent bonds are weaker than solute-solvent bonds

    C. Solute-solute and solvent-solvent bonds are stronger than solute-solvent bonds

    D. Solute-solute and solvent-solvent bond strengths are equal to solute-solvent bond strength

    E. The heat of solution does not support a conclusion about bond strength

6. Which statement supports that calcium fluoride is much less soluble in water than sodium fluoride?

> I. calcium fluoride is not used in toothpaste
>
> II. calcium fluoride is not used to fluoridate city water supplies
>
> III. sodium fluoride is not used to fluoridate city water supplies

A. I only

B. II only

C. I and II only

D. I and III only

E. I, II and III

7. The principle *like dissolves like* is NOT applicable for predicting solubility when the solute is a/an:

A. nonpolar liquid

B. polar gas

C. nonpolar gas

D. ionic compound

E. covalent compound

8. Which of the following is the most soluble in benzene (C_6H_6)?

A. glucose ($C_6H_{12}O_6$)

B. sodium benzoate

C. octane (C_8H_{18})

D. hydrobromic acid

E. dichloromethane (CH_2Cl_2)

9. Apply the *like dissolves like* rule to predict which liquids are immiscible in water.

A. ethanol, CH_3CH_2OH

B. acetone, C_3H_6O

C. acetic acid, CH_3COOH

D. formaldehyde, H_2CO

E. none of the above

10. What is the equilibrium constant expression (K_{sp}) for slightly soluble silver sulfate in an aqueous solution: $Ag_2SO_4 (s) \leftrightarrow 2\, Ag^+ (aq) + SO_4^{2-} (aq)$?

A. $K_{sp} = [Ag^+] \cdot [SO_4^{2-}]^2$

B. $K_{sp} = [Ag^+]^2 \cdot [SO_4^{2-}] / [Ag_2SO_4]$

C. $K_{sp} = [Ag^+] \cdot [SO_4^{2-}]$

D. $K_{sp} = [Ag^+] \cdot [SO_4^{2-}]^2 / [Ag_2SO_4]$

E. $K_{sp} = [Ag^+]^2 [SO_4^{2-}]$

11. In the reaction KHS (aq) + HCl (aq) → KCl (aq) + H_2S (g), which ions are spectator ions?

A. H^+ and HS^-

B. K^+ and HS^-

C. K^+ and Cl^-

D. K^+ and H^+

E. HS^- and Cl^-

12. What is the mass of a 10.0% blood plasma sample containing 2.50 g of dissolved solute?

A. 21.5 g

B. 25.0 g

C. 0.215 g

D. 0.430 g

E. 12.5 g

13. What is the molarity of KCl in seawater if KCl is 12.5% (m/m) and the density of seawater is 1.06 g/mL?

A. 1.78 M

B. 17.8 M

C. 2.78 M

D. 0.845 M

E. 27.8 M

14. If 25.0 mL of urine has a mass of 25.5 g and contains 1.8 g of solute, what is the mass/mass percent concentration of solute in the urine sample?

A. 12.48%

B. 17.42%

C. 7.06%

D. 3.82%

E. 47.4%

15. With increasing temperature, many solvents expand to occupy greater volumes. What happens to the concentration of a solution made with such a solvent as temperature increases?

A. Decreases because the solution has a greater ability to dissolve more solute at a higher temperature

B. Increases because the solution has a greater ability to dissolve more solute at a higher temperature

C. Decreases as the volume increases since concentration depends on the mass dissolved in a given volume

D. Increases as the solute fits into the new spaces between the molecules

E. Increases as volume increases because concentration depends on mass dissolved in a given volume

Notes for active learning

Detailed Explanations

1. A is correct.

Some solutions are colored (e.g., Kool-Aid powder dissolved in water).

The color of chemicals is a physical property of chemicals from (most common) the excitation of electrons due to the chemical's absorption of energy.

The observer sees not the absorbed color but the wavelength that is reflected.

Most simple inorganic (e.g., sodium chloride) and organic compounds (e.g., ethanol) are colorless.

Transition metal compounds are often colored due to transitions of electrons of *d*-orbitals of different energy.

Organic compounds tend to be colored when there is extensive conjugation (i.e., alternating double and single bonds), causing the energy gap between the *HOMO* (i.e., highest occupied molecular orbital) and *LUMO* (i.e., lowest occupied molecular orbital) to decrease.

This moves the absorption band from the UV to the visible region.

Color is due to the energy absorbed by compounds when electrons transition from the *HOMO* to the *LUMO*.

2. C is correct.

A *saturated solution* contains the maximum dissolved material in the solvent under normal conditions. Increased heat allows for a solution to become supersaturated.

Supersaturated refers to the vapor of a compound with a higher partial pressure than the vapor pressure of that compound.

A saturated solution forms a precipitate as more solute is added to the solution.

3. D is correct.

A *hygroscopic substance* (e.g., honey, glycerin) readily attracts water from its surroundings through adsorption or absorption.

Adsorption is the process in which atoms, ions, or molecules from a substance (e.g., gas, liquid, dissolved solid) adhere to a surface of the *adsorbent.*

4. C is correct.

Suppose the attraction between solute molecules is less than between solvent molecules. In that case, a solute has (a virtually) infinite solubility because solvent-solute attraction is stronger than solute-solute attraction.

5. B is correct.

Bond formation releases energy: if the heat of the solution is negative, then energy is released.

The *heat of the solution* is the net of *enthalpy* changes for making and breaking bonds.

Solvent-solvent and solute-solute bonds break during solution formation while solute-solvent bonds form.

The *breaking of bonds* absorbs energy, while the formation of bonds releases energy.

6. C is correct.

Fluorides (e.g., BaF_2, MgF_2 PbF_2) are frequently *insoluble*.

7. D is correct.

Like dissolves like means polar substances dissolve in polar solvents, and nonpolar substances in nonpolar solvents.

Molecules that can form hydrogen bonds with water are soluble.

Ionic compounds form anions and cations that bond with the polar water molecule and are soluble.

8. C is correct.

Like dissolves like means polar substances dissolve in polar solvents, and nonpolar substances in nonpolar solvents.

Benzene is a nonpolar molecule, and octane is also nonpolar.

Thus, nonpolar octane is soluble in nonpolar benzene.

9. E is correct.

The "like dissolves like" rule applies when a solvent is miscible (soluble) with a solute of similar properties.

A polar solute is miscible with a polar solvent.

All the molecules are polar and therefore are miscible in water.

10. E is correct.

The *solubility constant* (K_{sp}) calculation is similar to the equilibrium constant.

For K_{sp}, only aqueous species are included in the calculation.

The concentration of each species is raised to their coefficients' power and multiplied with each other.

Therefore,

$$K_{sp} = [Ag^+]^2 \cdot [SO_4^{2-}]$$

11. C is correct.

Spectator ions do not participate in the chemical reaction and solution before and after the reaction.

Ionic equation of the reaction:

$$K^+ (aq) + HS^- (aq) + H^+ (aq) + Cl^- (aq) \rightarrow K^+ (aq) + Cl^- (aq) + H_2S (g)$$

The *net ionic equation* only shows elements, compounds, and ions directly involved in the chemical reaction and is written without the spectator ions.

Net ionic equation of the reaction:

$$HS^- (aq) + H^+ (aq) \rightarrow H_2S (g)$$

12. B is correct.

$$\text{Mass \%} = \text{mass of solute / mass of solution}$$

Rearrange that equation to solve for mass of solution:

$$\text{mass of solution} = \text{mass of solute / mass \%}$$

$$\text{mass of solution} = 2.50 \text{ g} / 10.0\%$$

$$\text{mass of solution} = 2.50 \text{ g} / 0.1$$

$$\text{mass of solution} = 25.0 \text{ g}$$

13. A is correct.

Assume 1 L of seawater.

Molarity is the number of moles in 1 L of solution, so starting with 1 L makes the calculation easier.

>Mass of seawater = volume × density

>Mass of seawater = (1 L × 1,000 L/mL) × 1.06 g/mL

>Mass of seawater = 1,060 g

Use the mass of the seawater to determine the mass of KCl:

>Mass of KCl = mass % of KCl × mass of seawater

>Mass of KCl = 12.5% × 1,060 g

>Mass of KCl = 132.5 g

Calculate the number of moles:

>Moles of KCl = mass of KCl / molar mass of KCl

>Moles of KCl = 132.5 g / (39.1 g/mol + 35.45 g/mol)

>Moles of KCl = 1.78 moles

Divide moles by volume to calculate molarity:

>Molarity of KCl = moles of KCl / volume of KCl

>Molarity of KCl = 1.78 moles / 1 L

>Molarity of KCl = 1.78 M

14. C is correct.

>Mass % of solute = (mass of solute / total urine mass) × 100%

>Mass % of solute = (1.8 g / 25.5 g) × 100%

>Mass % of solute = 7.06%

15. C is correct.

As the volume of a solution increases, the concentration of a solution decreases

because concentration depends on how much mass is dissolved in a given volume.

Notes for active learning

Notes for active learning

CHAPTER 6

Kinetics and Equilibrium

- Reaction Rates

- Rate Dependence on Concentrations of Reactants

- Rate Order

- Reaction Order

- Rate Mechanisms

- Reaction Rate Depends on Temperature

- Kinetic *vs.* Thermodynamic Control of Reactions

- Catalysts and Enzymes

- Equilibrium in Reversible Reactions

- Mass Action

- Equilibrium Constants

- Equilibrium Variables

- Equilibrium Constant and $\Delta G°$ Relationship

- Practice Questions & Detailed Explanations

Reaction Rates

Measuring reaction rate

The *reaction rate* is the rate of change in the concentration of reactants or products (i.e., how fast a reactant is consumed and how fast a product is formed).

The *reaction rates*:

$$rate \ of \ reaction = \frac{change \ in \ concentration \ of \ reactant \ or \ product}{change \ in \ time}$$

or

$$rate \ of \ reaction = \frac{\Delta \ concentration}{\Delta \ time}$$

The rate is concentration divided by time; the unit is molarity per second (M/s).

Expressing reaction rates

The *rate of reaction* can be written for the forward and reverse directions of a chemical reaction.

A subscript "fwd" or "rev" specifies the direction of the reaction rate.

Forward rate of reaction:

$$rate_{fwd} = \frac{-\Delta reactant}{\Delta time} \qquad \textit{reactant disappears}$$

$$rate_{fwd} = \frac{\Delta product}{\Delta time} \qquad \textit{product forms}$$

Reverse rate of reaction:

$$rate_{rev} = \frac{-\Delta product}{\Delta time} \qquad \textit{product disappears}$$

$$rate_{rev} = \frac{\Delta reactant}{\Delta time} \qquad \textit{reactant reforms}$$

Collision theory

A reaction between two molecules occurs *if* the molecules

1) have sufficient *kinetic energy,* and

2) molecules are *appropriately oriented* to start a reaction in the image below.

The *collision theory* focuses on gas-phase chemical reactions.

Three *collision-related* factors affect the rate of a chemical reaction:

1. *Collision **frequency**:* Increasing the occurrence of reactants colliding increases the reaction rate. The more collisions, the higher the probability that a collision produces the product.

2. *Collision **energy**:* Reactants must collide with enough energy to *form new bonds* for a reaction to occur.

3. *Collision **orientation**:* Reactants need the correct orientation (i.e., properly aligned) for products to form.

A reaction forms a product from two molecules if the molecules have:

1) sufficient kinetic energy and

2) appropriately oriented molecules.

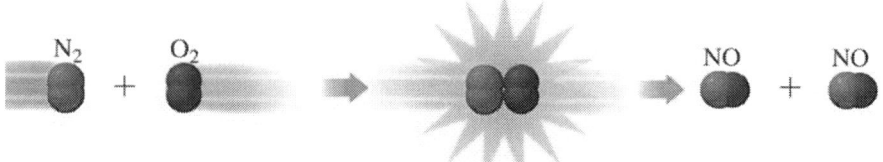

The two requirements of sufficient kinetic energy and appropriately oriented molecules are satisfied when $N_2 + O_2$ transforms into NO

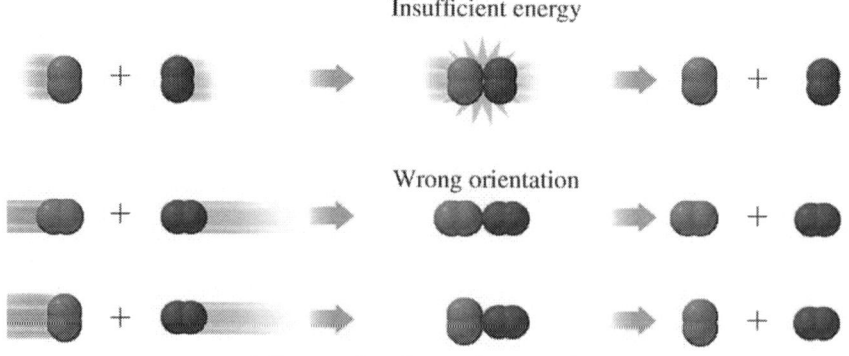

Collisions that do not form products

Rate Dependence on Concentration of Reactants

Rate law

The *rate law expression* describes how the reaction rate changes with concentrations of reactants and products.

For example, consider the general expression for a homogeneous reaction:

$$a \, \text{A} + b \, \text{B} \longrightarrow c \, \text{C} + d \, \text{D}$$

Its *rate law* expression is:

$$rate = -\Delta[A] \, / \, \Delta t$$

$$-\Delta[A] \, / \, \Delta t = k[A]^x \cdot [B]^y$$

where $[A]$ and $[B]$ are initial concentrations of the reactants, x and y are partial rate orders of the reaction determined experimentally, and k is the rate constant

Rate constant *k*

The rate constant (k) is *empirically* derived and quantifies a chemical reaction.

It accounts for factors influencing reaction rate, including *temperature* and *solvent*.

For example, consider the general expression for a homogeneous reaction:

$$a \, \text{A} + b \, \text{B} \longrightarrow c \, \text{C} + d \, \text{D}$$

The *rate constant* (k) expression is:

$$rate = k(\text{T})[A]^x \cdot [B]^y$$

$$k(\text{T}) = [A]^x \cdot [B]^y \, / \, rate$$

where $[A]$ and $[B]$ are initial concentrations of the reactants, x and y are partial rate orders of the reaction determined experimentally, and $k(\text{T})$ is the rate constant dependent on the reaction mechanism

Therefore, the constant is specific to the *experimental conditions* (i.e., *empirically* derived).

There is *no correlation* between the stoichiometric coefficients (a, b) and the rate exponents (x, y).

The *empirically derived exponents* are often small integers, fractions, or zero.

Relative reactant and product concentrations

If the reactants' initial concentration increases, there are more reactant molecules; thus, molecules are closer and collide more frequently.

A *higher collision frequency* results in a *higher rate* of reaction.

As the reaction proceeds, the concentration of *reactant* molecules decreases, and the *forward rate decreases*.

Thus, the concentration of *product* molecules increases, and the *reverse reaction rate increases*.

Rate Order

Partial rate order

The *partial rate order* is the exponent for each concentration term.

- For a reaction that is *zero-order* for a reactant, the rate is constant and does *not* depend on the reactant's concentration.

  The reactant's exponent is zero.

  Thus, the term equals one (since anything to the zeroth power is one).

  Since it equals one, it is excluded from the rate law expression, as the reaction rate does not depend on a zero-order reactant.

- For a reaction that is *first-order* for a reactant, the reaction rate is directly proportional to the concentration of that reactant.

  The exponent is one.

- For a reaction that is *second-order* for a reactant, the reaction rate is directly proportional to the square of the concentration of that reactant.

  The exponent is two.

Differential rate laws

The *overall order of the reaction* is the sum of the partial orders.

Most differential rate laws are zero-, first-, or second-order overall:

- *Zero-order overall*:

  The differential rate law is generally $r = k$

  The units of the rate constant k are mole $\times L^{-1} \times \sec^{-1}$ (or M/s)

- *First-order overall*:

  The differential rate law is generally $r = k[A]$

  The units of the rate constant k are $\sec^{-1}$

- *Second-order overall*:

  The differential rate law is generally $r = k[A]^2$

  The units of the rate constant k are $L \times mole^{-1} \times \sec^{-1}$ (or $M^{-1} \cdot s^{-1}$)

Notes for active learning

Reaction Order

Overall order of reaction

The example below illustrates how to sum the partial orders to obtain an equation's overall order.

For example, consider the reaction:

$$2 \text{ NO } (g) + O_2 (g) \leftrightarrow 2 \text{ NO}_2 (g)$$

The *rate law* was experimentally determined to be:

$$rate\ law = k[NO]^2 \cdot [O_2]$$

In the rate law, the partial order for NO is 2, and the partial order for O_2 is 1.

The overall order is three; this reaction is *third order overall.*

Note that the orders match the reaction coefficients, but rate orders and coefficients are *not necessarily* correlated. The coefficients only equal rate orders for elementary-step reactions. It is unknown if the reaction is an elementary or multi-step reaction in most problems unless it is explicitly stated.

Molecularity

In defining the overall order of a reaction, *molecularity* is important.

A reaction mechanism's molecularity is determined by the number of molecules (typically the coefficients) or ions participating in the rate-determining step.

A *unimolecular* reaction has a single reactant in the transition state.

A *bimolecular* reaction is a mechanism involving two reacting species.

A *termolecular* reaction is a mechanism when three independent molecules collide and react. Termolecular reactions are rare as the necessary factors are so specific that they are unlikely.

Reaction Type	Overall Reaction Order	Rate Law(s)
Unimolecular	1	$r = k[A]$
Bimolecular	2	$r = k[A]^2$, $r = k[A] \cdot [B]$
Termolecular	3	$r = k[A]^3$, $r = k[A]^2 \cdot [B]$, $r = k[A] \cdot [B] \cdot [C]$
Zero order	0	$r = k$

Differential *vs.* integrated rate law

The *differential rate law* and *integrated rate law* express specific rate laws.

> *Differential rate* laws (above) express the rate of reaction *vs.* concentration.

> *Integrated rate* laws express the change in concentration of components *vs.* time.

The *integrated rate* laws are:

> Integrated rate law for 0th order

> $$[A]_t = -kt + [A]_0$$

> Integrated rate law for 1st order

> $$\ln[A]_t = -kt + \ln[A]_0$$

> Integrated rate law for 2nd order

> $$1 / [A] = 1 / [A]_0 + kt$$

Initial experimental rate method

The *method of initial rates* determines the rate law using experimental data to determine the rate order for reactants.

For example, the results of an experiment show the reaction rates relative to the initial concentrations of reactants:

Concentrations (M)

$[O_2]$		Rate (M/s)	initial $[NO]$
0.10	Trial 1	1.2×10^{-8}	0.10
0.20	Trial 2	2.4×10^{-8}	0.10
0.10	Trial 3	1.08×10^{-7}	0.30

Based on the data above, determine:

> a) reaction orders of the reactants

> b) value and units of k

Rate law by concentrations

To determine the rate law, compare two trials in which the concentration of one reactant changes while the other reactant's concentration is held constant.

It is necessary to determine the reactant's partial order for changed concentrations.

Keep the other reactant concentration constant, so the change in rate is due to one reactant's concentration change.

Consider the results for Trial 1 and Trial 2.

Comparing these two trials shows that NO concentration is constant while O_2 is doubled.

To determine the partial order of O_2, note that doubling the concentration of O_2 doubles the reaction rate.

Expressed as:

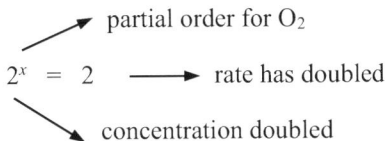

For equality, x equals one.

Therefore, the partial order for O_2 is first-order.

To determine the partial order of NO, select two trials with the concentration of O_2 constant while [NO] varies.

These criteria are satisfied with Trial 1 and Trial 3.

The concentration of O_2 is held constant, while the concentration of NO has tripled.

Under these conditions, the rate of the reaction increased nine-fold.

Expressed as:

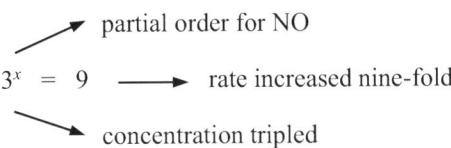

For equality, x equals two.

Therefore, the partial order for NO is second-order.

Calculating rate order experimentally

The calculated rate orders are substituted into the rate law:

$$rate\ law = k[A]^x \cdot [B]^y$$

$$rate\ law = k[NO]^2 \cdot [O_2]$$

Partial first orders are typically omitted, so $[O_2]$ does not have an exponent; the exponent is assumed as one.

With partial rate orders, it is possible to solve the rate constant, k.

For example, choose an experimental trial, and substitute the values of *rate* and *concentrations* of NO and O_2.

Using data from Trial 1:

$$rate = k[NO]^2 \cdot [O_2]$$

$$k = rate\ /\ [NO]^2 \cdot [O_2]$$

$$k = 1.2 \times 10^{-8}\ M^{-2} \cdot s^{-1}\ /\ [(0.10\ M)^2 \cdot (0.10\ M)]$$

$$k = 1.2 \times 10^{-5}\ M^{-2} \cdot s^{-1}$$

Thus, the rate law becomes:

$$rate = 1.2 \times 10^{-5}\ M^{-2} \cdot s^{-1}[NO]^2 \cdot [O_2]$$

The reaction is third order overall, and the units are $M^{-2} \times s^{-1}$.

Notice that the units of k cancel to yield M/s for the reaction rate.

Rate Mechanisms

Activation energy

The reactants must have enough energy to *break existing* and *create new bonds* for a reaction to proceed.

This minimum energy threshold is the *activation energy* (E_a).

The *activation energy barrier* (E_a as the hill on the reaction profile energy *vs.* reaction progress graph) is the difference between the reactants' energy and the highest point on the reaction profile toward product formation.

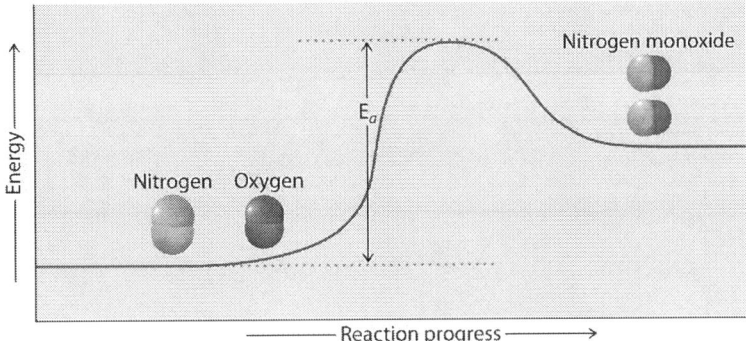

The reacting molecules need sufficient energy to proceed over the energy barrier to transform into products.

For the *forward* reaction:

> the activation energy E_a is the difference between the reactants' energy and the top of the reaction profile (i.e., transition state).

For the *reverse* reaction:

> the activation energy E_a is the difference between the products' energy and the top of the reaction profile (i.e., transition state).

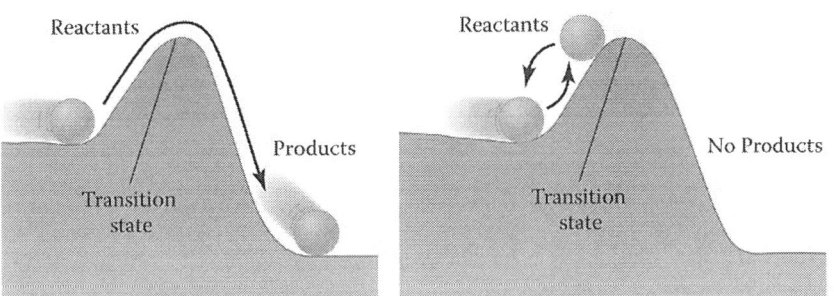

Exergonic reaction forms product   *Exergonic reaction without TS*

Unimolecular mechanisms

Unimolecular mechanisms involve *intermediates* (e.g., carbocation, radicals) during the reaction profile (two peaks on the reaction progress diagram).

The formation of an intermediate of the first step is often the rate-determining (i.e., slow) step.

The intermediate is consumed as the reaction forms products (via the second transition state).

Bimolecular mechanisms

The left image above is a completed bimolecular (E_2 or Sn_2) reaction.

The right image above shows a bimolecular reaction whereby the activation energy (energy barrier) is not overcome: no transition state (TS) forms and no products are generated.

Reaction mechanism influences rates

Most chemical reactions occur in several steps, known as a *reaction mechanism*.

The *reaction mechanism* is the sequence as reactant molecules transform into products.

Each step in a multiple-step mechanism is an *elementary step* with an individual rate constant and rate law.

For example, the following reaction steps were experimentally observed:

$$(CH_3)_3CCl\ (aq) + OH^-\ (aq) \rightarrow (CH_3)_3COH\ (aq) + Cl^-\ (aq)$$

Step 1:

$$(CH_3)_3CCl\ (aq) \rightarrow (CH_3)_3C^+\ (aq) + Cl^-\ (aq)\quad \text{(slow)}$$

$$r_1 = k[(CH_3)_3CCl]$$

Step 2:

$$(CH_3)_3C^+\ (aq) + OH^-\ (aq) \rightarrow (CH_3)_3COH\ (aq)\ \text{(fast)}$$

$$r_2 = k[(CH_3)_3C^+]\cdot[OH^-]$$

Rate-determining step

The *rate-determining step* has the slowest rate and most significant contributor to the reaction's overall rate.

In most cases, the overall reaction rate is almost equal to the rate-determining step's rate because the other steps are fast (almost instantaneous) and do not significantly affect the overall rate.

For example, the first step, the formation of the charged carbonium ion $(CH_3)_3C^+$, is slow compared to the second step, where the carbocation immediately reacts with the OH^- ion to form the neutral product $(CH_3)_3COH$.

$$(CH_3)_3CCl\ (aq) \rightarrow (CH_3)_3C^+\ (aq) + Cl^-\ (aq) \qquad \text{(slow – first step)}$$

$$(CH_3)_3C^+\ (aq) + OH^-\ (aq) \rightarrow (CH_3)_3COH\ (aq) \qquad \text{(fast – second step)}$$

Therefore, the first step is the rate-determining (slow) step.

The rate law of this (slow) step determines the overall rate:

$$r_1 = k[(CH_3)_3CCl]$$

There are reactions when the *rate-determining step* (or *slowest step*) is *not* the first step in the reaction mechanism.

In these examples, an intermediate may appear in the rate-determining step.

The intermediate must be substituted so that it is not in the rate law, requiring a detailed approach for determining the reaction mechanism's rate law.

Notes for active learning

Reaction Rate Depends on Temperature

Kinetic energy influences rate

Increasing temperature increases the colliding molecules' *average kinetic energy*; the molecules move faster.

These faster molecules are more likely to possess the energy necessary to overcome the activation energy and successfully collide and react.

Faster molecules (i.e., higher kinetic energy) collide at an increased frequency.

Therefore, using collision theory, increasing temperature increases the reaction rate.

However, some reactions might slow or stop as the temperature increases. This is common among reactions in biological systems because enzyme proteins involved in the reaction may be damaged by higher temperatures, reducing the enzyme's capability to participate in reactions.

Transition state as energy of activation

The *transition state* (or *activated complex*) is the energy peak (*y*-axis) of the energy profile when bonds begin to *break* (reactants) and other bonds begin to *form* (products).

Transition states form if the reactants have enough energy to achieve the needed activation energy (highest energy location on the reaction progress graph).

A molecule in the transition state *either* reverts to reactants or forms products.

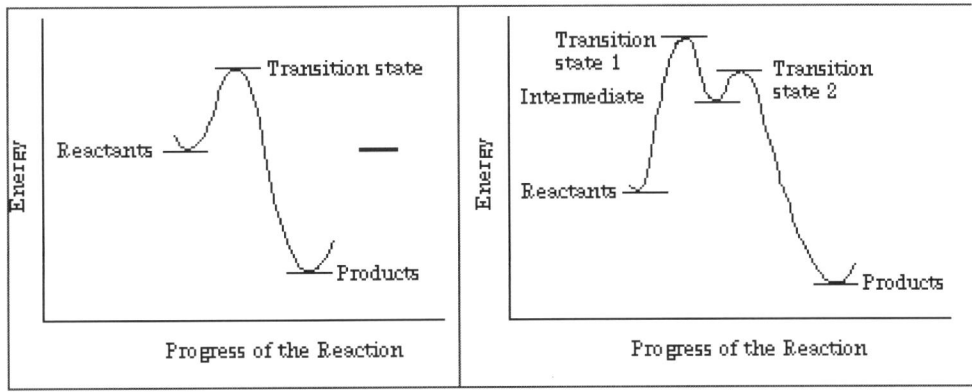

Bimolecular reactions   *Unimolecular reactions*

Molecules in the transition state *cannot be isolated.*

Transition state bond making and breaking events

Transition states signify bond-breaking, and bond-making events, as shown below.

Transition states are denoted by square brackets [ ] and a double dagger ($\ddagger$) around the fleeting (i.e., unable to be isolated) molecule.

Dashed lines symbolize temporary bonds simultaneously *formed and broken*, shown in brackets.

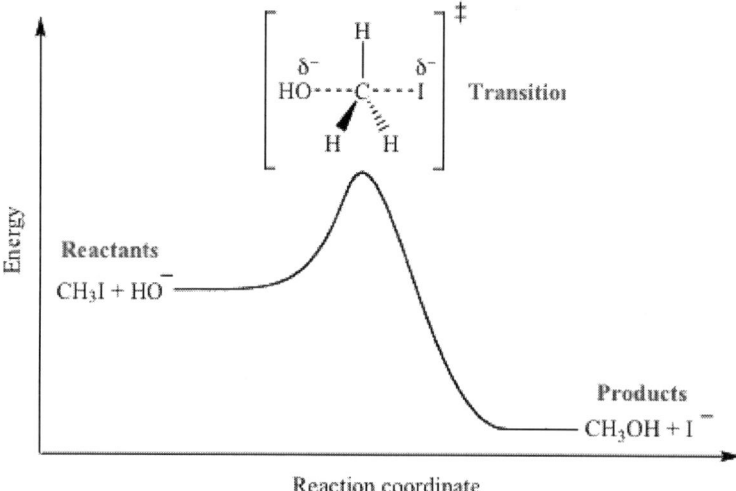

Energy profiles for activation energy E_{ac}

In a reaction, the products and reactants often have different energy levels.

The energy differential is released ($-\Delta H$) or absorbed ($+\Delta H$) from the system as heat (enthalpy or ΔH).

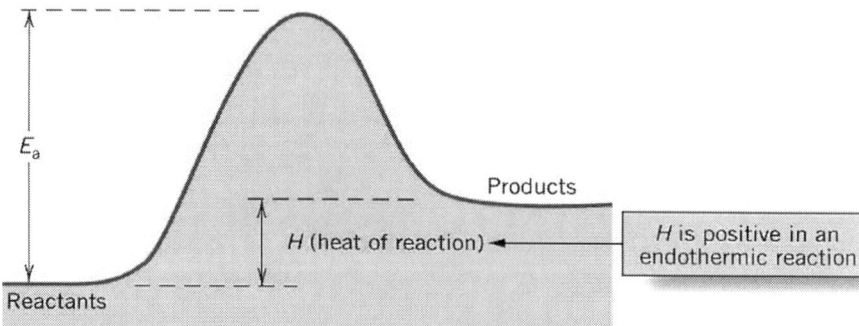

The graph's high point indicates the energy of the activation E_{ac} barrier during the reaction

Reactions are *endothermic* or *exothermic,* based on enthalpy.

In *endothermic* reactions,

> the products have higher potential energy than reactants ($+\Delta H$), favoring the *reverse reaction.*

In *exothermic* reactions,

> the products are lower potential energy than reactants ($-\Delta H$), favoring the *forward reaction.*

The forward reaction has a $-\Delta H$ (release energy) in the forward direction and a $+\Delta H$ in the reverse direction.

Enthalpy (ΔH)

In *endothermic reactions*, the potential energy of products is *higher* (i.e., less stable) than for reactants ($+\Delta H$).

> Heat is required as a reactant to drive the reaction forward.

> Heat is *absorbed* by the system from its surroundings to *break bonds.*

> This is an *energetically unfavorable* reaction, and the temperature of the surroundings *decreases*.

In *exothermic reactions*, the potential energy of products is *lower* (i.e., more stable) than for reactants ($-\Delta H$).

> Excess heat formed by the reaction is released by the system towards its surroundings.

> Energy is *released* when by the system to *form bonds.*

This is an *energetically favorable* reaction, and the temperature of the surroundings *increases*.

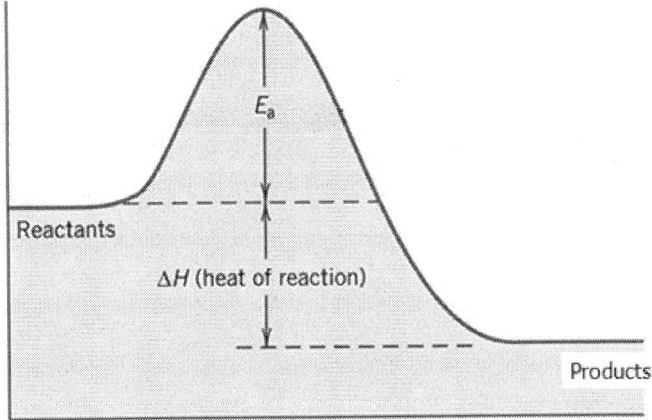

Exothermic reactions form products with lower energy than reactants

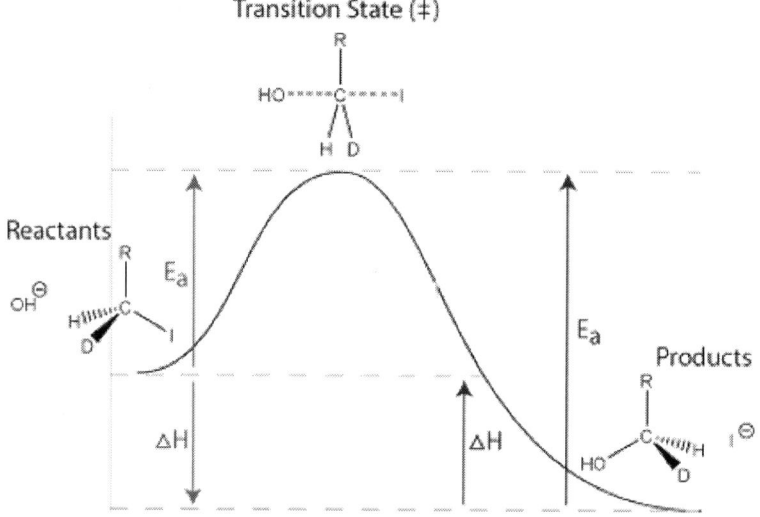

Energy profile diagrams summarize the reaction energies (ΔH and E_{ac})

Hammond postulate

Hammond's postulate (or *axiom*) states that *the transition state resembles either the reactants or the products, to whichever it is closer in energy*.

In an *exothermic* reaction, the transition state (TS) is *closer to the reactants* than to the products in energy.

Therefore, according to Hammond's postulate, the transition state *resembles the reactants*.

In *endothermic* reactions, the transition state is *closer to the products* in energy than to reactants.

Therefore, according to Hammond's postulate, the transition state *resembles the products*.

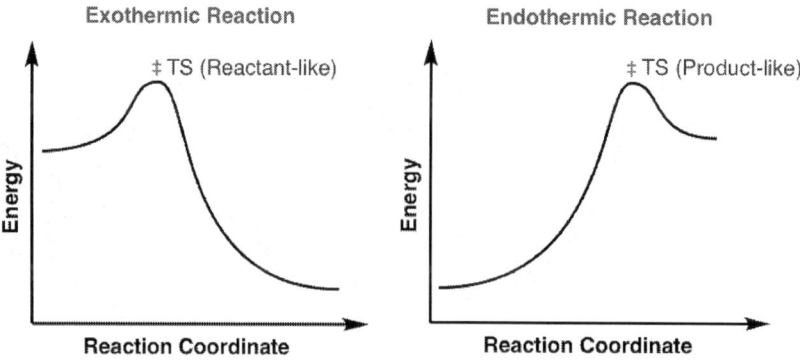

The Hammond postulate states that the transition state resembles the species closer in energy

The E_{ac} is lower (measured from the energy of reactants) to the peak of the energy barrier compared to the E_{ac} in the reverse reaction (measured from the energy of the products).

Arrhenius equation described temperature dependence of reaction rate

In 1889, Swedish Nobel laureate Svante Arrhenius (1859-1927) described the *temperature dependence* of a reaction rate whereby chemical reactions generally occur rapidly at higher temperatures.

Arrhenius equation uses the relationship between rate constant (k), temperature (T), and activation energy (E_a):

$$k = Ae^{-E_{ac}/RT}$$

where k is the rate constant, R is the ideal gas constant ($R = 8.314$ J mol^{-1} K^{-1}), E_{ac} is the activation energy, T is the temperature in Kelvin, and A is the *Arrhenius frequency factor*

The *Arrhenius frequency factor* is the pre-exponential factor (or collision factor) representing collision frequency.

For the Arrhenius equation, ensure that the quantities' units are proper. T is in Kelvin, E_{ac} is in J/mol, and A has the same units as the rate constant k

Point-slope form for the Arrhenius equation

Taking the natural logarithm of each side of the Arrhenius equation yields the point-slope form for the equation:

$$\ln k = -(E_a / RT) + \ln A$$

A *straight-line slope* is produced if data is graphed, where $\ln k$ is on the y-axis, and $1 / T$ is on the x-axis.

On the $\ln k$ *vs.* $1 / T$ plot, the line's slope is E_a / R, and the y-intercept corresponds to $\ln A$.

Determine the slope of the line mathematically to calculate E_a, whereby the slope of a line is the change in y over the change in x (i.e., rise over run).

By extrapolation, the y-intercept and $\ln A$ collision factor are calculated.

The reaction rate constant at two temperatures is required to calculate the point-slope form.

The rate constants relationship at two temperatures is expressed by:

$$\ln k_2 - \ln k_1 = \left(\ln A - \frac{E_a}{RT_2}\right)\left(\ln A - \frac{E_a}{RT_1}\right) = \frac{E_a}{R}\left(\frac{1}{T_1} - \frac{1}{T_2}\right)$$

The equation calculates the rate constant k of a reaction at two temperatures (in Kelvin).

The y-intercept ($\ln A$) can be ignored because it is a constant and not part of the slope calculation.

Eliminate the $\ln A$ term by subtracting the expressions for the two $\ln$-k.

Simplified Arrhenius equation expresses activation energy

Since the slope embodies the activation energy (E_{ac}), the simplified Arrhenius equation determines the E_{ac}.

Rearrange the expression to isolate activation energy.

$$E_a = \frac{R \ln \frac{k_2}{k_1}}{\frac{1}{T_1} - \frac{1}{T_2}}$$

For example, a reaction rate for a chemical process is investigated at two temperatures. The reaction rate at 25 °C is 1.55×10^{-4} s^{-1}. At 50 °C, the reaction rate is 3.88×10^{-4} s^{-1}. Based on this data, what is the energy of activation for the chemical process expressed in J/mol? (Use R = 8.314 J/mol·K)

Inspect the units. The units for rates, s^{-1}, cancel. Since the activation energy is to be expressed in J/mol, the gas constant R is expressed as 8.314 J/mol·K. This choice of the gas constant R dictates temperature units.

The R constant has temperature units expressed in K; temperatures are converted to Kelvin:

$T_1 = 25 °C + 273 = 298$ K

$T_2 = 50 °C + 273 = 323$ K

$$E_a = \frac{R \ln \frac{k_2}{k_1}}{\frac{1}{T_1} - \frac{1}{T_2}}$$

Substitute the rate constants and temperatures into the Arrhenius equation:

$$E_{ac} = \frac{8.314 \text{ J/mol·K} \times \ln(3.88 \times 10^{-4} \text{ s}^{-1} / 1.55 \times 10^{-4} \text{ s}^{-1})}{(1/298 \text{ K}) - (1/323 \text{ K})}$$

$$E_{ac} = \frac{8.314 \text{ J/mol·K} \times \ln(2.503)}{(2.6 \times 10^{-4} \text{ K})}$$

$$E_{ac} = 2.93 \times 10^4 \text{ J/mol}$$

Van 't Hoff equation for K_{eq}

The Van 't Hoff equation relates the change in the equilibrium constant (K_{eq}) of a chemical reaction to the change in temperature.

The Van 't Hoff equation calculates the *equilibrium constant K* (uppercase *K* for equilibrium; lowercase *k* denotes rate) of a reaction at *two Kelvin temperatures*.

The Van 't Hoff equation modifies the Arrhenius equation by replacing E_{ac} with ΔH.

This modified Arrhenius equation is the Van 't Hoff equation:

$$\ln\left(\frac{K_{T_2}}{K_{T_1}}\right) = \frac{\Delta H^\circ}{R}\left(\frac{1}{T_1} - \frac{1}{T_2}\right)$$

The above equation is *not* the Arrhenius equation.

Notes for active learning

Kinetic *vs.* Thermodynamic Control of Reactions

Reaction selectivity

Several chemical reactions have two possible products, and the *reaction conditions* affect selectivity.

Thermodynamics explains how ΔG and other variables affect if a reaction occurs spontaneously.

Kinetics describes how fast a reaction occurs (based on the activation energy), and kinetics has no impact on the spontaneity or thermodynamics of the reaction.

The *kinetic product* requires lower activation energy and is formed preferentially (according to the transition state being of lower energy) at a lower temperature.

The *thermodynamic product* has a greater negative ΔG (favorable) formed preferentially at a higher temperature.

The energy profile diagram displays a reaction under kinetic *vs.* thermodynamic control.

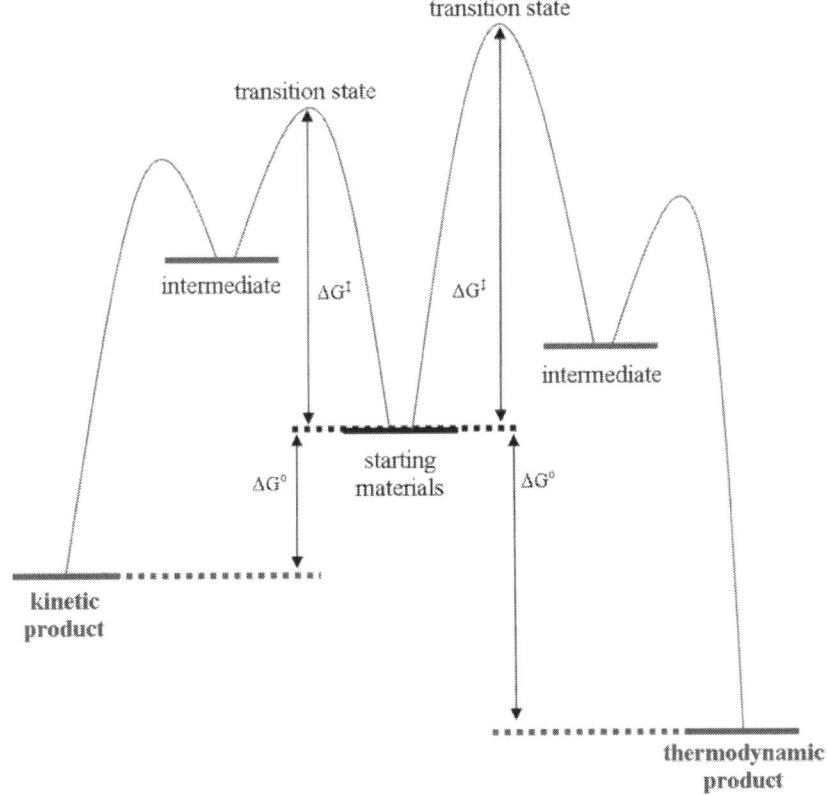

Kinetic vs. thermodynamic control for products depends on energy of transition state

The reactants (i.e., starting material) are in the center of the graph above.

On the left is an intermediate with lower energy than the intermediate on the right.

The intermediate with lower energy forms faster (I.e., kinetic control).

However, the final product from this lower energy intermediate yields a higher energy product than the more stable product on the right (thermodynamic product).

The product on the right on the graph above is the thermodynamic product (overall more stable than the kinetic product). However, it requires a higher (more unstable) transition state before forming a product.

> *Kinetic* products form *fast* because of a *lower energy* transition state.

> *Thermodynamic* products form *slowly* because of a *higher energy* transition state.

Reaction mechanism drives product formation

However, the ΔG, the transition states, and the intermediates and transition states' energy are different.

The product is different, depending on whether the rxn is under kinetic control (top) or thermodynamic (bottom).

When strong acids are added to certain conjugated dienes (e.g., butadiene), two products form the 1,2- product and the 1,4-product.

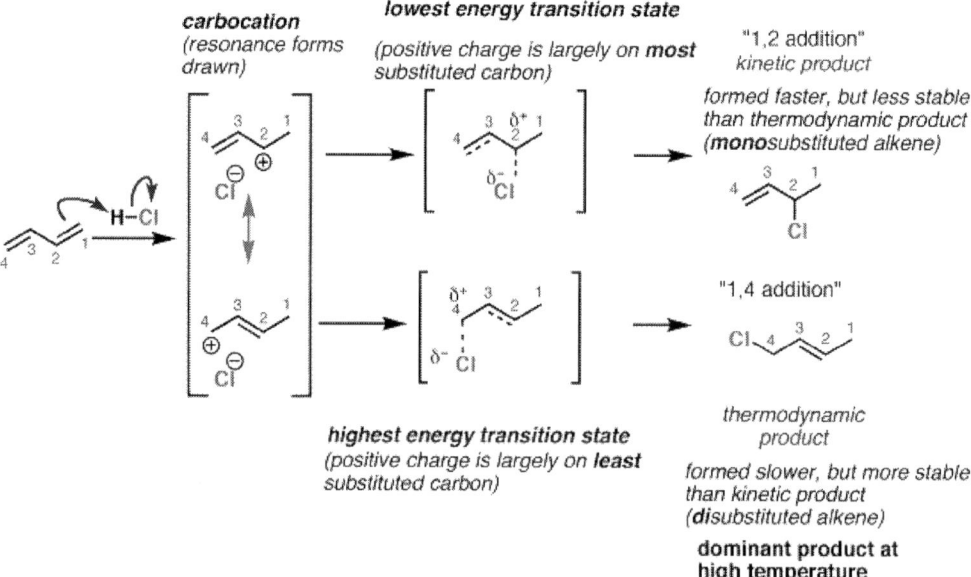

1,2 vs. 1,4 addition of conjugated dienes form kinetic (less stable) and thermodynamic (more stable) products

In the top transition state, the carbocation's positive charge is localized on the most substituted (2°) carbon, and the transition state has lower energy.

The lower potential energy transition state indicates that the reaction has lower activation energy.

Kinetic (1,2) product

The 1,2 *kinetic product* is preferentially formed at *lower temperatures*.

The bottom transition state is higher in energy because the carbocation's positive charge is localized on the less substituted (1°) carbon.

Therefore, this transition state has higher energy. This intermediate indicates that the bottom reaction has higher activation energy.

Thermodynamic (1,4) product

The 1,4 *thermodynamic product* is preferentially formed at higher temperatures needed to proceed with the less stable (2°) carbon intermediate.

The 1,4-product is more stable than the 1,2-product because it is an internal alkene instead of a terminal alkene.

The 1,4-addition produces the thermodynamically stable product (i.e., lower potential energy) but proceeds slower through the transition state of the higher energy (less substituted carbocation).

The 1,4-addition product is the major product formed at higher temperatures needed to proceed through the less stable transition state.

Terminal *vs.* internal alkene

The energy profile diagram for an E_1 *vs.* E_2 reaction is shown:

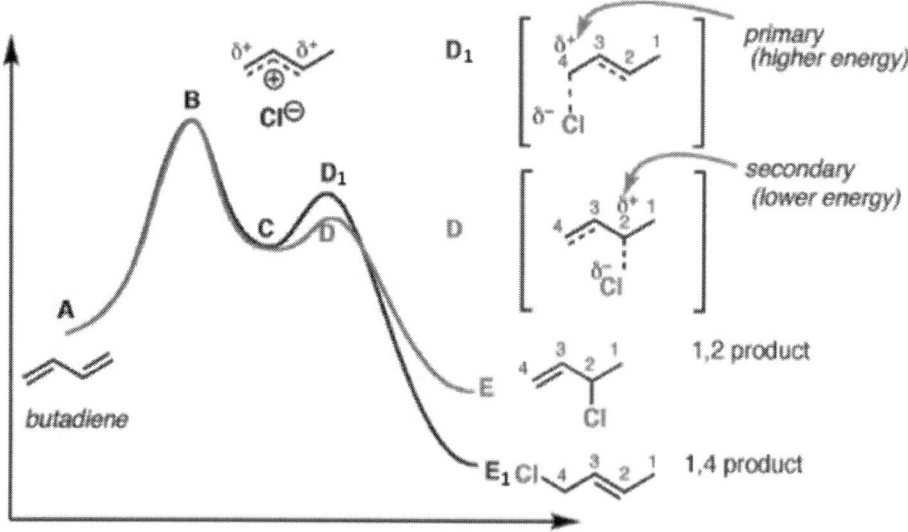

The energy of HCl addition to 1,2-addition vs. 1,4-addition to butadiene

The height of transition states D and D_1 determines the reaction rate once the carbocation C forms.

Lower energy transition states (D *vs.* D_1) proceed faster to products.

Therefore, E (terminal alkene) forms faster from the secondary carbocation intermediate C since the energy of transition is less for D (secondary carbocation) than D_1 (primary carbocation).

The energy of E (terminal alkene) and E_1 (internal alkene) is related to greater stability for the 1,4-addition compared to the 1,2-addition.

E_1 has a more substituted (internal) double bond than E (terminal alkene), which is more stable.

E_1 represents the thermodynamic (more stable) product that is *slower* to form (higher energy transition state) but is *more stable* than the kinetic product E, which formed faster (lower energy transition state) but is less stable than E_1.

Catalysts and Enzymes

Catalysts lower energy of activation

Catalysts are substances in a chemical reaction that increase the reaction rate by lowering the activation energy (often through an alternative pathway) needed to proceed.

A catalyst participates in a chemical reaction but remains chemically unchanged and used repeatedly.

For a reaction mechanism separated into elementary steps, a catalyst appears at the start and reappears at the end of the reaction. A catalyst is neither a reactant nor a product in the reaction.

Acids, bases, and metal ions often act as catalysts.

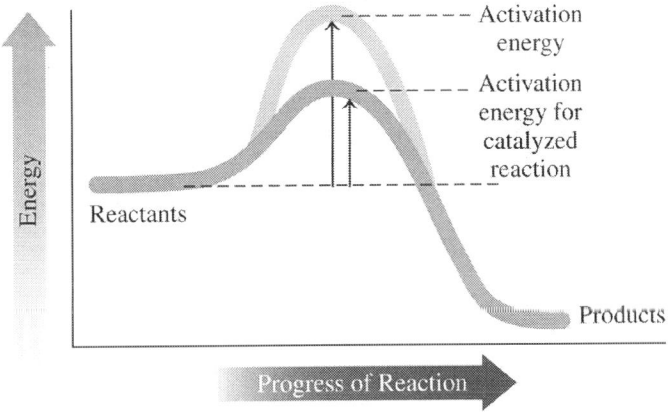

Homogeneous *vs.* heterogeneous catalysts

There are homogeneous and heterogeneous catalysts.

Homogeneous catalysts are in the same phase as the reactants.

Heterogeneous catalysts are in another phase than the reactants. These catalysts often immobilize reactants near each other, increasing the probability of a collision and, therefore, the corresponding reaction rate.

An example of a heterogeneous catalyst is a dissolved acid that catalyzes esters' hydrolysis in an aqueous solution.

Most heterogeneous catalysts are solids, and the reactants are often gases or liquids.

For example, iron catalyzes NH_3 synthesis from gaseous N_2 and H_2.

Modern cars have *catalytic converters* in their exhaust systems as heterogeneous catalysts providing a solid surface for exhaust molecules to bind and react. Catalytic converters reduce pollutants such as carbon monoxide (CO), nitrogen oxide (NO), and hydrocarbons (e.g., octane as C_8H_{18}).

The catalyst usually consists of solid particles, including platinum (Pt) and palladium (Pd).

For example, the reaction:

$$2 \text{ NO } (g) \rightarrow N_2 (g) + O_2 (g)$$

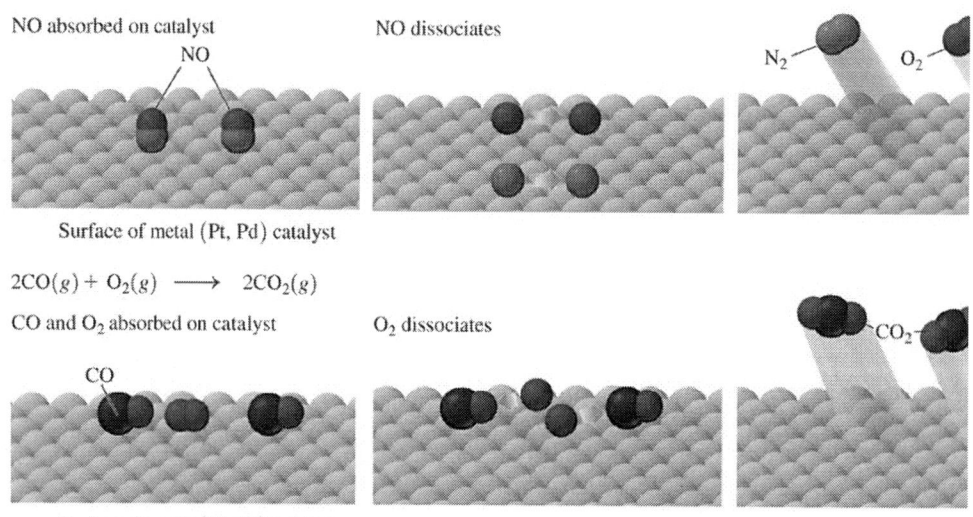

Biological enzymes

Enzymes (e.g., proteins) are *biological catalysts* that increase the reaction rates of biochemical reactions.

Biological enzymes are more efficient than chemical catalysts; enzymes can increase a reaction rate by a factor of more than ten million (1×10^7).

Enzymes are highly specific; they form enzyme-substrate complexes with reactants.

A *substrate* is a molecule on which the enzymes act.

For example, when hydrogen peroxide (H_2O_2) is applied to a wound, oxygen (O_2) gas is produced by decomposing hydrogen peroxide by the catalase enzyme in the blood.

Decomposition of hydrogen peroxide by catalase:

$$2 \text{ H}_2O_2 (l) \xrightarrow{catalase} 2 \text{ H}_2O (l) + O_2 (g)$$

Catalysts are written above or below the arrow in the reaction.

Catalyzed *vs.* non-catalyzed reactions

Changing variables (i.e., reactant concentration, temperature) could increase the reaction rate.

Catalyst has a similar effect on the reaction rate but uses a different method.

Factor	Result
Increasing reactant concentration	More collisions
Increasing temperature	More collisions exceeding the energy of activation
Adding a catalyst	Lowers energy of activation

Notes for active learning

Equilibrium in Reversible Reactions

Reversible reactions

Some chemical reactions are reversible; they proceed in either direction, resulting in an equilibrium mixture of reactants and products.

If a chemical bond forms, it can be broken with sufficient energy input.

Reversible chemical reactions are indicated by a double-headed arrow or two arrows facing opposite directions.

$$A + B \leftrightarrow AB$$

$$A + B \leftrightarrows AB$$

The forward reaction occurs when A and B form AB.

When AB (product) accumulates, the reverse reaction occurs from the decomposition of AB, and A and B (reactants) are reformed.

When the forward and the reverse reactions have the *same rate*, the reaction is in chemical *equilibrium*.

Equilibrium

Equilibrium does not mean equal amounts of reactants and products but is the point where the number of reactants and products *remain constant* because the *forward and reverse reactions proceed at equal rates*.

For example, consider the following reaction:

$$2\,SO_2\,(g) + O_2\,(g) \rightleftarrows 2\,SO_3\,(g)$$

The reaction proceeds forward with SO_2 and O_2, and SO_3 products form until equilibrium (i.e., the equal rate for the forward and reverse reaction).

From the SO_3 products, the reaction proceeds in the reverse direction and forms SO_2 and O_2 until equilibrium.

As shown, the equilibrium concentrations of SO_2, O_2, and SO_3 are the same for the forward and reverse reactions.

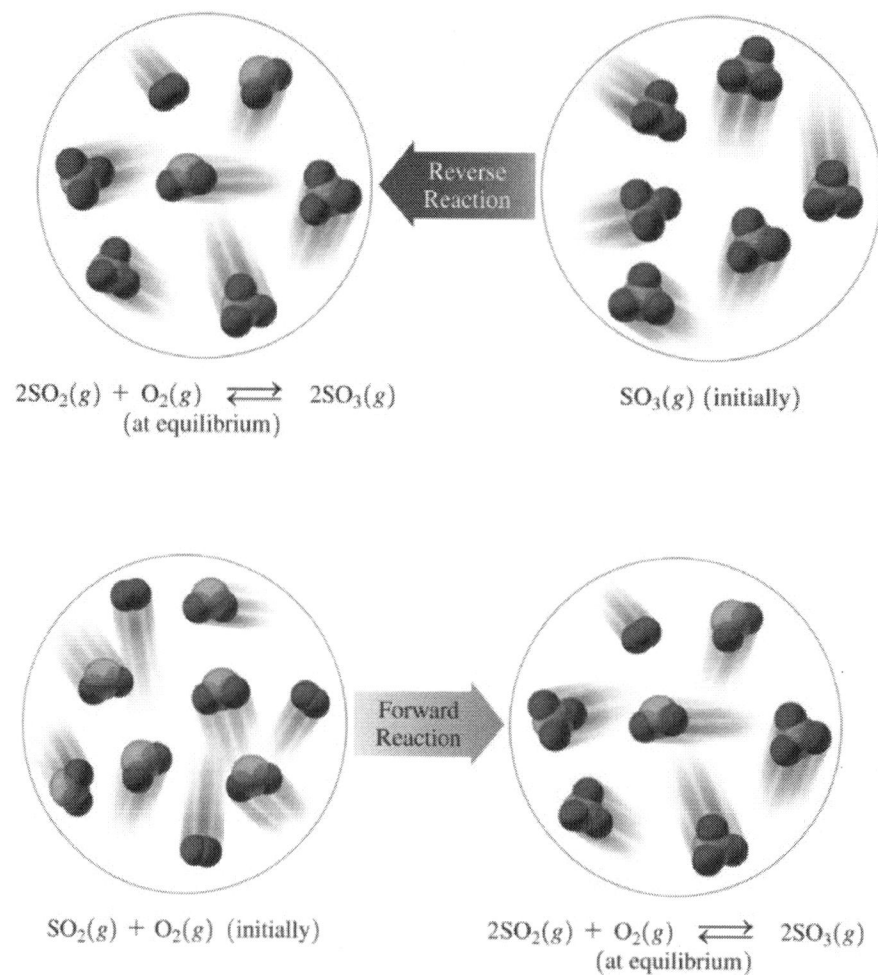

$2SO_2(g) + O_2(g) \rightleftharpoons 2SO_3(g)$
(at equilibrium)

$SO_3(g)$ (initially)

$SO_2(g) + O_2(g)$ (initially)

$2SO_2(g) + O_2(g) \rightleftharpoons 2SO_3(g)$
(at equilibrium)

Experimental evidence for equilibrium

For example, consider the reversible reaction for the evaporation and condensation of water. Suppose water is placed over a gas-filled chamber, separated by a moving piston as diagramed.

Work *on* or *by* the system is observed by measuring the gas's compression or expansion.

As water evaporates, the water's mass decreases, and the gas expands.

When water vapor condenses, it increases water weight and compresses the gas.

Once the system reaches equilibrium, the liquid-to-vapor ratio remains constant. The gas neither compresses nor expands.

Therefore, no work is done *on* or *by* the system at equilibrium.

Water molecules constantly transform between liquid and gas phases, but the evaporation rate matches the condensation rate, so there is no noticeable change in composition.

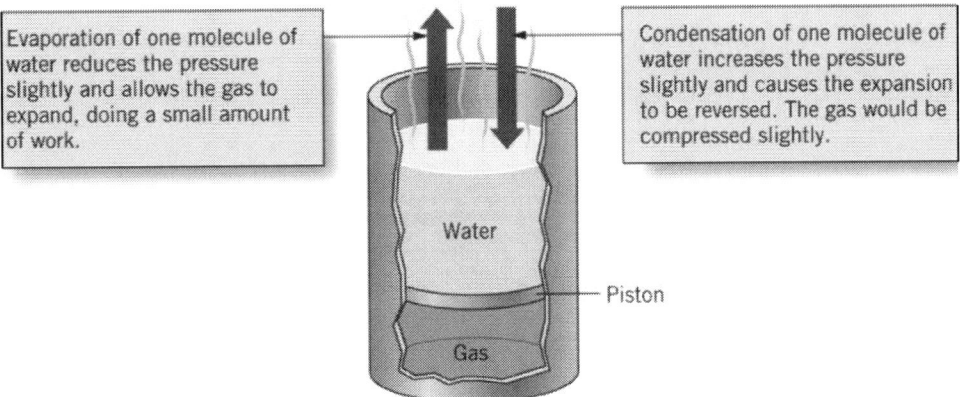

Evaporation of one molecule of water reduces the pressure slightly and allows the gas to expand, doing a small amount of work.

Condensation of one molecule of water increases the pressure slightly and causes the expansion to be reversed. The gas would be compressed slightly.

Water

Piston

Gas

Piston system detects equilibrium for water

Notes for active leaning

Mass Action

Law of Mass Action

The *Law of Mass Action* states that the *reaction rate depends on the concentration of the participating substances*.

$$aA + bB \rightleftarrows cC + dD$$

$$\text{rate}_{\text{rxn}} = \left(-\frac{1}{a}\right)\frac{\Delta A}{\Delta t} = \left(-\frac{1}{b}\right)\frac{\Delta B}{\Delta t} = \left(+\frac{1}{c}\right)\frac{\Delta C}{\Delta t} = \left(+\frac{1}{d}\right)\frac{\Delta D}{\Delta t}$$

Deriving the equilibrium constant

Law of Mass Action *derives the equilibrium constant* by setting the forward reaction rate equal to the reverse reaction rate (i.e., equilibrium).

For the reaction:

$$aA + bB \rightleftarrows cC + dD$$

The following relationships exist:

$$r_{\text{forward}} = r_{\text{reverse}}$$

$$k_{\text{forward}} \cdot [A]^a \cdot [B]^b = k_{\text{reverse}} \cdot [C]^c \cdot [D]^d$$

$$K_c = k_{\text{forward}} / k_{\text{reverse}} = [C]^c \cdot [D]^d / [A]^a \cdot [B]^b$$

Phases for Law of Mass Action

The Law of Mass Action is based on the equilibrium constant K_c, which is the ratio of the concentration of products to the concentration of reactants (i.e., [products] / [reactants]).

The mass action expression of equilibrium *only* includes the equilibrium components in the *same phase*.

Pure solids and liquids do *not* appear in the mass action expression.

The absence of pure solids and liquids from the equation is:

$$CaCO_3 \ (s) \ \rightleftarrows \ CaO \ (s) + CO_2 \ (g)$$

$$K_c = [CO_2]$$

For example, consider the following equation:

$$2\ HCl\ (aq) + CaCO_3\ (s) \leftrightarrows CaCl_2\ (aq) + CO_2\ (g) + 2\ H_2O\ (l)$$

The *net ionic equation* for the above reaction is:

$$2\ H^+\ (aq) + CaCO_3\ (s) \leftrightarrows Ca^{2+}\ (aq) + CO_2\ (g) + 2\ H_2O\ (l)$$

Cl^- ions were present on each side; therefore, they are spectator ions and *removed* from the net equation.

Mass Action equilibrium constants

There are two forms of *mass action equilibrium constants*.

K_{c1} is expressed as the concentrations of the aqueous species.

$$K_{c1} = [Ca^{2+}] / [H^+]^2$$

K_{c2} is expressed as the concentration of the gaseous species.

$$K_{c2} = [CO_2]$$

The K_{c1} and K_{c2} constants have different values.

The commonly reported K values are based on the most accessible phase to measure experimentally.

Do not mix phases, and do not include solids in either expression. When solving mass action problems, check the molecules' phases.

Equilibrium Constants

Equilibrium constant K_c

The *equilibrium constant* K_c is the ratio of the molar concentrations of products divided by reactants at equilibrium.

The K_c is abbreviated as K_{eq}, a general term for an equilibrium constant for concentration or partial pressures.

For the reaction $aA + bB \leftrightarrows cC + dD$, the equilibrium constant is:

$$K_c = \frac{[C]^c[D]^d}{[A]^a[B]^b} = \frac{[\text{products}]}{[\text{reactants}]}$$

The equilibrium constant K_c is a temperature-specific constant, and it has the same value, regardless of the molar concentrations, if the temperature remains constant.

The units of K_c depend on the reaction; K_c is without units.

Calculating K_c

Guidelines for calculating the K_c value:

1. State the given and needed qualities.

2. Write the K_c expression for the equilibrium.

3. Substitute equilibrium (molar) concentrations and calculate K_c.

For example, what is the value of K_c at 443 °C for the equilibrium concentrations:

$H_2(g) + I_2(g) \leftrightarrows 2\,HI(g)$

$[H_2] = 1.2$ mol/L

$[I_2] = 1.2$ mol/L

$[HI] = 0.35$ mol/L

Step 1: State the given and needed quantities

Given values		Needed quantities
Reactants	**Products**	
$[H_2] = 1.2$ mol/L	$[HI] = 0.35$ mol/L	
$[I_2] = 1.2$ mol/L		K_c

Step 2: Write the K_c expression for the equilibrium

$K_c = [HI]^2 / [H_2] \cdot [I_2]$

Step 3: Substitute equilibrium (molar) concentrations and calculate K_c

$K_c = [0.35]^2 / [1.2] \cdot [1.2]$

$K_c = 8.5 \times 10^{-2}$

K_c magnitude indicates reaction direction

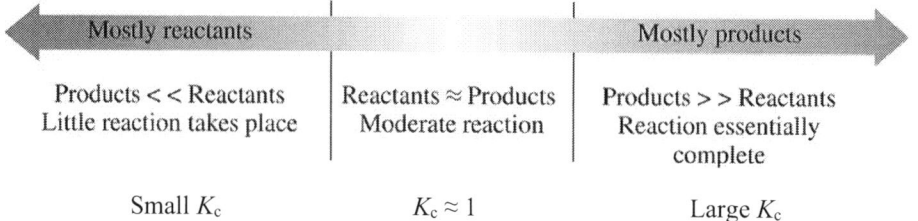

Mostly reactants		Mostly products
Products < < Reactants Little reaction takes place	Reactants ≈ Products Moderate reaction	Products > > Reactants Reaction essentially complete
Small K_c	$K_c \approx 1$	Large K_c

The K_c value depends on whether the equilibrium has a greater quantity of products or reactants.

However, the equilibrium constant's magnitude does not affect how fast equilibrium is reached.

Reactions with a large K_c have more products created from the forward reaction, and products predominate at equilibrium.

Reactions with small K_c have *reactants* predominate at equilibrium.

For example, the equilibrium for the reaction of SO_2 and O_2 has a large K_c.

$$2 \, SO_2 \, (g) + O_2 \, (g) \; \leftrightarrows \; 2 \, SO_3 \, (g)$$

At equilibrium, the reaction contains mostly *products*.

$K_c = [SO_3]^2 / [SO_2]^2 \cdot [O_2]$

$K_c = [\text{many products}] / [\text{few reactants}]$

$K_c = 3.4 \times 10^2$

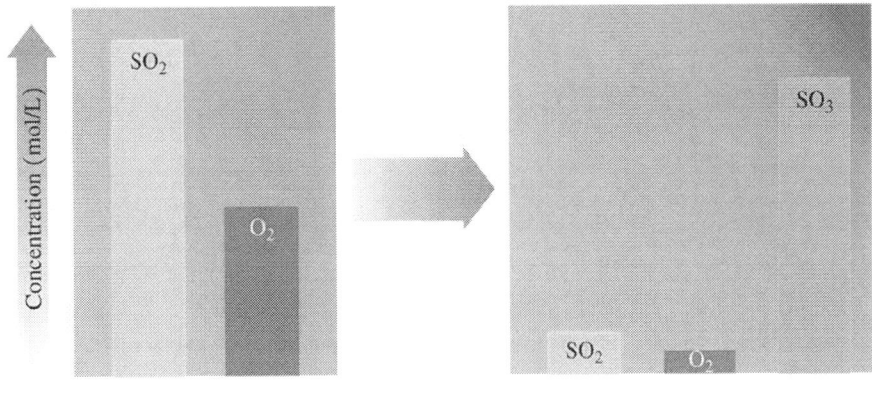

Initially   *At equilibrium*

For example,

$$2 SO_2 (g) + O_2 (g) \leftrightarrows 2 SO_3 (g)$$

Reactions with a small K_c have an equilibrium mixture with a *low concentration of products* and a *high concentration of reactants*.

The equilibrium constant for the reaction of N_2 and O_2 has a small K_c.

$$N_2 (g) + O_2 (g) \leftrightarrows 2 NO (g)$$

At equilibrium, the reaction mixture contains mostly *reactants*.

$$K_c = [NO] / [N_2]\cdot[O_2]$$

$$K_c = [few products] / [mostly reactants]$$

$$K_c = 2 \times 10^{-9}$$

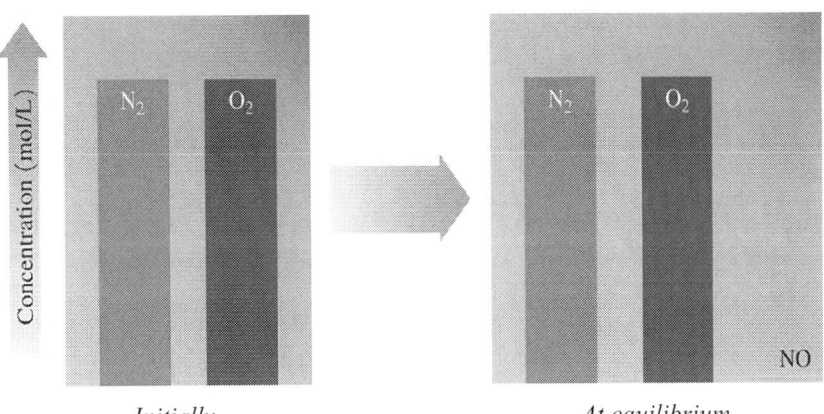

Initially   *At equilibrium*

Homogeneous *vs.* heterogeneous equilibrium

A *homogeneous equilibrium* is a reaction with products and reactants in the same physical state, such as:

$$2 \, SO_2 \, (g) + O_2 \, (g) \rightleftarrows 2 \, SO_3 \, (g)$$

$$K_c = [SO_3]^2 \, / \, [SO_2]^2 \cdot [O_2]$$

A *heterogeneous equilibrium* is a reaction with substances in different physical states:

$$C \, (s) + H_2O \, (g) \rightleftarrows CO \, (g) + H_2 \, (g)$$

$$K_c = [CO] \cdot [H_2] \, / \, [H_2O]$$

The concentrations of liquids and solids do not change and are omitted from the equilibrium constant K_c expression.

This omission of liquids and solids is like mass-action equilibrium.

Reaction quotient Q_c

The reaction quotient Q_c is calculated using the method for K_c, but Q_c calculation can be performed during the reaction.

Q_c is a "snapshot" of the system, representing the ratio between products and reactants at a given time.

Comparing Q_c and K_c determines the direction of the reaction:

If $Q_c > K_c$, the reaction creates more reactants

If $Q_c < K_c$, the reaction creates more products

If $Q_c = K_c$, the reaction is at equilibrium, $\Delta G = 0$

Equilibrium Variables

Le Châtelier's Principle for equilibrium

Le Châtelier's Principle states that when a reversible reaction at equilibrium is stressed by a change in a variable (e.g., concentration, pressure, volume, or temperature), equilibrium shifts to relieve the effect of the change.

For the equilibrium between colorless N_2O_4 and brown NO_2:

$$N_2O_4\ (g) \rightleftarrows 2\ NO_2\ (g)$$

Effect of concentration on equilibrium

If the N_2O_4 *reactant increases*, the reaction shifts to the right to produce more NO_2.

If the NO_2 *product increases*, the reaction shifts to the left to produce more N_2O_4.

Effect of pressure on equilibrium

Increasing pressure shifts the reaction to the side with fewer gas molecules in a gaseous equilibrium.

For the reaction:

$$N_2O_4\ (g) \rightleftarrows 2\ NO_2\ (g)$$

Increasing pressure shifts the reaction to the left, producing N_2O_4.

Changing pressure does *not* shift equilibrium for equal moles of gases as reactants and products.

Effect of decreasing volume on equilibrium

A change in the volume of a gas mixture at equilibrium changes the concentration of gases in the mixture.

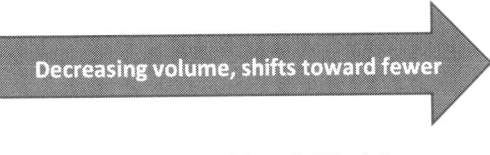

Decreasing volume, shifts toward fewer

$$2\ CO_2\ (g) + O_2\ (g) \rightleftarrows 2\ CO_2\ (g)$$

Decreasing volume increases the concentration of the gases as the system shifts toward the smaller number of moles from the decrease in volume.

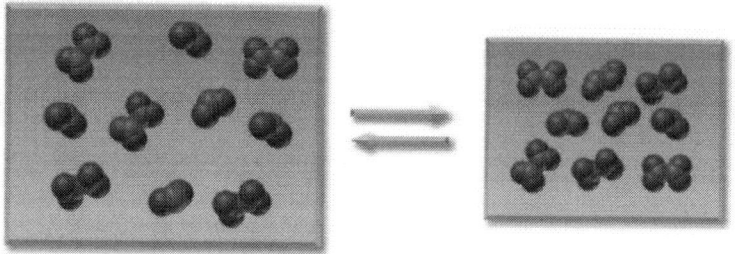

1.00 L at equilibrium   *0.750 L at equilibrium*

Effect of increasing volume on equilibrium

A change in the volume of a gas mixture changes its concentration.

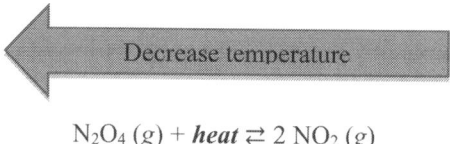

$$2\ CO\ (g) + O_2\ (g) \rightleftarrows 2\ CO_2\ (g)$$

Increasing the volume decreases gas concentration as the system shifts toward the larger number of moles to compensate.

Temperature effect equilibrium depends on the ΔH (i.e., enthalpy) of a reaction (i.e., endothermic or exothermic).

Endothermic reaction equilibrium and temperature

Decreasing temperature of an endothermic reaction ($+\Delta H$ with heat as a reactant) causes the system to shift the reaction toward heat; toward *reactants*, increasing heat in the system.

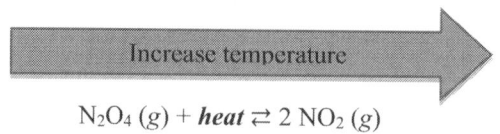

$$N_2O_4\ (g) + \textbf{\textit{heat}} \rightleftarrows 2\ NO_2\ (g)$$

Increasing temperature of an endothermic reaction ($+\Delta H$ with heat as a reactant) causes the system to shift the reaction to remove heat; toward *products* consuming the heat.

$$N_2O_4\ (g) + \textbf{\textit{heat}} \rightleftarrows 2\ NO_2\ (g)$$

Exothermic reaction equilibrium and temperature

Decreasing temperature of an exothermic reaction ($-\Delta H$ with heat as a product) causes the system to shift the reaction toward heat; it shifts the reaction toward *products*, increasing heat in the system.

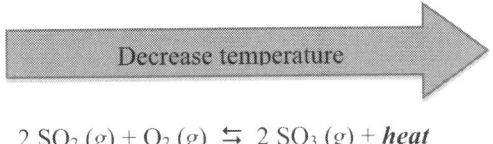

$$2\ SO_2\ (g) + O_2\ (g) \leftrightarrows 2\ SO_3\ (g) + \textbf{\textit{heat}}$$

Increasing temperature of an exothermic reaction ($-\Delta H$ with heat as a product) causes the system to shift the reaction toward removing heat; it shifts the reaction toward *reactants*, decreasing heat in the system.

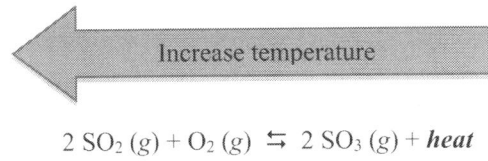

$$2\ SO_2\ (g) + O_2\ (g) \leftrightarrows 2\ SO_3\ (g) + \textbf{\textit{heat}}$$

Variables affecting equilibrium

Variable	Stress	Equilibrium shifts toward
Concentration	Add a reactant	Products (forward reaction)
	Remove a reactant	Reactants (reverse reaction)
	Add a product	Reactants (reverse reaction)
	Remove a product	Products (forward reaction)
Volume (container)	Decrease volume	Side with lower total coefficients (*aq* and *g*)*
	Increase volume	Side with higher total coefficients (*aq* and *g*)
Pressure	Decrease pressure	Side with higher total coefficients (*aq* and *g*)
	Increase pressure	Side with lower total coefficients (*aq* and *g*)
Temperature	**Endothermic Rxn**	
	Raise Temperature	Products (forward reaction to *remove heat*)
	Lower Temperature	Reactants (reverse reaction to *add heat*)
	Exothermic Rxn	
	Raise Temperature	Reactants (reverse reaction to *add heat*)
	Lower Temperature	Products (forward reaction to *remove heat*)
Catalyst	Increases rates equally	No effect

* Pure solids and pure liquids are not included in the equilibrium expression.

Biological equilibrium

Equilibrium is an important concept in many systems, including biological ones.

For example, O_2 transport uses an equilibrium between hemoglobin (Hb), oxygen, and oxyhemoglobin (HbO_2).

$$Hb\ (aq) + O_2\ (g) \rightleftarrows HbO_2\ (aq)$$

$$K_c = [\text{products}] / [\text{reactants}]$$

$$K_c = [HbO_2] / [Hb]{\cdot}[O_2]$$

If there is a high concentration of O_2 (reactant) in the lungs' alveoli, the reaction shifts to the right to make more oxyhemoglobin (product).

When the concentration of O_2 is low in tissues, the reverse reaction releases O_2 from oxyhemoglobin.

At normal atmospheric pressure, oxygen diffuses into the blood because the partial pressure of oxygen in the alveoli is higher than that in the blood.

At altitudes above 8,000 ft, a decrease in atmospheric pressure results in the lower partial pressure of O_2.

Hypoxia may occur at high altitudes where the $[O_2]$ is low. At an altitude of 18,000 ft., a person obtains 29% less oxygen and may experience hypoxia.

According to Le Châtelier's Principle, a decrease in O_2 shifts the equilibrium in the direction of the reactants and depletes the concentration of HbO_2, which causes hypoxia.

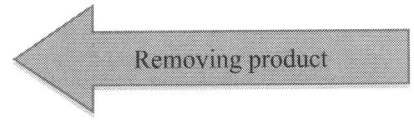

Removing product

$$Hb\ (aq) + O_2\ (g) \rightleftarrows \quad HbO_2\ (aq)$$

Another reaction that applies Le Châtelier's Principle follows.

For example, indicate the shift in equilibrium caused by each change:

$$2\ NO_2\ (g) + \text{heat} \leftrightarrows 2\ NO\ (g) + O_2\ (g)$$

The following changes cause the equilibrium to shift toward reactants or products (*answers below*)?

A. adding NO   C. removing O_2

B. lowering the temperature   D. increasing the volume

                                             E. removing NO

Answers:

A: reactants	C: products	
B: reactants	D: products	E: products

Equilibrium Constant and ΔG° Relationship

Direction and extent of reaction

Gibbs free energy ($\Delta G°$) is an expression for a system's free energy.

The equations relate Gibbs free energy to the equilibrium constant:

$$\Delta G° = -RT \ln K$$

Taking the antilog (e^x) of each side gives:

$$K = e^{-\Delta G° / RT}$$

The ΔG and K imply the direction and extent of a reaction, *not* the rate.

ΔG at equilibrium

If $\Delta G° = 0$, the reaction is at equilibrium; no net change to the system or surroundings.

$$\Delta G° = 0$$

The reaction appears spontaneous in either direction if some small change is made to the system.

The system remains at equilibrium if no heat is added or removed.

Phases exist indefinitely.

ΔG affects equilibrium

ΔG affects the position of equilibrium in the following ways:

$$\Delta G° > 0 \text{ (positive)}$$

If $\Delta G° > 0$, equilibrium is closer to reactants; reverse reaction (towards reactants) is spontaneous.

$$\Delta G° < 0 \text{ (negative)}$$

If $\Delta G° < 0$, equilibrium is closer to products; forward reaction (towards products) is spontaneous.

For example, consider the freezing of water at 0 °C:

$$H_2O \ (l) \rightarrow \ H_2O \ (s)$$

Below 0 °C:

$\Delta G < 0$ and freezing is spontaneous

Above 0 °C:

$\Delta G > 0$ and freezing is nonspontaneous

Energy profile for spontaneous reactions

The energy profile displays a reaction with a $+\Delta G$ (i.e., nonspontaneous).

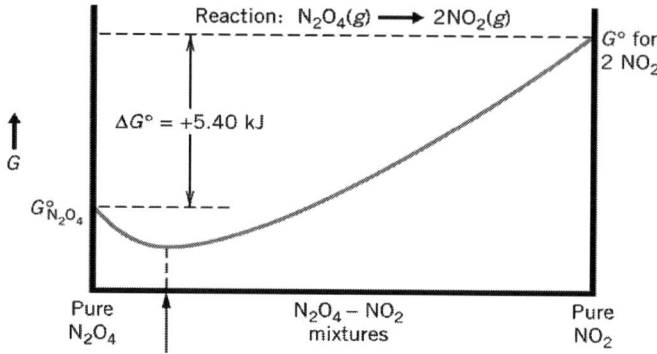

Energy profile for nonspontaneous reactions

The following energy profile represents a reaction with $-\Delta G$ (i.e., spontaneous).

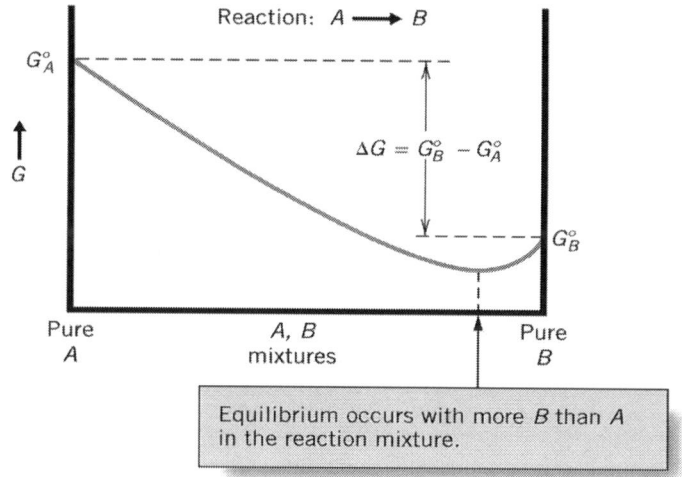

$\Delta G°$ and equilibrium constant K_c relationship

The equilibrium constant Kc is the ratio of the equilibrium concentrations of products over the equilibrium concentrations of reactants, each raised to the power of their stoichiometric coefficients.

$$Kc = [\text{products}]^x / [\text{reactants}]^y$$

where [ ] represents concentration and exponents x and y are stoichiometric coefficients

For example, calculate the equilibrium constant K_c at 25 °C for the decarboxylation (~CO_2) of liquid pyruvic acid ($CH_3COCOOH$) to form gaseous acetaldehyde (CH_3COH) and carbon dioxide (CO_2).

$$CH_3COCOOH \ (l) \rightleftarrows CH_3COH \ (g) + CO_2 \ (g)$$

Compound	$\Delta G°_f$ (kJ/mol)
CH_3COH	− 33.30
$CH_3COCOOH$	−463.38
CO_2	−394.36

Equilibrium constant K_{eq} from ΔG

K_{eq} is a measure of the ratio of the concentrations of products to the concentrations of reactants.

$\Delta G°$ is the standard change in free energy or the change in free energy *under standard conditions*.

$$\Delta G° = -RT \ln K_{eq}$$

where $\Delta G°$ is the standard change in free energy, K_{eq} is the ratio of the concentrations of products to the concentrations of reactants, R = 8.314 J mol^{-1} K^{-1} and T is temperature in Kelvin

Sum $\Delta G°_f$ for each reaction to obtain total $\Delta G°$.

$$\Delta G° = \Sigma \Delta G°_f$$

$$\Delta G° = \Delta G°_f \ (CH_3COH) + \Delta G°_f \ (CO_2) − \Delta G°_f \ (CH_3COCOOH)$$

$$\Delta G° = (−133.30 \ (kJ/mol) + (−394.36 \ (kJ/mol) − (−463.38(kJ/mol))$$

$$\Delta G° = −64.28 \ kJ/mol$$

Use the $\Delta G°$ value to calculate K_{eq}

$$K_{eq} = e^{-\Delta G° / RT}$$

where K is the equilibrium constant, R is the gas constant, and T is the temperature in Kelvin

Solve for the exponent

$$\Delta G° / RT = (-64.28 \text{ kJ/mol}) / [(8.314 \text{ J/K}) \times (298 \text{ K}) \times (1{,}000 \text{ J/kJ})]$$

$$\Delta G° / RT = -25.94$$

Substitute exponent value

$$K_{eq} = e^{-(-25.945)}$$

$$K_{eq} = e^{25.945}$$

$$K_{eq} = 1.85 \times 10^{11}$$

The value for K_{eq} means that the ratio for concentration of products to reactants is 1.85×10^{11}

Therefore, product formation is favored.

Practice Questions

1. Which statement is NOT a correct characterization for a catalyst?

    A. Catalysts are not consumed in a reaction

    B. Catalysts do not actively participate in a reaction

    C. Catalysts lower the activation energy for a reaction

    D. Catalysts can be either solids, liquids, or gases

    E. Catalysts do not alter the equilibrium of the reaction

2. What is the rate law for the following reaction that was found to be first-order in each of the two reactants and second-order overall?

$$2\ NO\ (g) + O_2\ (g) \rightarrow 2\ NO_2\ (g)$$

    A. rate $= k[NO]^2 \cdot [O_2]^2$

    B. rate $= k[NO_2]^2 \cdot [NO]^{-2} \cdot [O_2]^{-\frac{1}{2}}$

    C. rate $= k[NO] \cdot [O_2]$

    D. rate $= k[NO]^2$

    E. rate $= k([NO] \cdot [O_2])^2$

3. Which equilibrium constant applies to a reversible reaction involving a gaseous mixture at equilibrium?

    A. Ionization equilibrium constant, K_w

    B. General equilibrium constant, K_p

    C. Solubility product equilibrium constant, K_{sp}

    D. Ionization equilibrium constant, K_i

    E. None of the above

4. What is the equilibrium constant (K_{eq}) expression for the following reaction:

$$CaCO_3\ (s) \leftrightarrow CaO\ (s) + CO_2\ (g)$$

    A. $K_{eq} = [CO_2]$

    B. $K_{eq} = 1 / [CO_2]$

    C. $K_{eq} = [CaO] \cdot [CO_2] / [CaCO_3]$

    D. $K_{eq} = [CaO] \cdot [CO_2]$

    E. None of the above

5. Which of the following statements about catalysts is NOT true?

    A. A catalyst does not change the energy of the reactants or the products

    B. A catalyst increases the rate of slow reactions

    C. A catalyst is consumed in a reaction

    D. A catalyst alters the rate of a chemical reaction

    E. All the above are true

6. Increasing the temperature of a chemical reaction increases the rate of reaction because:

    A. both the collision frequency and collision energies of reactant molecules increase

    B. the collision frequency of reactant molecules increases

    C. the activation energy increases

    D. the activation energy decreases

    E. the stability of the products increases

7. If $K_{eq} = 6.1 \times 10^{-11}$, which statement is true?

    A. Slightly more products are present

    B. The number of reactants equals products

    C. Mostly products are present

    D. Mostly reactants are present

    E. Cannot be determined

8. Which is true before a reaction reaches chemical equilibrium?

    A. The amounts of reactants and products are equal

    B. The amounts of reactants and products are constant

    C. The amount of products is decreasing

    D. The amount of reactants is increasing

    E. The amount of products is increasing

9. Why might increasing temperature alter the rate of a chemical reaction?

    A. The molecules combine with other atoms at high temperature to save space

    B. The density decreases as a function of temperature that increases volume and decreases the reaction rate

    C. The molecules have higher kinetic energy and have more force when colliding

    D. The molecules are less reactive at higher temperatures

    E. None of the above

10. In the following reaction, how does increasing the pressure affect the equilibrium?

$$2 \, SO_2 \, (g) + O_2 \, (g) \leftrightarrow 2 \, SO_3 \, (g) + heat$$

    A. Remains unchanged, but the reaction mixture gets warmer

    B. Remains unchanged, but the reaction mixture gets cooler

    C. Shifts to the right towards products

    D. Shifts to the left towards reactants

    E. Pressure does not affect equilibrium

11. According to Le Châtelier's Principle, which shifts the equilibrium to the left for the following reactions?

$$N_2 (g) + 3 H_2 (g) \leftrightarrow 2 NH_3 (g) + heat$$

A. Decreasing the temperature

B. Increasing $[H_2]$

C. Increasing $[N_2]$ the pressure

D. Decreasing the pressure on the system

E. Increasing $[N_2]$ and $[H_2]$

12. What is the term for a dynamic state of a reversible reaction in which the rates of the forward and reverse reactions are equal?

A. dynamic equilibrium

B. rate equilibrium

C. reversible equilibrium

D. concentration equilibrium

E. chemical equilibrium

13. When a system is at equilibrium, the:

A. reaction rate of the forward reaction is low compared to the reverse

B. number of products and reactants are equal

C. reaction rate of the forward reaction equals the rate of the reverse

D. reaction rate of the reverse reaction is low compared to the forward

E. none of the above

Notes for active learning

Detailed Explanations

1. B is correct.

Enzymes (i.e., *biological catalysts*) bind to substrates to form an enzyme-substrate complex. While in this complex, enzymes often align reactive chemical groups and hold them closer.

Enzymes can induce structural changes that strain substrate bonds. Therefore, catalysts can actively participate in reactions.

Catalysts *lower the reaction's activation energy* (energy barrier), thus increasing its rate (k).

Although catalysts decrease the amount of energy required to reach the rate-limiting transition state, they do *not* decrease the relative energy of the products and reactants.

Therefore, a catalyst does *not* affect ΔG.

A catalyst provides an *alternative pathway* for the reaction to proceed to product formation. It lowers activation energy (i.e., relative energy between reactants and transition state) and, therefore, *speeds* the reaction rate.

Catalysts do not affect the Gibbs free energy (ΔG: stability of products *vs.* reactants) or the enthalpy (ΔH: bond breaking in reactants or bond making in products).

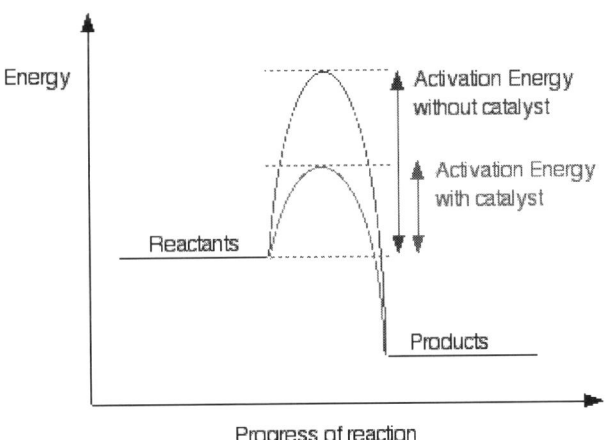

2. C is correct.

The *rate law* is calculated by comparing trials and determining how changes in the initial concentrations of the reactants affect the reaction rate.

$$\text{rate} = k[A]^x \cdot [B]^y$$

where k is the rate constant, and the exponents x and y are the partial reaction orders (i.e., determined experimentally). They are *not* equal to the stoichiometric coefficients

To write a complete rate law, each reactant's order is required.   *continued...*

The problem provided each reactant's order (first order in each).

The total order is a sum of the individual orders: $1 + 1 = 2$.

However, the total order is usually not expressed in the rate law.

3. B is correct.

To determine the amount of each compound at equilibrium, consider the chemical reaction written in the form:

$$aA + bB \leftrightarrow cC + dD$$

The *equilibrium constant* is:

$$K_{eq} = ([C]^c \times [D]^d / ([A]^a \times [B]^b)$$

K_w is the self-ionization (i.e., autoionization) constant for water.

K_{sp} is used for "solubility" and is only applicable for equilibriums for solids dissolving in liquids.

K_i is used for "ionization" and only applies to equilibrium systems involving liquids.

For equilibria in the gas phase, the equilibrium equation (K_p) is a function of the partial pressures (P) of the reactants and products.

Therefore:

$$K_p = ([P_C]^c \times [P_D]^d) / ([P_A]^a \times [P_B]^b)$$

where P represents partial pressure, usually in atm

4. A is correct.

General formula for the *equilibrium constant* of a reaction:

$$aA + bB \leftrightarrow cC + dD$$

$$K_{eq} = ([C]^c \times [D]^d) / ([A]^a \times [B]^b)$$

For equilibrium constant calculation, only include species in *aqueous* or *gas phases*:

$$K_{eq} = [CO_2]$$

5. C is correct.

The primary function of catalysts is lowering the reaction's activation energy (energy barrier), thus increasing its rate (k).

Although catalysts decrease the amount of energy required to reach the rate-limiting transition state, they do *not* decrease the relative energy of the products and reactants.

Therefore, a catalyst does *not* affect ΔG.

A catalyst is never consumed in a reaction; reagents are consumed.

A catalyst lowers the energy of the high-energy transition state (i.e., the activation energy), but it does not change the energy of the reactants or the products.

Catalysts *increase the rate* of chemical reactions.

6. A is correct.

Temperature measures the molecules' average kinetic energy (KE).

Increasing the temperature increases the collision frequency (i.e., due to an increased probability of molecules striking each other) *and* collision energies (kinetic energy = $\frac{1}{2}mv^2$) of reactant molecules.

7. D is correct.

For the general equation:

$$a\mathrm{A} + b\mathrm{B} \leftrightarrow c\mathrm{C} + d\mathrm{D}$$

The equilibrium constant is:

$$K_{eq} = ([\mathrm{C}]^c \times [\mathrm{D}]^d) / ([\mathrm{A}]^a \times [\mathrm{B}]^b)$$

or

$$K_{eq} = [\text{products}] / [\text{reactants}]$$

If K_{eq} is less than 1 (e.g., 6.1×10^{-11}), then the numerator (i.e., products) is smaller than the denominator (i.e., reactants), and fewer products have formed relative to the reactants.

If the reaction favors reactants compared to products, the equilibrium lies to the left.

8. E is correct.

A reaction proceeds with the formation of products.

Therefore, the forward reaction rate is greater than the reverse reaction rate until the reaction achieves equilibrium.

At equilibrium, the rate (not relative concentrations of products and reactants) decreases for the forward reaction while the reverse reaction rate increases.

At equilibrium, the forward reaction rate equals the reverse reaction rate.

An *endergonic reaction* (i.e., $+\Delta G$) has fewer products forming than reactants remaining (i.e., the reaction is nonspontaneous) with higher energy than the reactants.

An *exergonic reaction* (i.e., $-\Delta G$) has more products formed than reactants remaining (i.e., the reaction is spontaneous), with the products lower in energy than the reactants.

Catalysts lower the reaction's activation energy (energy barrier), thus increasing its rate (k).

Although catalysts decrease the amount of energy required to reach the rate-limiting transition state, they do *not* decrease the relative energy of the products and reactants.

Therefore, a catalyst does not affect ΔG.

A catalyst provides an alternative pathway for the reaction to proceed to product formation.

Catalysts lower the energy of activation (i.e., relative energy between reactants and transition state) and, therefore, speed the reaction rate.

Catalysts do *not* affect the Gibbs free energy (ΔG: stability of products *vs.* reactants) or the enthalpy (ΔH: bond breaking in reactants or bond making in products).

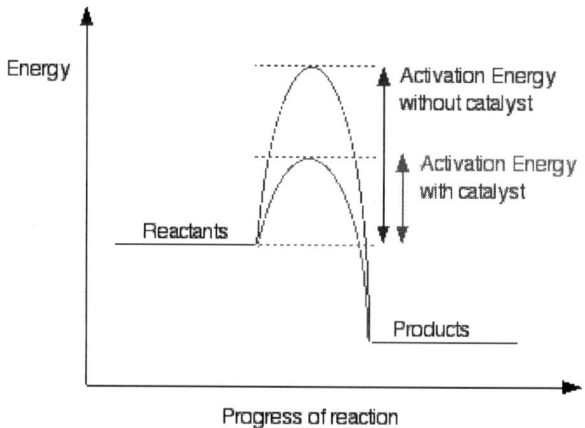

9. C is correct.

As the average kinetic energy (i.e., $KE = \frac{1}{2}mv^2$) *increases*, the particles move faster, collide more frequently per unit time, and possess greater energy when they collide.

This *increases* the *reaction rate*. Hence the *reaction rate* of most *reactions increases* with *increasing temperature*.

10. C is correct.

When changing the reaction conditions, Le Châtelier's Principle states that the position of equilibrium shifts to counteract the change.

If the reaction temperature, pressure, or volume change, the position of equilibrium changes.

In general, increasing the pressure tends to favor the side of the reaction with a lower molar coefficient sum (i.e., toward the products in this example).

11. D is correct.

When changing the reaction conditions, Le Châtelier's Principle states that the position of equilibrium shifts to counteract the change.

If the reaction temperature, pressure, or volume changes, the equilibrium changes.

Removing reactants or adding products shifts the equilibrium to the left.

In this reaction, reducing the pressure of the system favors the reactants because of their respective molar concentrations of reactants (i.e., 1 + 3) and products (i.e., 2).

All other modifications listed would shift the equilibrium toward products (i.e., to the right).

12. E is correct.

Equilibrium refers to the state when the forward reaction rate equals the reverse reaction rate.

Equilibrium does not describe the state when the relative energy of the reactants and products is the same, nor does it describe the state when the relative amounts of reactants and products are the same.

13. C is correct.

Chemical equilibrium refers to a dynamic process whereby the rate of reactant molecules transforming into products equals the rate for a product molecule to be transformed into a reactant.

Therefore, the reaction rate of the forward reaction equals the reverse rate.

Notes for active learning

Notes for active learning

Notes for active learning

CHAPTER 7

Acids and Bases

- Acid–Base Nomenclature

- Acidic and Basic Solutions

- Acid–Base Equilibria

- Acid–Base Ionization

- Periodic Trends of Acids and Bases

- Percent Dissociation

- Salts of Acid–Base Reactions

- pH calculations for Acid–Base Solutions

- Buffered Solutions

- Ionic Equations and Neutralization

- Titration

- Indicators

- Titration Curves

- Practice Questions & Detailed Explanations

Acid–Base Nomenclature

Characteristics of acids and bases

Before the twentieth century, acids, bases, and salts were characterized by properties such as taste and the ability to change litmus's color (a water-soluble mixture of organic dyes).

Acids taste *sour* (e.g., lemon juice), bases taste *bitter* (e.g., mustard), and salts, as their name suggests, taste *salty* (e.g., sodium chloride as table salt).

Acids cause blue litmus paper to turn *red*, bases turn red litmus paper *blue*, while neutral compounds do not affect litmus paper's color.

Bases are recognized by their slippery feel (e.g., soap).

The position on the pH scale categorically classifies acids and bases.

Some complex metabolic processes in the human body are controlled by physiological pH ($\approx$ 7.4); a small change may lead to severe illness and death.

The chemistry of acids and bases has a significant role in nature and industry processes. For example, soil acidity is essential for plant growth.

Acids and bases are particularly essential in manufacturing industries. Sulfuric acid (H_2SO_4) is the most widely produced chemical, and it is needed to produce fertilizers, polymers, steel, and many other materials.

The extensive use of sulfuric acid has led to environmental problems, such as the phenomenon of acid rain.

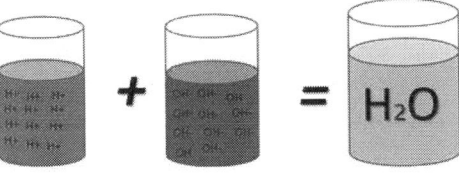

Acids containing   *Bases containing*   *Acids and bases combine*
protons (H^+)   *hydroxides (^-OH)*   *to produce water (H_2O)*

Binary acids and oxoacids

There are *two types of acids*:

> *binary acids* (acids that do not contain oxygen atoms)

> *oxoacids* (acids containing oxygen atoms).

Acids have slightly different names from the ionic naming rules discussed, while bases follow the ionic naming rules discussed.

Naming binary acids

Binary acids start with *hydro-*, then the first syllable of the anion's name, and end with *~ic*:

HF : *hydro*fluor*ic* acid

HCl : *hydro*chlor*ic* acid

HI : *hydro*iod*ic* acid

HBr : *hydro*brom*ic* acid

H$_2$S : *hydro*sulfur*ic* acid

HCN : *hydro*cyan*ic* acid

Oxoacid's names derive from the name of the oxyanion of the acid.

Naming associated anions

For anions ending with *~ate*, the acid's name starts with the first syllable of the anion name and ends with *~ic*.

For anions ending with *~ite*, the name of acid starts with the first syllable of the anion's name and ends with *-ous*.

Anion	Name of Anion	Acid	Name of Acid
NO$_3^-$	nitrate ion	HNO$_3$	nitric acid
NO$_2^-$	nitrite ion	HNO$_2$	nitrous acid
SO$_4^{2-}$	sulfate ion	H$_2$SO$_4$	sulfuric acid
SO$_3^{2-}$	sulfite ion	H$_2$SO$_3$	sulfurous acid
PO$_4^{3-}$	phosphate ion	H$_3$PO$_4$	phosphoric acid
C$_2$H$_3$O$_2^-$	acetate ion	HC$_2$H$_3$O$_2$	acetic acid
ClO$^-$	hypochlorite	HClO	hypochlorous acid
ClO$_2^-$	chlorite	HClO$_2$	chlorous acid
ClO$_3^-$	chlorate	HClO$_3$	chloric acid
ClO$_4^-$	perchlorate	HClO$_4$	perchloric acid

Arrhenius acids and bases

In 1884, Swedish chemist Svante Arrhenius defined acids and bases:

> *Acids* dissociate in water to produce *hydrogen ions* (H^+)

> *Bases* dissociate to in water produce *hydroxide ions* (^-OH)

For example, HCl is an *acid*:

$$HCl\ (aq) \rightarrow H^+\ (aq) + Cl^-\ (aq)$$

For example, NaOH is a *base*:

$$NaOH\ (aq) \rightarrow Na^+\ (aq) + {}^-OH\ (aq)$$

Arrhenius acids increase the hydronium ion concentration $[H_3O^+]$ in an aqueous solution.

Arrhenius bases increase the hydroxide ion concentration $[^-OH]$ in an aqueous solution.

Acids, according to the Arrhenius concept:

1. $HCl\ (aq) + H_2O \rightarrow H_3O^+\ (aq) + Cl^-\ (aq)$

2. $HNO_3\ (aq) + H_2O \rightarrow H_3O^+\ (aq) + NO_3^-\ (aq)$

3. $CH_3COOH\ (aq) + H_2O \rightleftarrows H_3O^+\ (aq) + CH_3COO^-\ (aq)$

Bases, according to the Arrhenius concept:

1. $NaOH\ (aq) \rightarrow Na^+\ (aq) + OH^-\ (aq)$

2. $Ba(OH)_2\ (aq) \rightarrow Ba^{2+}\ (aq) + 2\ OH^-\ (aq)$

3. $NH_3\ (aq) + H_2O \rightleftarrows NH_4^+\ (aq) + OH^-\ (aq)$

Lewis acids and bases

A *Lewis base* provides *a lone pair of electrons* to another reactant to form a coordinate covalent bond.

A *Lewis acid* has an empty orbital to accept *a pair of electrons* from another reactant to form a coordinate covalent bond.

In 1923, Gilbert Newton Lewis (1875-1946) proposed a definition that Lewis acids (e.g., H^+) accept a pair of nonbonding electrons.

> Lewis acids are electron-pair *acceptors*.

> Bases are electron-pair *donors*.

H+ is the acid while water and ammonia are the Lewis bases in the following reactions:

$$H^+ \; + \; H_2O \; \rightarrow \; H_3O^+ \qquad\qquad H^+ \; + \; NH_3 \; \rightarrow \; NH_4^+$$

Lewis	Lewis		Lewis	Lewis
acid	base		acid	base

In reactions forming new covalent bonds, the species with an incomplete octet (i.e., an electron-deficient molecule) may act as Lewis acids; a lone pair of electrons may act as Lewis bases.

In the following reactions,

BF_3, $AlCl_3$, and $FeBr_3$ are Lewis acids

while

NH_3, Cl^- and Br^- are Lewis bases

$$BF_3 + NH_3 \rightarrow F_3B{:}NH_3 \qquad AlCl_3 + Cl^- \rightarrow AlCl_4^- \qquad FeBr_3 + Br^- \rightarrow FeBr_4^-$$

In the formation of complex ions, the positive ions act as Lewis acids, and the ligands (anions or small molecules) act as Lewis bases:

$$Cu^{2+}(aq) \quad + \quad 4\,NH_3\,(aq) \; \rightleftarrows \; Cu(NH_3)_4^{2+}\,(aq)$$

Lewis acid   Lewis base

$$Al^{3+}(aq) \quad + \quad 6\,H_2O \; \rightleftarrows \; [Al(H_2O)_6]^{3+}\,(aq)$$

Lewis acid   Lewis base

The *ionizable hydrogen* in oxoacids is bonded to the oxygen.

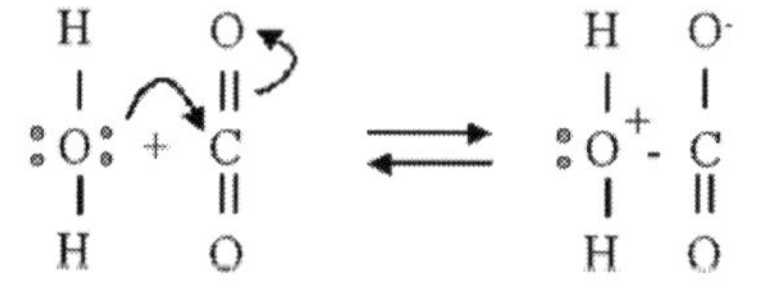

Lewis base   Lewis acid

Brønsted–Lowry acids and bases

The *Brønsted-Lowry theory* was proposed, in 1923, independently by Danish chemist Johannes Nicolaus Brønsted (1879-1947) and British chemist Martin Lowry (1874-1936).

The Brønsted-Lowry theory states that:

an acid is a substance that acts as a proton donor, and

a base is a substance that acts as a proton acceptor.

Brønsted–Lowry conjugate acids and conjugate bases

From the exchange of protons,

acids form its *conjugate base*

bases form its *conjugate acid*

Brønsted-Lowry acid-base reactions can be represented as follows:

$$HA \; + \; B \; \rightleftarrows \; BH^+ \; + \; A^-$$
acid   base   conjugate   conjugate
                acid   base

Brønsted-Lowry acids, bases, conjugate acids, and conjugate bases

$$HCl \; + \; H_2O \; \rightarrow \; H_3O^+ (aq) \; + \; Cl^- (aq)$$
acid   base   conjugate   conjugate
                       acid   base

$$HC_2H_3O_2 + \; H_2O \; \rightleftarrows \; H_3O^+ (aq) \; + \; C_2H_3O_2^- (aq)$$
acid   base   conjugate   conjugate
                          acid   base

$$NH_3 + H_2O \; \rightleftarrows \; NH_4^+ (aq) + OH^- (aq)$$
base   acid   conjugate   conjugate
                   acid   base

The transfer of protons in a Brønsted-Lowery acid-base reaction

$$HF \, (aq) \quad + \quad NH_3 \quad \rightleftarrows \quad F^- \quad + \quad {}^+NH_4 \, (aq)$$

$HF (aq)$	NH_3	F^-	$^+NH_4 (aq)$
Acid donates H$^+$ to NH$_3$	Base accepts H$^+$ from HF	Conjugate base accepts H$^+$ / $^+$NH$_4$	Conjugate acid donates H$^+$ to F$^-$

Amphoteric substances

Water is an *amphoteric* substance because it can *act as an acid or a base*. It acts as a base by accepting a hydrogen ion; another water molecule acts as an acid.

Therefore, the ions H_3O^+ *(aq)* + OH^- *(aq)* form in pure water.

As ions form, they react to produce water again with the following equilibrium:

$$H_2O \;+\; H_2O \;\rightleftarrows\; H_3O^+\,(aq) \;+ OH^-\,(aq)$$

    acid   base   conjugate   conjugate

                              acid   base

Autoionization of water

In the autoionization of water, a proton is transferred from one water molecule to another, producing a hydronium ion (H_3O^+) and a hydroxide ion (OH^-).

Ions formed in the auto-ionization of water are:

$$H_2O \,(l) + H_2O \,(l) \;\rightleftarrows\; H_3O^+\,(aq) + OH^-\,(aq)$$

This process of exchange is the *autoionization of water*.

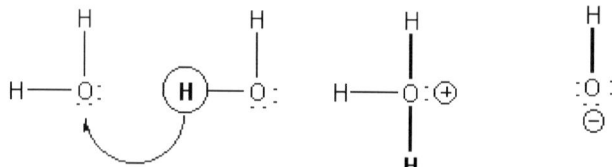

H^+ acceptor base   *H^+ donor acid*   *Conjugate acid*   *Conjugate base*

Equilibrium constant K_w for autoionization

The equilibrium constant K_w:

$$K_w = [H_3O^+]\cdot[^-OH]$$

where K_w is the autoionization constant for water

At standard temperature and pressure (25 °C, 1 atm), the equilibrium constant K_w (*ion-product constant*) for water:

$$K_w = [H_3O^+]\cdot[^-OH]$$

$$K_w = 1.0 \times 10^{-14}$$

For example,

If $[H_3O^+]$ increases ($> 1.0 \times 10^{-7}$ *M*)

If $[^-OH]$ decreases ($< 1.0 \times 10^{-7}$ M)

If $[H_3O^+] = [^-OH] = 1.0 \times 10^{-7}$ M, neutral solution (e.g., pure water)

If $[H_3O^+] > 1.0 \times 10^{-7}$ M, $[^-OH] < 1.0 \times 10^{-7}$ M, solution is acid ($[H^+] > [^-OH]$)

If $[H_3O^+] < 1.0 \times 10^{-7}$ M, $[^-OH] > 1.0 \times 10^{-7}$ M, solution is basic ($[H^+] < [^-OH]$)

Notes active learning

Acidic and Basic Solutions

pH and pOH

The pH scale measures acidity or basicity, especially when the hydrogen ion (H^+) concentration is extremely low.

$$pH = -\log[H^+]$$

$$pOH = -\log[^-OH]$$

Neutral solutions

$$[H^+] = 1.0 \times 10^{-7} \text{ M}$$

$$pH = -\log(1.0 \times 10^{-7})$$

$$pH = 7.00$$

Neutral solutions contain

$$[^-OH] = 1.0 \times 10^{-7} \text{ M}$$

$$pOH = -\log(1.0 \times 10^{-7})$$

$$pOH = 7.00$$

Acidic solutions

$$[H^+] > 1.0 \times 10^{-7} \text{ M}$$

$$pH < 7.00$$

Basic solutions

$$[H^+] < 1.0 \times 10^{-7} \text{ M}$$

$$pH > 7.00$$

pH standards

$$pH = 7 \rightarrow \text{a neutral solution}$$

$$pH < 7 \rightarrow \text{an acidic solution}$$

$$pH > 7 \rightarrow \text{a basic solution}$$

Note

$$K_w = [H^+] \cdot [^-OH]$$

$$K_w = 1.0 \times 10^{-14}$$

$$pK_w = -\log(K_w)$$

$$pK_w = -\log(1.0 \times 10^{-14})$$

$$pK_w = 14.00$$

but

$$pK_w = pH + pOH = 14.00$$

$$pOH = 14.00 - pH$$

$$pH = 14.00 - pOH$$

if

$$pH = 7$$

$$pOH = 7$$

if

$$pH < 7$$

$$pOH > 7$$

if

$$pH > 7$$

$$pOH < 7$$

thus

$$pH < 7$$

$$[H^+] > [^-OH]$$

and

$$pH > 7$$

$$[^-OH] > [H^+]$$

For example. if

$$[H^+] = 1.0 \times 10^{-4}\ M$$

$$pH = -\log(1.0 \times 10^{-4})$$

$$pH = 4.00$$

when

$$[^-OH] = 1.0 \times 10^{-4} \text{ M}$$

$$[H^+] = 1.0 \times 10^{-14} / 1.0 \times 10^{-4} \text{ M}$$

$$[H^+] = 1.0 \times 10^{-10} \text{ M}$$

$$pH = -\log(1.0 \times 10^{-10})$$

$$pH = 10.00$$

Alternatively,

$$[^-OH] = 1.0 \times 10^{-4} \text{ M}$$

$$pOH = -\log[^-OH]$$

$$pOH = -\log(1.0 \times 10^{-4} \text{ M})$$

$$pOH = 4.00$$

$$pH = 14.00 - 4.00$$

$$pH = 10.00$$

Reference values for [H$^+$] and pH

[H$^+$], M	pH	[H$^+$], M	pH
1.0×10^{-1}	1.00	1.0×10^{-8}	8.00
1.0×10^{-2}	2.00	1.0×10^{-9}	9.00
1.0×10^{-3}	3.00	1.0×10^{-10}	10.00
1.0×10^{-4}	4.00	1.0×10^{-11}	11.00
1.0×10^{-5}	5.00	1.0×10^{-12}	12.00
1.0×10^{-6}	6.00	1.0×10^{-13}	13.00
1.0×10^{-7}	7.00	1.0×10^{-14}	14.00

pH of an acidic solution

To calculate the pH of an acidic solution, use the expression:

$$pH = -\log[H_3O^+]$$

For example, if

$$[H_3O^+] = 1.0 \times 10^{-2} \text{ M}$$

$$pH = -\log(1.0 \times 10^{-2})$$

$$pH = -(-2.00)$$

$$pH = 2.00 \; (\rightarrow \text{acidic})$$

pH of a basic solution

To calculate the pH of a basic solution, use the expression:

$$pOH = -\log[^-OH]$$

If a solution has

$$[OH^-] = 1.0 \times 10^{-2} \text{ M}$$

$$pOH = -\log(1.0 \times 10^{-2})$$

$$pOH = -(2.00)$$

$$pOH = 2.00 \; (\rightarrow \text{basic})$$

Since at 25 °C

$$K_w = [H_3O^+]\cdot[OH^-]$$

$$K_w = 1.0 \times 10^{-14}$$

$$pK_w = -\log(K_w)$$

$$pK_w = -\log[H_3O^+] + (-\log[^-OH])$$

$$pK_w = -\log(1.0 \times 10^{-14})$$

$$pK_w = -(-14.00)$$

$$pK_w = 14.00$$

$$pK_w = pH + pOH$$

$$pK_w = 14.00$$

$$pOH = 14.00 - pH$$

Thus, in aqueous solutions

$$pH = 2$$

$$pOH = 12$$

and

$$pOH = 2$$

$$pH = 12$$

Notes for active learning

Acid–Base Equilibria

Equilibrium constants K_a and K_b

Equilibrium constants K_a and K_b measure the extent to which an acid or base dissociates (dissociation constants).

The strength of an acid is defined by its dissociation (ionization) in an *aqueous* (*aq*) solution.

$$HX \ (aq) + H_2O \rightleftarrows H_3O^+ \ (aq) + X^- \ (aq)$$

The equilibrium constant, K_a, for the acid ionization:

$$K_a = [H_3O^+] \cdot [X^-] / [HX]$$

For K_a, the products (conjugate acid H_3O^+ and conjugate base X^-) of the dissociation are the numerator, while the parent acid (HX) is the denominator.

The strength of a base is defined by its dissociation (ionization) in an *aqueous* (*aq*) solution.

$$X^- \ (aq) + H_2O \rightleftarrows OH^- \ (aq) + HX \ (aq)$$

$$K_b = [HX^+] \cdot [^-OH] / [X]$$

For K_b, the products (conjugate acid HX^+ and conjugate base OH^-) of the dissociation are the numerator, while the parent base (X) is the denominator.

Acid dissociation

The value of K_a measures the extent of *acid dissociation*, hence the acid's relative strength.

For strong acids (e.g., $HClO_4$, HCl, H_2SO_4, HNO_3), K_a is exceptionally large (not in the reported K_a values).

For weak acids, $K_a \ll 10^{-1}$.

> *Strong bases* have larger K_b values.

> *Weak bases* have exceedingly small K_b values.

If the K_a value for a conjugate acid-base pair is known, the K_b can be calculated (and vice versa) using the following relationship:

$$K_a K_b = K_w$$

Sodium acetate dissolves in water dissociating into sodium (Na^+) and acetate ions ($C_2H_3O_2^-$):

$$NaC_2H_3O_2 \ (aq) \rightarrow Na^+ \ (aq) + C_2H_3O_2^- \ (aq)$$

If the K_a is 1.8×10^{-5} (provided in a reference table), what is the K_b?

The acetate ion reacts with water with the following equilibrium:

$$C_2H_3O_2^- \, (aq) + H_2O \rightleftharpoons HC_2H_3O_2 \, (aq) + OH^- \, (aq)$$

$$K_b = [HC_2H_3O_2] \cdot [OH^-] \, / \, [C_2H_3O_2^-]$$

For the dissociation of acetic acid:

$$HC_2H_3O_2 \, (aq) + H_2O \rightleftharpoons H_3O^+ \, (aq) + C_2H_3O_2^- \, (aq)$$

$$K_a = [H_3O^+] \cdot [C_2H_3O_2^-] \, / \, [CH_3COOH]$$

$$K_a \times K_b = [H_3O^+] \cdot [C_2H_3O_2^-] \, / \, [CH_3COOH] \times [HC_2H_3O_2] \cdot [OH^-] \, / \, [C_2H_3O_2^-]$$

$$K_w = [H_3O^+] \cdot [OH^-]$$

$$K_w = 1.0 \times 10^{-14}$$

Thus, for acetate ion $C_2H_3O_2^-$ in solution

$$K_b = K_w / K_a \, (\text{for } HC_2H_3O_2)$$

$$K_b = (1.0 \times 10^{-14}) / (1.8 \times 10^{-5})$$

$$K_b = 5.6 \times 10^{-10} \, (> K_w)$$

An aqueous solution of 0.10 M $NaC_2H_3O_2$ has $[OH^-] \sim 7.5 \times 10^{-6}$ M and pH ~ 8.9.

pK_a and pK_b measure acidity and basicity

The operator "p" means "take the negative logarithm of."

Therefore

$$pK_a = -\log K_a$$

As K_a gets larger, pK_a gets smaller.

The *smaller* the pK_a, the *stronger* the acid (e.g., pK_a for $H_2SO_4 = 1.92$).

$$pK_b = -\log K_b$$

As K_b gets larger, pK_b gets smaller.

The *smaller* the value of pK_b, the *stronger* the base.

Conjugate acids and bases

A *conjugate base* is the species remaining from the acid after it *loses* a proton.

A *conjugate acid* is the species remaining from the base after *gaining* a proton.

The *conjugate acid-base pairs* (acid₁–conjugate base₁ and acid₂–conjugate base₂) are related substances by the loss or gain of a single proton (H^+).

H_2O and H_3O^+, and H_2O and OH^- are conjugate acid-base pairs, but H_3O^+ and OH^- are not.

Strong acids have *weak conjugate bases*.

Weak acids have *strong conjugate bases*.

The *weaker the acid*, the stronger its conjugate base.

Strong bases have *weak conjugate acids*.

Weak bases have *strong conjugate acids*.

The *weaker the base*, the stronger the conjugate acid it produces.

HCl is a strong acid, and Cl^- is a weak base.

HF is a weak acid, and F^- is a strong conjugate.

A Brønsted-Lowry acid-base reaction involves a competition between two bases for a proton, in which the stronger base is the most protonated at equilibrium.

$$HCl + H_2O \rightarrow H_3O^+ (aq) + Cl^- (aq)$$

H_2O is a stronger base than the Cl^-.

At equilibrium, the HCl solution contains mainly H_3O^+ and Cl^- ions.

$$HC_2H_3O_2 (aq) + H_2O \rightleftarrows H_3O^+ (aq) + C_2H_3O_2^- (aq)$$

$C_2H_3O_2^-$ is the stronger base.

At equilibrium, acetic acid contains mainly $HC_2H_3O_2$ and few H_3O^+ and $C_2H_3O_2^-$ ions.

$$NH_3 (aq) + H_2O \rightleftarrows NH_4^+ (aq) + \ ^-OH (aq)$$

H_2O is an acid. Competition for protons occurs between NH_3 and ^-OH, where ^-OH is the stronger base, and the equilibrium favors the reactants. An aqueous ammonia solution contains *mostly NH₃ molecules* and smaller amounts of NH_4OH, NH_4^+, and ^-OH.

According to Brønsted-Lowry, the net acid-base reactions favor *strong acid-strong base* combinations to *weak acid-weak base* combinations.

Acid-base reaction equilibrium

The following acid-base reactions proceed in the forward direction:

$$HCl \, (aq) + NH_3 \, (aq) \rightarrow NH_4^+ \, (aq) + Cl^- \, (aq)$$

$$HSO_4^- \, (aq) + CN^- \, (aq) \rightarrow HCN \, (aq) + SO_4^{2-} \, (aq)$$

HSO_4^- ($pK_a = 6.91$) is a stronger acid than HCN ($pK_a = 9.21$)

Many acid-base reactions reach a state of equilibrium.

For the following acid-base reactions, the equilibrium may favor the products or the reactants, depending on the relative strength of the acid:

$$H_2PO_4^- \, (aq) + C_2H_3O_2^- \, (aq) \rightleftarrows HC_2H_3O_2 \, (aq) + HPO_4^{2-} \, (aq)$$

Equilibrium shifts to the *left* (reactants):

$HC_2H_3O_2$ ($pK_a = 4.75$) is the stronger acid

HPO_4^{2-} ($pK_a = 7.21$) is the stronger base

$$HNO_2 \, (aq) + C_2H_3O_2^- \, (aq) \rightleftarrows HC_2H_3O_2 \, (aq) + NO_2^- \, (aq)$$

Equilibrium shifts to the *right* (products).

HNO_2 ($pK_a = 3.39$) is the stronger acid

$C_2H_3O_2^-$ ($pK_a = 4.75$) is the stronger base

Acid-Base Ionization

Strength of acids and bases

A *strong acid* ionizes entirely in aqueous solutions.

Strong acids include $HClO_4$, HCl, H_2SO_4, HNO_3, HBr, and HI. For these strong acids, the equilibrium lies *far to the right*.

Weak acids partially ionize with ionization equilibriums *far to the left*. Weak acids include:

$HC_2H_3O_2$, HNO_2, H_2SO_3, H_3PO_4, and HClO.

For example, consider the reversible process when an acid dissolves in water:

$$HA\ (aq) + H_2O \leftrightarrow H_3O^+\ (aq) + A^-\ (aq)$$

In the forward reaction above, the acid HA donates a proton to the water molecule to form a hydronium ion (H_3O^+) and conjugate base (A^-).

Water acts as a Brønsted-Lowry base, and the acid strength is measured by its degree of ionization (or dissociation) in water. If an acid ionizes completely, it is a strong acid.

A *strong acid* has a *weak conjugate base* (i.e., the conjugate base loses its proton to water readily). Cl^-, Br^-, I^-, ClO_4^-, HSO_4^-, and NO_3^- are weak conjugate bases.

A strong acid is less able to compete with water for a proton.

Ionization constant K_a

The ionization constant K_a for a *strong acid is exceptionally large*, and the equilibrium shifts far to the right (i.e., dissociated proton and stable anion).

A weak acid does not readily give up its proton to water, and it has a strong conjugate base.

Strong conjugate bases include:

$C_2H_3O_2^-$, F^-, CN^-, NO_2^-, HSO_3^-, SO_3^{2-}, $H_2PO_4^-$, HPO_4^{2-}, PO_4^{3-}.

The weaker the acid, the stronger its conjugate base.

The ionization equilibrium for weak acids shifts far to the left.

An aqueous solution of a strong acid has hydronium ions (H_3O^+) and the acid's conjugate base.

For example, there are H_3O^+ and Cl^- ions in an aqueous HCl solution, but virtually no HCl molecules.

An aqueous solution of a weak acid, acetic acid, contains mainly undissociated molecules, $HC_2H_3O_2$, with about ~1% H_3O^+ and $C_2H_3O^-$ ions.

Bond strength and polarity for acids

Factors determine the strength of acids, such as the *strength* and *polarity* of the X−H bond in the molecule and the hydration energy of the ionic species in an aqueous solution.

For inorganic binary acids, H−X bond strength decreases down the group:

HF, HCl, HBr, HI),

The weaker the bond, the easier it ionizes in an aqueous solution.

The stronger bond is the acid; strength increases down the group:

HF < HCl < HBr < HI

Relative ionization strength

Among the hydrohalic acids, HF is the weak acid; the others are strong acids. The relative strength of HCl, HBr and HI cannot be differentiated in an aqueous solution because each of them dissociates almost completely.

Less polar solvents are used to determine their relative strength.

> *Strong bases*, like strong acids (e.g., NaOH, KOH, and Ba(OH)₂), ionize completely when dissolved in water. A strong base tends to accept a proton.

> *Weak bases* do not ionize completely when dissolved in water and show a minimal tendency to accept a proton.

HCl, HBr, and HI ionize partially in acetone or methanol, which have a weaker ionizing strength than water.

The ionization of HCl in acetone:

$(CH_3)_2CO$ (*l*) (*acetone*) + HCl ⇌ $(CH_3)_2COH^+$ + Cl⁻

The degree of ionization in acetone increases as:

HCl < HBr < HI

Oxoacids produce H+

Oxoacids contain one or more ~OH groups covalently bonded to a central atom, a metal, or a nonmetal.

The ~OH group ionizes entirely or partially in an aqueous solution, producing hydrogen ions.

Examples of oxoacids: H_2CO_3, HNO_3, H_3PO_4, H_2SO_4, $HClO_4$, $HC_2H_3O_2$

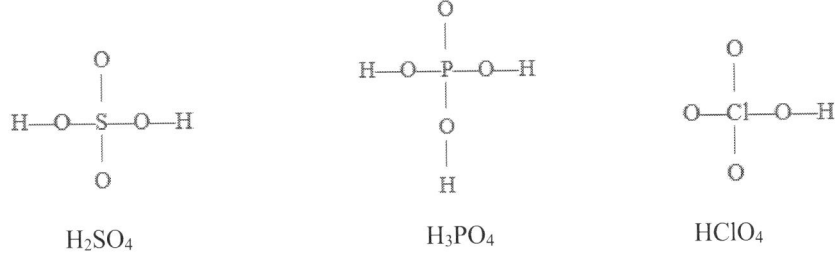

| H_2SO_4 | H_3PO_4 | $HClO_4$ |

Notes for active learning

Periodic Trends of Acids and Bases

Periodic trends for acidity

The acidity of hydrogen halides increases down a group; trend of relative acidity for hydrides of Group 15:

$$H_2O < H_2S < H_2Se < H_2Te$$

For the same period hydrides, relative acidity increases from left to right:

$$CH_4 < NH_3 < H_2O < HF$$

$$PH_3 < H_2S << HCl$$

Water is a stronger acid than ammonia, and in an acid-base reaction, H_2O acts as a Brønsted-Lowry acid, which donates a proton to NH_3:

$$H_2O + NH_3\ (aq) \rightleftarrows NH_4^+\ (aq) + OH^-\ (aq)$$

In reaction with HF, water acts as a Brønsted-Lowry base, which accepts a proton:

$$HF\ (aq) + H_2O \rightleftarrows H_3O^+\ (aq) + F^-\ (aq)$$

Electronegativity for acid strength

Relative acid strength depends on the central atoms' *electronegativity*. The more electronegative, the more polarized the O–H bond, and the *more readily it ionizes* in an aqueous solution to release the H^+ ion.

For example, N, S, and Cl are more electronegative than P; HNO_3, H_2SO_4, and $HClO_4$ are stronger acids, whereas H_3PO_4 is weak.

The order of electronegativity for relative acid strength:

$$Cl \approx N > S > P$$

Acid strength:

$$HClO_4 > HNO_3 > H_2SO_4 > H_3PO_4$$

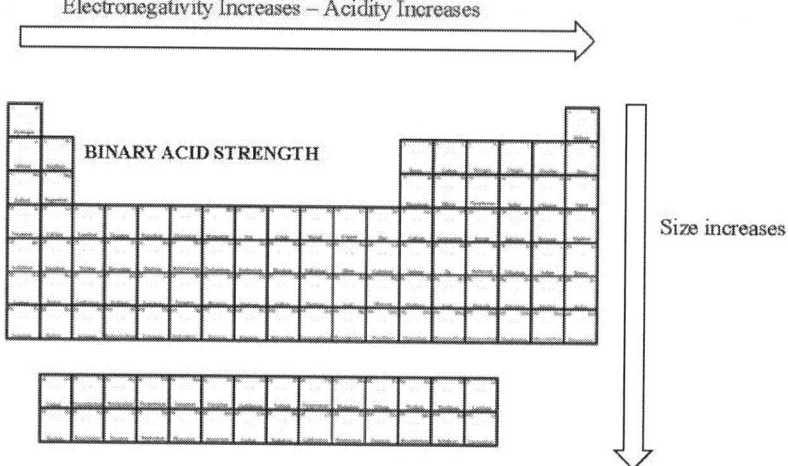

Acidity increases with increases in electronegativity and atomic size

Induction and acid strength

For oxoacids that have the central atoms with elements of the same group in the periodic table, relative strength decreases from top to bottom (as the electronegativity of the central atom decreases):

$$HOCl > HOBr > HOI \qquad HClO_2 > HBrO_2 > HIO_2 \qquad HClO_4 > HBrO_4 > HIO_4$$

For oxoacids containing identical central atoms, acidity increases as oxygen atoms bond.

Acidity increases as follows:

$$HOCl < HClO_2 < HClO_3 < HClO_4$$

$$H_2SO_3 < H_2SO_4$$

$$HNO_2 < HNO_3$$

When oxygen atoms bond to the central atom, the O–H bond in the molecule becomes highly polarized (due to the inductive electronegative effect).

The O–H bond ionizes readily to release an H^+ ion.

Acetic acid (CH_3COOH) is an organic acid containing the carboxyl (~COOH) group. In an aqueous solution, ionization of an acetic acid involves breaking the carboxyl group's O–H bond, but not the C–H bonds in the methyl group (CH_3).

However, if one or more of the hydrogen atoms in the methyl group is substituted with a more electronegative atom, the inductive effect causes the electron cloud to be drawn away from the carbonyl group. The O–H bond becomes more polarized and ionizes readily, increasing the acidity.

The following K_a values illustrate the effect on acetic acid's acidity and its derivatives when methyl hydrogens are substituted with electronegative atoms.

The stronger the acid, the higher the K_a value:

$$CH_3COOH\ (aq)\ (acetic\ acid) + H_2O \rightleftarrows CH_3COO^-\ (aq) + H_3O^+\ (aq)$$

$$K_a = \mathbf{1.8 \times 10^{-5}}$$

$$ClCH_2COOH\ (aq)\ (chloroacetic\ acid) + H_2O \rightleftarrows ClCH_2COO^-\ (aq) + H_3O^+\ (aq)$$

$$K_a = \mathbf{1.4 \times 10^{-3}}$$

$$FCH_2COOH\ (aq)\ (fluoroacetic\ acid) + H_2O \rightleftarrows FCH_2COO^-\ (aq) + H_3O^+\ (aq)$$

$$K_a = \mathbf{2.6 \times 10^{-3}}$$

$$CCl_3COOH\ (aq)\ (trichloroacetic\ acid) + H_2O \rightleftarrows CCl_3COO^-\ (aq) + H_3O^+\ (aq)$$

$$K_a = \mathbf{3.0 \times 10^{-1}}$$

Monoprotic and polyprotic acids

The *monoprotic acids* include HCl, HF, HOCl, HNO_2, and $HC_2H_3O_2$ because each molecule contains a single ionizable hydrogen ion.

The *polyprotic acids* contain more than one ionizable hydrogen, and examples include:

$$H_2SO_4,\ H_2SO_3,\ H_2C_2O_4\ and\ H_3PO_4$$

The hydrogen ionizes in stages with ionization constants, such as for H_3PO_4.

$$H_3PO_4\ (aq) \rightleftarrows H^+\ (aq) + H_2PO_4^-\ (aq);\qquad K_{a1} = 7.5 \times 10^{-3}$$

$$H_2PO_4^-\ (aq) \rightleftarrows H^+\ (aq) + HPO_4^{2-}\ (aq);\qquad K_{a2} = 6.2 \times 10^{-8}$$

$$HPO_4^{2-}\ (aq) \rightleftarrows H^+\ (aq) + PO_4^{3-}\ (aq);\qquad K_{a3} = 4.8 \times 10^{-13}$$

From above, acid strength decreases in order:

$$H_3PO_4 \gg H_2PO_4^- \gg HPO_4^{2-}$$

Ionization of polyprotic acids

Sulfuric acid (H_2SO_4) is a strong acid, but only the first hydrogen ionizes completely:

$$H_2SO_4\,(aq) \rightarrow H^+\,(aq) + HSO_4^-\,(aq) \qquad K_{a1} = \text{exceptionally large}$$

The second H does not often dissociate, and HSO_4^- is a weak acid:

$$HSO_4^-\,(aq) \rightleftarrows H^+\,(aq) + SO_4^{2-}\,(aq) \qquad K_{a2} = 1.2 \times 10^{-2}$$

Hydroxide bases

Weak bases include NH_3 (or NH_4OH), NH_2OH, $Mg(OH)_2$, and hydroxides and oxides as slightly soluble in water.

Hydroxides of Group I metals (LiOH, NaOH, KOH, ROH, and CsOH) are strong bases, but only NaOH and KOH are commercially important and commonly used laboratory bases.

Hydroxide bases are *soluble* in water, and they dissociate entirely in an aqueous solution, producing a high concentration of hydroxide ions.

A moderately dilute solution of NaOH contains [$^-$OH], [Na+], and [NaOH].

$Ba(OH)_2$ is a relatively strong base among the alkaline earth metal hydroxides.

The other metal hydroxides are sparingly soluble in water, limiting their basicity.

A saturated solution of these hydroxides contains a low concentration of $^-$OH.

These hydroxides react with strong acids:

$$Na_2O\,(s) + H_2O \rightarrow 2\,NaOH\,(aq)$$

$$BaO\,(s) + H_2O \rightarrow Ba(OH)_2\,(aq)$$

$$MgO\,(s) + 2\,HCl\,(aq) \rightarrow MgCl_2\,(aq) + H_2O$$

Hydroxides of some metals (e.g., $Al(OH)_3$, $Cr(OH)_3$, $Zn(OH)_2$, $Sn(OH)_2$ and $Pb(OH)_2$) exhibit amphoteric properties (i.e., act as an acid or a base).

For example:

$$Al(OH)_3\,(s) \ + \ OH^-\,(aq) \rightleftarrows Al(OH)_4^-\,(aq)$$

$$Al(OH)_3\,(s) \ + \ 3\,H_3O^+\,(aq) \rightleftarrows [Al(H_2O)_6]^{3+}\,(aq)$$

Hydride bases

Hydrides of reactive metals (e.g., NaH, MgH_2, CaH_2) form strongly basic solutions if dissolved in water.

The hydride ion reacts with water to produce hydroxide ions and hydrogen gas:

$$H^- \, (aq) + H_2O \rightarrow H_2 \, (g) + OH^- \, (aq)$$

The oxide ion O^{2-} has an extraordinarily strong affinity for protons (H^+), reacting with H_2O to form hydroxide ions.

$$O^{2-} \, (aq) + H_2O \rightarrow 2 \, OH^- \, (aq)$$

Oxides of nonmetals are acidic, forming acidic solutions when dissolved in water.

$$CO_2 \, (g) + H_2O \rightleftharpoons H_2CO_3 \, (aq) \rightleftharpoons H^+ \, (aq) + HCO_3^- \, (aq)$$

$$SO_2 \, (g) + H_2O \rightleftharpoons H_2SO_3 \, (aq) \rightleftharpoons H^+ \, (aq) + HSO_3^- \, (aq)$$

Ammonia is the only weak base that is of commercial importance.

Ammonia does not contain hydroxide ions, but it reacts with water and ionizes as follows:

$$NH_3 \, (aq) + H_2O \rightleftharpoons NH_4^+ \, (aq) + OH^- \, (aq)$$

The base dissociation constant K_b is given by the expression:

$$K_b = [NH_4^+] \cdot [^-OH] \, / \, [NH_3]$$

$$K_b = 1.8 \times 10^{-5}$$

Notes for active learning

Percent Dissociation

Degree of ionization

The *degree of ionization* (or *percent dissociation*) of a weak acid is:

Percent dissociation = [acid ionized] / [initial acid] × 100%

For strong acids, the percent dissociation at equilibrium is ~ 100%.

The percent dissociation for weak acids depends on the acid's K_a and initial concentration.

For example, the percent dissociation of acetic acid (Use $HC_2H_3O_2$ $K_a = 1.8 × 10^{-5}$) at 0.10 M concentration is:

$(1.3 × 10^{-3}$ M / 0.10 M$) × 100\% = 1.3$ %

The *stronger* the acid, the larger is the K_a, and the *greater* is the percent ionization.

The *percent ionization of a weak acid* depends on the K_a and the extent of dilution.

The *more dilute* an acid solution, the *higher* is the percentage of ionization.

For example, consider a 0.010 M acetic acid solution and its ionization products.

Use an ICE table (**I**nitial, **C**hange, **E**quilibrium) to simplify calculations in reversible equilibrium reactions.

Once the equilibrium row is completed (by summing the initial and change rows), its contents can be substituted into the equilibrium constant expression to solve for K_a.

$$CH_3COOH\ (aq) \rightleftarrows H^+\ (aq) + CH_3COO^-\ (aq)$$

Initial [ ], M:	0.010	0.00	0.00
Change, Δ[ ], M:	$-x$	$+x$	$+x$
Equilibrium [ ], M :	$(0.010 - x)$	x	x

Acid ionization constant K_a

The acid ionization constant, K_a, is given by:

$K_a = [H_3O^+]·[C_2H_3O_2^-] / [CH_3COOH]$

$K_a = x^2 / (0.010 - x)$

$K_a = 1.8 × 10^{-5}$

Since $K_a \ll 0.010$, approximate that $x \ll 0.010$, and $(0.010 - x) \sim 0.010$

Then

$$K_a = x^2 / (0.010 - x)$$

$$K_a \approx x^2 / 0.010$$

$$K_a = 1.8 \times 10^{-5}$$

$$x^2 = 1.8 \times 10^{-7}$$

and

$$x = \sqrt{(1.8 \times 10^{-7})}$$

$$x = 4.2 \times 10^{-4}$$

$$x = [H_3O^+]$$

$$x = 4.2 \times 10^{-4} \text{ M}$$

The *degree of ionization* of acetic acid at this concentration is

$$(4.2 \times 10^{-4} \text{ M} / 0.010 \text{ M}) \times 100\% = 4.2\%$$

The degree of ionization of the acid increases as the solution is diluted. For 0.1 M acetic acid, the degree of ionization is 0.42%, 10-fold lower than in 0.010 M acid solution.

The percent dissociation of weak bases depends on the base solution's K_b value and dilution.

Larger K_b and greater dilution result in a higher percent dissociation.

Hydrolysis of salts

The presence of salt affects the dissociation of acids and bases due to salts' *hydrolysis*.

For example, CH_3COOH dissociates less in a solution containing CH_3COONa salt, while NH_4OH dissociates less in a solution containing NH_4Cl salt.

When salts (or ionic compounds) dissolve in water, it is assumed that they dissociate entirely into separate ions.

Some of these ions react with water and behave as acids or bases.

The acidic or basic nature of a salt solution depends on whether it is a product of a:

> 1) strong acid-strong base reaction,
>
> 2) a weak acid-strong base reaction,
>
> 3) a strong acid-weak base reaction, or
>
> 4) a weak acid-weak base reaction.

Salts of Acid–Base Reactions

Salts

Salts are the products of acid-base reactions.

The general reaction to produce salt:

Acid + Base → Salt + Water

The acid and base are neutralized; H^+ and OH^- form water.

The acid's nonmetallic ions and metal ions of the base form the salt.

For example, NaCl is a product of the following acid-base reaction:

$HCl\ (aq) + NaOH\ (aq) \rightarrow NaCl\ (aq) + H_2O$

In the chemical formula of salt, the cation is contributed by the base (e.g., Na^+), while the acid contributes the anion (e.g., Cl^-).

Four types of salts

The four types of salts are:

- **Salts of strong acid-strong base** reactions (e.g., $NaCl$, KNO_3, $NaClO_4$)

- **Salts of weak acid-strong base** reactions (e.g., $NaC_2H_3O_3$, K_2CO_3, KCN, $NaCHO_2$)

- **Salts of strong acid-weak base** reactions (e.g., NH_4Cl, NH_4NO_3, $HONH_3Cl$)

- **Salts of weak acid-weak base** reactions (e.g., $NH_4C_2H_3O_2$, NH_4CN, NH_4HS)

When dissolved in water, these salts produce acidic, basic, or neutral solutions.

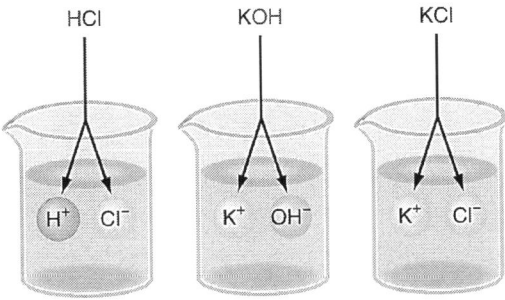

The acidic dissociation constant increases with stable anions

*Salts of **Strong Acid-Strong Base** Reactions*: (e.g., NaCl, NaNO$_3$, KBr)

- Salts form a neutral solution because neither the cation nor anion reacts with water and offsets the equilibrium concentrations of H$_3$O$^+$ and OH$^-$ in the solution.

*Salts of **Weak Acid-Strong Base** Reactions*: (e.g., NaF, NaNO$_2$, NaC$_2$H$_3$O$_2$)

- Salts form basic solutions when dissolved in H$_2$O.

  The anions of such salts react with H$_3$O increasing [$^-$OH].

- Sodium acetate (NaC$_2$H$_3$O$_2$) is a product of the reaction between acetic acid (HC$_2$H$_3$O$_2$), which is a weak acid, and a strong base (NaOH).

$$HC_2H_3O_2 \ (aq) + NaOH \ (aq) \rightarrow NaC_2H_3O_2 \ (aq) + H_2O$$

*Salts of **Strong Acid-Weak Base** Reactions*: (e.g., NH$_4$Cl, NH$_4$NO$_3$, (CH$_3$)$_2$NH$_2$Cl, C$_5$H$_5$NHCl)

- Aqueous solutions of salts are acidic.

- The cations react with H$_2$O and increase [H$_3$O$^+$].

- Example: NH$_4$Cl is produced when HCl (strong acid) reacts with NH$_3$ (a weak base):

$$HCl \ (aq) + NH_3 \ (aq) \rightarrow NH_4Cl \ (aq) \rightarrow NH_4^+ \ (aq) + Cl^- \ (aq)$$

In an aqueous solution, NH$_4^+$ equilibrium increases [H$_3$O$^+$] and creates an acidic solution:

$$NH_4^+ \ (aq) + H_2O \rightleftarrows H_3O^+ \ (aq) + NH_3 \ (aq)$$

$$K_a = [H_3O^+] \cdot [NH_3] / [NH_4^+]$$

While in an NH$_3$ solution, the following equilibrium occurs:

$$NH_3 \ (aq) + H_2O \rightleftarrows NH_4^+ \ (aq) + OH^- \ (aq)$$

$$K_b = [NH_4^+] \cdot [^-OH] / [NH_3]$$

$$K_a \times K_b = \{[H_3O^+] \cdot [NH_3] / [NH_4^+]\} \times \{[NH_4^+] \cdot [^-OH] / [NH_3]\}$$

$$K_a \times K_b = K_w = [H_3O^+] \cdot [^-OH]$$

$$K_w = 1.0 \times 10^{-14}$$

For NH_4^+ :

$K_a = K_w / K_b$ (for NH_3)

$K_a = (1.0 \ x \ 10^{-14}) / (1.8 \ x \ 10^{-5})$

$K_a = 5.6 \times 10^{-10}$

An aqueous solution of NH_4^+ has a $K_a = 5.6 \times 10^{-10}$ at 25 °C (which is $> K_w$).

A 0.10 M solution of NH_4Cl or NH_4NO_3 has $[H_3O^+] \approx 7.5 \times 10^{-6}$ M and pH ≈ 5.1

*Salts of **Weak Acid-Weak Base** Reactions*: (e.g., $NH_4C_2H_3O_2$, NH_4CN, NH_4NO_2)

Aqueous salts can be neutral, acidic, or basic, depending on the relative magnitude of the K_a of the weak acid and the K_b of the weak base.

- If $K_a \approx K_b$, salt forms an approximately neutral solution.

For example:

K_a of $HC_2H_3O_2 = 1.8 \times 10^{-5}$

K_b of $NH_3 = 1.8 \times 10^{-5}$

When $NH_4C_2H_3O_2$ dissolves in water and dissociates, the following equilibria exist:

$NH_4C_2H_3O_2 \, (aq) \rightarrow NH_4^+ \, (aq) + C_2H_3CO_2^- \, (aq)$

$NH_4^+ \, (aq) + H_2O \rightleftarrows H_3O^+ \, (aq) + NH_3 \, (aq)$ $\qquad\qquad K_a = 5.6 \times 10^{-10}$

$C_2H_3O_2^- \, (aq) + H_2O \rightleftarrows HC_2H_3O_2 \, (aq) + OH^- \, (aq)$ $\qquad\qquad K_b = 5.6 \times 10^{-10}$

K_a (for NH_4^+) = K_b (for $C_2H_3O_2^-$), at equilibrium $[H_3O^+] = [^-OH]$ and the $NH_4C_2H_3O_2$ solution is neutral.

If $\boldsymbol{K_a > K_b}$, the salt solution is acidic.

For NH_4NO_2:

K_a (HNO_2) = 4.0×10^{-4}

K_b (NH_3) = 1.8×10^{-5}

$K_a > K_b \rightarrow$ acidic solution because hydrolysis forms a solution with $[H_3O^+] > [^-OH]$.

If $\boldsymbol{K_a < K_b}$, the salt solution is basic.

For NH_4CN in solution,

K_a (HCN) = 6.2×10^{-10}

K_b (NH$_3$) = 1.8×10^{-5}

The following equilibria exist:

NH_4CN (*aq*) $\rightarrow$ NH_4^+ (*aq*) + CN^- (*aq*)

NH_4^+ (*aq*) + H_2O $\rightleftarrows$ H_3O^+ (*aq*) + NH_3 (*aq*)

K_a = 5.6×10^{-10}

CN^- (*aq*) + H_2O $\rightleftarrows$ HCN (*aq*) + OH^- (*aq*)

K_b = 1.6×10^{-5}

K_b (CN^-) > K_b (NH_4^+), at equilibrium [^-OH] > [H_3O^+]; an aqueous solution of NH_4CN is basic.

pH Calculations for Acid–Base Solutions

Calculating pH of strong acid and strong base solutions

Strong acids are assumed to ionize completely in an aqueous solution.

For monoprotic acids (i.e., acids with a single ionizable hydrogen) such as HCl and HNO_3, the [hydronium ion; H_3O^+] in solution is the same as the molar concentration of the acid:

$$[H_3O^+] = [HX]$$

For example, consider 0.10 M HCl (*aq*)

$$[H_3O^+] = 0.10 \text{ M}$$

$$pH = -\log(0.10)$$

$$pH = 1.00$$

A strong base such as NaOH has [⁻OH] equal to the molar concentration of dissolved NaOH.

A solution of 0.10 M NaOH (*aq*) has:

$$[^-OH] = 0.10 \text{ M}$$

$$pOH = -\log[^-OH]$$

$$-\log[^-OH] = -\log(0.10)$$

$$pOH = 1.00$$

$$pH = 14.00 - 1.00$$

$$pH = 13.00$$

A strong base such as $Ba(OH)_2$ produces twice the concentration of ⁻OH as the molar concentration of $Ba(OH)_2$ in solution:

$$Ba(OH)_2 \, (aq) \rightarrow Ba^{2+} \, (aq) + 2 \, OH^- \, (aq)$$

$$[^-OH] = 2 \times [Ba(OH)_2]$$

In a solution of 0.010 M $Ba(OH)_2$

$$[^-OH] = 0.020 \text{ M}$$

$$pOH = 1.70, \text{ and } pH = 12.30$$

Calculating pH of weak acid solutions

Unlike strong acids, weak acids do not ionize completely.

At equilibrium, $[H^+]$ is less than the concentration of the acid.

The concentration of H^+ in a weak acid solution depends on the initial acid concentration and the K_a of the acid.

Determine $[H^+]$ of a weak acid using the "ICE" table as follows:

For example, consider a 0.10 M acetic acid solution and its ionization products.

$$HC_3H_3O_2\,(aq) \rightleftarrows H^+\,(aq) + C_2H_3O_2^-\,(aq)$$

Initial [ ], M:	0.10	0.00	0.00
Change, Δ[ ], M:	$-x$	$+x$	$+x$
Equilibrium [ ], M :	$(0.10 - x)$	x	x

The acid ionization constant, K_a, is given by the expression:

$$K_a = [H_3O^+]\cdot[C_2H_3O_2^-] / [CH_3COOH]$$

$$K_a = x^2 / (0.010 - x)$$

$$K_a = 1.8 \times 10^{-5}$$

Since,

$$K_a \ll 0.10$$

approximate that $x \ll 0.10$, and $(0.10 - x) \sim 0.10$

$$K_a = x^2 / (0.10 - x)$$

$$K_a \approx x^2 / 0.10$$

$$K_a = 1.8 \times 10^{-5}$$

$$x^2 = 1.8 \ x \ 10^{-6}$$

$$x = \sqrt{(1.8 \ x \ 10^{-6})}$$

$$x = 1.3 \times 10^{-3}$$

Note that

$$x = [H_3O^+]$$

$$[H_3O^+] = 1.3 \times 10^{-3} \text{ M}$$

$$pH = -\log(1.3 \times 10^{-3})$$

$$pH = 2.89$$

Calculating pH of weak base solutions

The concentration of $^-$OH in a weak base, such as NH_3 (aq), depends on its K_b value and the initial concentration of the base.

To determine [$^-$OH] and pH of 0.10 M NH_3 (aq), set the following "ICE" table:

$$NH_3 \,(aq) + H_2O \rightleftarrows NH_4^+ \,(aq) + {}^-OH \,(aq)$$

Initial [ ], M:	0.10	0.00	0.00
Change, Δ[ ], M:	$-x$	$+x$	$+x$
Equilibrium [ ], M:	$(0.10 - x)$	x	x

$$K_b = [NH_4^+] \cdot [{}^-OH] / [NH_3]$$

$$K_b = x^2 / (0.10 - x)$$

$$K_b = 1.8 \times 10^{-5}$$

Using the approximation method

$$K_b = x^2 / (0.10 - x)$$

$$K_b \approx x^2 / 0.10$$

$$K_b = 1.8 \times 10^{-5}$$

$$x^2 = 1.8 \times 10^{-6}$$

$$x = \sqrt{(1.8 \times 10^{-6})}$$

$$x = 1.3 \times 10^{-3}$$

where

$$x = [^-OH]$$

$$x = 1.3 \times 10^{-3} \, M$$

$$pOH = -\log(1.3 \times 10^{-3})$$

$$pOH = 2.87$$

$$pH = 11.13$$

Calculating pH of basic or acidic salt solutions

1.   For example, consider a solution of 0.050 M sodium acetate, which dissociates completely and establishes the following equilibrium:

$$NaC_2H_3O_2 \, (aq) \rightarrow Na^+ \, (aq) + C_2H_3O_2^- \, (aq)$$

The acetate ion establishes the equilibrium in aq solution:

$$C_2H_3O_2^- \, (aq) + H_2O \rightleftarrows HC_2H_3O_2 \, (aq) + {}^-OH \, (aq)$$

$$K_b = [HC_2H_3O_2] \cdot [^-OH] / [C_2H_3O_2^-]$$

$$K_b = 5.6 \times 10^{-10}$$

By approximation

$$[^-OH] = \sqrt{(K_b[C_2H_3O_2^-])}$$

$$[^-OH] = \sqrt{\{(5.6 \times 10^{-10}) \cdot (0.050)\}}$$

$$[^-OH] = 5.3 \times 10^{-6} \, M$$

$$pOH = -\log(5.3 \times 10^{-6})$$

$$pOH = 5.28$$

$$pH = 8.72 \text{ (solution is basic)}$$

2.   For example, consider a solution of 0.050 M NH_4Cl, which dissociates and establishes the following equilibrium:

$$NH_4Cl \, (aq) \rightarrow NH_4^+ \, (aq) + Cl^- \, (aq)$$

$$NH_4^+ \, (aq) + H_2O \rightleftarrows H_3O^+ \, (aq) + NH_3 \, (aq)$$

$$K_a = [H_3O^+] \cdot [NH_3] / [NH_4^+]$$

$K_a = 5.6 \times 10^{-10}$

By approximation

$[H_3O^+] = \sqrt{(K_a[NH_4^+])}$

$[H_3O^+] = \sqrt{\{(5.6 \times 10^{-10}) \cdot (0.050)\}}$

$[H_3O^+] = 5.3 \times 10^{-6}$ M

$pH = -\log(5.3 \times 10^{-6})$

$pH = 5.28$, (solution is acidic)

Notes for active learning

Buffered Solutions

Buffers

A *buffer* is a solution that maintains pH (little or no change) even when a small amount of strong acid or strong base is added. A buffer solution contains a weak acid and the "salt" of its conjugate base or a weak base and the "salt" of its conjugate acid.

The *buffering capacity* is the amount of H^+ or OH^- ion the buffer absorbs without significantly altering the pH.

A buffer with large concentrations of buffering components absorbs significant quantities of strong acid or strong base, with little change in its pH, has a large buffering capacity.

A buffer is effective within a pH range, typically about ±1 of the pK_a of its acid component.

Some *common buffer systems*:

Buffer	pK_a	pH Range
$HCHO_2 - NaCHO_2$	3.74	2.75–4.75
$HC_2H_3O_2 - NaC_2H_3O_2$	4.74	3.75–5.75
$KH_2PO_4 - K_2HPO_4$	7.21	6.20–8.20 (a blood buffer)
$CO_2/H_2O - NaHCO_3$	6.37	5.40–7.40 (a blood buffer)
$NH_3 - NH_4Cl$	9.25	8.25–10.25

When 0.01 mol of HCl is added to 1 L of pure water, $[H^+]$ increases from 10^{-7} to 10^{-2} M, and the pH changes from about 7 to 2. This change in pH indicates that water is not a buffer.

When the same HCl amount is added to a solution containing a mixture of 1 M acetic acid ($HC_2H_3O_2$) and 1 M sodium acetate ($NaC_2H_3O_2$), the solution's pH changes little – it goes from 4.74 to 4.66. A solution composed of acetic acid and sodium acetate is a buffer solution.

For example, consider a buffered solution composed of KH_2PO_4 and K_2HPO_4. The species present in the solution are primarily K^+, $H_2PO_4^-$ and HPO_4^{2-}. K^+ is a spectator ion and not involved in the buffering reaction.

If a small amount of strong acid is added to this solution, the H^+ ions from the acid react with the base component of the buffer (HPO_4^{2-}):

$$H^+(aq) + HPO_4^{2-}(aq) \rightarrow H_2PO_4^-(aq) \ldots \text{(buffering reaction 1)}$$

When a strong base such as NaOH is added, the $^-$OH reacts with the buffer's acid component ($HC_2H_3O_2$).

$^-$OH (aq) + $HC_2H_3O_2$ (aq) → H_2O + $C_2H_3O_2^-$ (aq) . . . (buffering reaction 2)

Reactions (1) and (2) are critical buffering reactions that maintain the solution's pH.

Biological buffers

Buffered solutions are vital to living organisms. Metabolic reactions are controlled or accelerated by biological catalysts called *enzymes*, often proteins that function within a narrow pH range.

The human body fluid must be at a specific (narrow) pH range. Human blood is maintained at the pH range of 7.30 - 7.40. A drop below pH 7 or rise above pH 7.5 can be fatal.

Two buffer systems – the phosphate buffer ($H_2PO_4^-$-HPO_4^{2-}) and carbonic acid-bicarbonate buffer (H_2CO_3-HCO_3^-) are essential to maintaining the normal blood pH.

The buffering reactions of bicarbonate buffer are:

H^+ (aq) + HCO_3^- (aq) → H_2O + CO_2 (aq)

$^-$OH (aq) + CO_2 (aq) + H_2O → 2 HCO_3^- (aq)

Buffers resist changes in pH

The *capacity of a buffer* is determined by the sizes of [AB] and [B$^-$].

Consider a buffer made up of acetic acid and sodium acetate, in which the significant species present in the solution are $HC_2H_3O_2$ and $C_2H_3O_2^-$.

For example, if a small amount of HCl (aq) is added to this solution, most H^+ (from HCl) is absorbed by the conjugate base, $C_2H_3O_2^-$.

H^+ (aq) + $C_2H_3O_2^-$ (aq) → $HC_2H_3O_2^-$ (aq)

Since $C_2H_3O_2^-$ is present in a larger quantity than the added H^+, the reaction shifts almost entirely to the right. This buffering reaction prevents a significant increase in [H^+] and minimizes the pH change.

For example, if a strong base, such as NaOH (aq), is added, most $^-$OH ions (from NaOH) react with the buffer's acidic component.

$^-$OH (aq) + $HC_2H_3O_2$ (aq) → H_2O + $C_2H_3O_2^-$ (aq)

Because of the larger concentration of $HC_2H_3O_2$ compared to $^-$OH, this reaction goes almost to completion. This buffering reaction prevents a significant increase in [$^-$OH] and minimizes pH changes.

For example, for a buffer containing the weak acid HB and the salt NaB, such that B^- is the conjugate base to the acid, the concentration $[H^+]$ and pH of the buffer depending on the dissociation constant, K_a, of the acid component, and the concentration ratio $[B^-]$ / $[HB]$ in the buffer solution.

For example, consider the equilibrium:

$$HB\ (aq) \leftrightharpoons H^+\ (aq) + B^-\ (aq)$$

$$K_a = [H^+] \cdot [B^-] / [HB]$$

Rearranging the expression:

$$[H^+] = K_a \times ([HB] / [B^-])$$

$$pH = pK_a + \log([B^-] / [HB])$$

Calculating pH of buffered solutions

The *pH of a buffered solution* is determined by the ratio $[B^-]$ / $[HB]$.

The following example illustrates how to calculate the change in pH and the buffering capacity of a buffered solution after a strong acid is added.

Calculate the change in pH when 0.010 mol of HCl adds to 1.0 L of each of the buffers:

Buffer A: 1.0 M $HC_2H_3O_2$ + 1.0 M $NaC_2H_3O_2$

Buffer B: 0.020 M $HC_2H_3O_2$ + 0.020 M $NaC_2H_3O_2$

Henderson-Hasselbalch equation

The expression $pH = pK_a + \log([B^-] / [HB])$ is the *Henderson-Hasselbalch equation*, which is useful for calculating the pH of solutions when the K_a and the ratio $[B^-]$ / $[HB]$ are known.

For buffers, the *Henderson-Hasselbalch equation* calculates pH:

$$pH = pK_a + \log([B^-] / [HB])$$

$$pH = pK_a + \log([C_2H_3O_2^-] / [HC_2H_3O_2])$$

$$pH = -\log(1.8 \times 10^{-5}) + \log(1)$$

$$pH = 4.74$$

For example, 0.010 mol HCl is added to Buffer A with the following reaction:

$$H^+(aq) + C_2H_3O_2^-(aq) \rightarrow HC_2H_3O_2(aq)$$

[ ] before reaction: 0.010 M   1.0 M   1.0 M

[ ] after reaction: 0   0.99 M   1.01 M

The new pH:

$pH = pK_a + \log([B^-] / [HB])$

$pH = 4.74 + \log(0.99 / 1.01)$

$pH = 4.74 - 0.010$

$pH = 4.73$ (pH is changed $\approx 0.21\%$)

0.010 mol HCl is added to Buffer B with the following reaction:

$$H^+(aq) + C_2H_3O_2^-(aq) \rightarrow HC_2H_3O_2(aq)$$

[ ] before rxn: 0.010 M   0.020 M   0.020 M

[ ] after rxn: 0   0.010 M   0.030 M

The expression to calculate pH:

$pH = pK_a + \log([B^-] / [HB])$

$pH = 4.74 + \log(0.010 / 0.030)$

$pH = 4.74 - 0.48$

$pH = 4.26$ (pH decreases by 10%)

From above, Buffer A, which contains larger quantities of buffering components, has a higher buffering capacity than Buffer B.

For Buffer A to decrease its pH by 0.48 units (or 10%), it must absorb the equivalent of 0.50 mol of HCl.

Summary of buffer solutions

1. The solution contains a weak acid HX and its conjugate base X^-, or a weak base B and its conjugate acid BH^+ in appreciable amounts.

2. A buffer solution maintains its pH by absorbing H^+ or OH^- produced by a strong acid or strong base, so these ions do not accumulate.

3. The buffering reaction involves H^+ with the conjugate base **X** in the buffer, or the reaction of ^-OH with the acid component (**HX**) of the buffer:

    $$H^+ (aq) + \mathbf{X}^- (aq) \rightarrow HX (aq)$$

    $$^-OH (aq) + \mathbf{HX} (aq) \rightarrow H_2O + X^-$$

    These two reactions prevent a significant increase in $[H^+]$ or $[^-OH]$ in the solution.

4. The *buffering capacity* of a solution implies the amount of H^+ or ^-OH it absorbs without significantly changing its pH. This depends on the concentration of the weak acid and its conjugate base in the solution.

5. The *buffering range* of a solution depends on the pK_a of the buffer's acid component.

    A buffer is most effective when the pH range $= pK_a \pm 1$.

Notes for active learning

Ionic Equations and Neutralization

Molecular, ionic and net ionic equations

Examples of molecular, complete (or total) ionic, and net ionic equations for strong acid-strong base reactions:

Molecular:

$$HCl\ (aq) + NaOH\ (aq) \rightarrow H_2O\ (l) + NaCl\ (aq)$$

Complete ionic:

$$H^+\ (aq) + Cl^-\ (aq) + Na^+\ (aq) + OH^-\ (aq) \rightarrow H_2O\ (l) + Na^+\ (aq) + Cl^-\ (aq)$$

Net ionic:

$$H^+\ (aq) + OH^-\ (aq) \rightarrow H_2O\ (l)$$

Spectator ions:

$$Na^+ \text{ and } Cl^-$$

Weak acids and weak bases ionize partially, and most weak acids and bases remain in the molecular form in solution.

Therefore, weak acids and bases should NOT be written in ionized forms when writing the ionic equations.

The three equations for the reaction between acetic acid (a weak acid) and sodium hydroxide (a strong base) are:

Molecular:

$$HC_2H_3O_2\ (aq) + NaOH\ (aq) \rightarrow H_2O\ (l) + NaC_2H_3O_2\ (aq)$$

Complete ionic:

$$HC_2H_3O_2\ (aq) + Na^+\ (aq) + OH^-\ (aq) \rightarrow H_2O\ (l) + Na^+\ (aq) + C_2H_3O_2^-\ (aq)$$

Net ionic:

$$HC_2H_3O_2\ (aq) + OH^-\ (aq) \rightarrow H_2O\ (l) + C_2H_3O_2^-\ (aq)$$

Spectator ion:

$$Na^+$$

Neutralization

Neutralization is a chemical reaction where an acid and a base react quantitatively, resulting in no excess of hydrogen (H^+) or hydroxide ions (OH^-) in the aqueous solution.

The products of acid-base reactions are salt and water.

$$HCl\ (aq) + NaOH\ (aq) \rightarrow H_2O\ (l) + NaCl\ (aq)$$

$$HClO_4\ (aq) + KOH\ (aq) \rightarrow H_2O\ (l) + KclO_4\ (aq)$$

$$HC_2H_3O_2\ (aq) + NaOH\ (aq) \rightarrow H_2O\ (l) + NaC_2H_3O_2\ (aq)$$

Note that the substances involved are subject to dissociation.

Stoichiometry of acid-base neutralization

For example, how many mL of 0.1725 M NaOH (*aq*) are needed to neutralize 25.00 mL of 0.2040 M HCl (*aq*)? According to the equation:

$$HCl\ (aq) + NaOH\ (aq) \rightarrow H_2O\ (l) + NaCl\ (aq)$$

Moles of NaOH needed = Moles of HCl present

(Liters NaOH × Molarity NaOH) = (Liters HCl × Molarity HCl)

(Liters of NaOH × 0.1725 mol/L) = (0.02500 L × 0.2040 mol/L)

Divide each side by 0.1725 mol/L:

$$\text{Liters of NaOH} = \frac{0.02500\ L \times 0.2040\ mol/L}{0.1725\ mol/L}$$

Liters of NaOH = 0.02957 L = 29.57 mL

For example, if 10.00 mL of acetic acid $HC_2H_3O_2$ of an unknown concentration requires 38.64 mL of 0.2250 M KOH to neutralize, what is the molarity of the acetic acid?

From the reaction:

$$HC_2H_3O_2\ (aq) + NaOH\ (aq) \rightarrow H_2O\ (l) + NaC_2H_3O_2\ (aq)$$

Moles of acetic acid = Moles of NaOH

(Liter of acid × Molarity of acid) = (Liter of base × Molarity of base)

(0.01000 L × Molarity $HC_2H_3O_2$) = (0.03864 L × 0.2250 M NaOH)

Divide each side by 0.01000 L:

$$\text{Molarity of } HC_2H_3O_2 = \frac{0.03864 \text{ L} \times 0.2250 \text{ M}}{0.01000 \text{ L}}$$

$HC_2H_3O_2 = 0.8694$ M

Notes for active learning

Titration

Molar concentration determination

Titration (or *volumetric analysis*) is a standard laboratory procedure to determine an *unknown concentration*.

Titration (*volumetric analysis*) is an essential application of neutralization reactions. A reagent of known concentration is slowly added to a sample of unknown concentration until the neutralization reaction (i.e., a reaction between an acid and a base) is complete.

Titration determines the molar concentration of a solution (analyte) using the volume and concentration of another (titrant) by adding an exact amount of the *titrant* from a buret to another reactant (*analyte*) in a flask.

The *analyte* is the substance (i.e., sample) of an unknown concentration being analyzed.

The *titrant* is the analytical reagent of known concentration added to the sample. It is carefully added from a buret until the *equivalence point* (i.e., the point of neutralization) is reached.

If the titrant's volume and concentration are known, its number of moles can be calculated.

An *indicator* is added to the reaction to determine the endpoint of a redox titration (i.e., when sample molecules have been depleted).

The *endpoint* is marked by an indicator's color change when a solution transitions from acidic to slightly basic. A redox indicator undergoes a color change at the neutralization reaction's equivalence point.

In a useful titration, the *equivalence point* should match the *endpoint*.

The number of moles of the analyte and its concentration can be determined from reaction stoichiometry.

Titration experimental factors

1. The reaction between titrant and analyte occurs rapidly.

2. The balanced equation is known.

3. The endpoint occurs precisely at or close to the equivalence point.

4. The volume of titrant to reach the equivalence point is accurately measurable.

Stoichiometry determines unknown concentrations

An acid (e.g., HCl (*aq*)) is titrated with aqueous NaOH as follows:

$$HCl\ (aq) + NaOH\ (aq) \rightarrow NaCl\ (aq) + H_2O$$

For example, the stoichiometric ratio of HCl to NaOH is 1 mole HCl to 1 mole NaOH. The standard solution's volume and concentration (acid or base) are known in titration. However, only the volume of the other solution (acid or base), whose concentration is to be determined, is known.

The above stoichiometry enables calculating unknown concentrations.

For example, suppose that 25.00 mL of 0.2250 M HCl (*aq*) is required to neutralize 27.45 mL of aqueous NaOH solution, whose concentration is unknown. Calculate the moles of each reactant and the concentration of NaOH.

No. of mol of HCl reacted:

$$25.0\ mL \times (1\ L/1{,}000\ mL) \times (0.2250\ mol/L) = 0.005625\ mol$$

Since HCl and NaOH react in a 1:1 ratio,

No. of mol of NaOH = No. of mol of HCl

No. of mol of NaOH = 0.005625 mol

Molarity of NaOH = (0.005625 mol / 0.02745 L)

Molarity of NaOH = 0.2049 M

The stoichiometric ratio may not be 1 to 1, as below.

For example, suppose 20.0 mL H_2SO_4 of unknown concentration requires 32.0 mL of 0.205 M NaOH. Calculate the concentration of the H_2SO_4 solution.

$$H_2SO_4\ (aq) + 2\ NaOH\ (aq) \rightarrow Na_2SO_4\ (aq) + 2\ H_2O$$

The stoichiometric ratio is 1 mole of H_2SO_4 to 2 moles of NaOH.

No. of mol of NaOH reacted = $\dfrac{0.205\ mol\ NaOH}{1\ L\ solution} \times 32.0\ mL \times \dfrac{1\ L}{1{,}000\ mL}$

No. of mol of NaOH reacted = 0.00656 mol

No. of mol of H_2SO_4 = Mol NaOH × (1 mol H_2SO_4 / 2 mol NaOH)

No. of mol of H_2SO_4 = 0.00656 mol NaOH × (1 mol H_2SO_4 / 2 mol NaOH)

No. of mol of H_2SO_4 = 0.00328 mol

Molarity of H_2SO_4 = $\dfrac{0.00328 \text{ mol } H_2SO_4}{0.0200 \text{ L}}$

Molarity of H_2SO_4 = 0.164 M

Strong acid–strong base titrations

For example, consider the strong acid-strong base titration of 20.0 mL of 0.100 M HCl (*aq*) with 0.100 M NaOH (*aq*) solution. Calculate the pH of the acid solution:

(a) before NaOH is added,

(b) after 15.0 mL of NaOH is added,

(c) after 19.5 mL of NaOH is added,

(d) after 20.0 mL of NaOH is added,

(e) after 21.0 mL of NaOH is added and

(f) after 25.0 mL of NaOH is added.

(a) Before titration:

$[H^+] = 0.100$ M

pH = 1.000

(b) When 15.0 mL of NaOH is added, the following reaction occurs:

H^+ (*aq*) + OH^- (*aq*) $\rightarrow H_2O$

[ ] before mixing:

      0.100 M   0.100 M

[ ] after mixing, but before reaction:

      0.0571 M   0.0429 M

[ ] after reaction:

      0.0142 M   0 M

$[H^+] = 0.0142$ M

pH = $-\log(0.0142)$

pH = 1.848

(c) After 19.5 mL of NaOH is added, the calculation of $[H^+]$ is:

$$H^+(aq) + {}^-OH(aq) \rightarrow H_2O$$

[ ] before mixing:

    0.100 M   0.100 M

[ ] after mixing, but before reaction:

    0.0506 M   0.0494 M

[ ] after reaction:

    0.0012 M   0 M

$[H^+] = 0.0012$ M

$pH = -\log(0.0012)$

$pH = 2.92$

Before the equivalent point, $[H^+]$ can be calculated:

$$[H^+] = \frac{\text{(initial mol of } H^+ - \text{mol of } {}^-OH \text{ added)}}{\text{(L of HCl titrated + L of NaOH added)}}$$

$$[H^+] = \frac{(0.00200 \text{ mol } H^+ - 0.00195 \text{ mol } {}^-OH)}{(0.0200 \text{ L of HCl} + 0.0195 \text{ L NaOH})}$$

$[H^+] = (0.000050 \text{ mol} / 0.0395 \text{ L})$

$[H^+] = 0.0013$ M

$pH = 2.90$

(d) When 20.0 mL of 0.100 M NaOH has been added,

$$H^+(aq) + {}^-OH(aq) \rightarrow H_2O$$

[ ] after mixing, but before reaction:

    0.0500 M   0.0500 M

[ ] after reaction:

     0   0

This point of the titration is the equivalence point, whereby only Na^+ and Cl^- occur in the solution.

Since neither reacts with water, the solution has a pH = 7.00.

(e) When 21.0 mL of 0.100 M NaOH has been added, there is excess OH^-:

$$H^+ (aq) + {}^-OH (aq) \rightarrow H_2O$$

[ ] after mixing, but before reaction:

        0.0488 M   0.0512 M

[ ] after reaction:

        0   0.0024 M

[$^-$OH] = 0.0024 M

pOH = 2.62

pH = 11.38

(f) When 25.0 mL of NaOH is added,

$$H^+ (aq) + {}^-OH (aq) \rightarrow H_2O$$

[ ] after mixing, but before reaction:

        0.0444 M   0.0556 M

[ ] after reaction:

        0   0.0112 M

[$^-$OH] = 0.0112 M

pOH = 1.953

pH = 12.047

In strong acid-strong base titrations, an abrupt change from about pH 3 to 11 occurs within ±0.5 mL of NaOH, added near the equivalent point.

Weak acid–strong base titrations

An example of a weak acid-strong base titration follows.

General reaction:

$$HA \ (aq) + {}^{-}OH \ (aq) \ \rightarrow \ H_2O + A^{-} \ (aq)$$

When acetic acid (weak acid) is titrated with sodium hydroxide (strong acid), the net reaction is:

$$HC_2H_3O_2 \ (aq) + {}^{-}OH \ (aq) \ \rightarrow \ C_2H_3O_2{}^{-} \ (aq) + H_2O$$

For example, consider the titration of 20.0 mL of 0.100 M $HC_2H_3O_2$ (aq) with 0.100 M NaOH (aq) solution.

Calculate the pH of the solution:

(a) before the NaOH is added;

(b) after 10.0 mL of NaOH is added;

(c) after 15.0 mL of NaOH is added;

(d) after 20.0 mL of NaOH is added;

(e) after 25.0 mL of NaOH is added.

(a) Before titration:

$$[H^{+}] = \sqrt{(0.100 \ M) \cdot (1.8 \times 10^{-5})}$$

$$[H^{+}] = 1.34 \times 10^{-3} \ M$$

$$pH = -\log(1.34 \times 10^{-3})$$

$$pH = 2.873$$

(b) After adding 10.0 mL of 0.100 M NaOH, $[H^{+}]$ is calculated as follows:

$$HC_2H_3O_2 \ (aq) + \ OH^{-} \ (aq) \ \rightarrow \ C_2H_3O_2{}^{-} \ (aq) + H_2O$$

Before mixing:

    0.100 M   0.1000 M   0.0000 M

After mixing (before rxn):

    0.0667 M   0.0333 M   0.0000 M

After reaction:

    0.0333 M   0.0000 M   0.0333 M

Using the Henderson-Hasselbalch equation,

$$pH = pK_a + \log([B^-] / [HB])$$

$$[H^+] = K_a \times \frac{[HC_2H_3O_2]}{[C_2H_3O_2^-]}$$

$$[H^+] = 1.8 \times 10^{-5} \times (0.0333 \text{ M} / 0.0333 \text{ M})$$

$$[H^+] = 1.8 \times 10^{-5} \text{ M}$$

$$pH = pK_a + \log([C_2H_3O_2^-] / [HC_2H_3O_2])$$

$$pH = 4.74 + \log(1)$$

$$pH = 4.74$$

When a weak acid is half-neutralized, 50% of the acid is converted to its conjugate base. Halfway to the equivalence point,

$$[C_2H_3O_2^-] = [HC_2H_3O_2]$$

Under this condition,

$$[H^+] = K_a, \text{ and } pH = pK_a$$

(c) After adding 15.0 mL of 0.100 M NaOH, $[H^+]$ is calculated as follows:

$$HC_2H_3O_2\,(aq) + {}^-OH\,(aq) \rightarrow C_2H_3O_2^-\,(aq) + H_2O$$

Before mixing:	0.100 M	0.100 M	0.000 M
After mixing (before rxn):	0.0571 M	0.0429 M	0.000
After reaction:	0.0142 M	0.000 M	0.0429 M

Using the Henderson-Hasselbalch equation:

$$pH = pK_a + \log([B^-] / [HB])$$

$$[H^+] = K_a \times [HC_2H_3O_2] / [C_2H_3O_2^-]$$

$$[H^+] = 1.8 \times 10^{-5} \times (0.0142 \text{ M} / 0.0429 \text{ M})$$

$$[H^+] = 6.0 \times 10^{-6} \text{ M}$$

$$pH = pK_a + \log([C_2H_3O_2^-] / [HC_2H_3O_2])$$

$$pH = 4.74 + \log(3.02)$$

$$pH = 4.74 + 0.48$$

$$pH = 5.22$$

(d) At the equivalent point, when 20.0 mL of 0.100 M NaOH has been added, acid has been reacted and converted to its conjugate base.

The latter undergoes hydrolysis (reacts with water) as follows:

$$C_2H_3O_2^- \,(aq) + H_2O \,(l) \leftrightarrows HC_2H_3O_2 \,(aq) + {}^-OH \,(aq)$$

Initial [ ], M:

0.0500	0.000	0.000

Change, Δ[ ], M:

$-x$	$+x$	$+x$

Equilibrium [ ], M:

$(0.0500 - x)$	x	x

By approximation:

$$[{}^-OH] = x = \sqrt{(K_b \times [C_2H_3O_2^-]_0)}$$

$$[{}^-OH] = \sqrt{(5.6 \times 10^{-10}) \cdot (0.0500)}$$

$$[{}^-OH] = 5.3 \times 10^{-6} \, M$$

$$pOH = -\log(5.3 \times 10^{-6})$$

$$pOH = 5.28$$

$$pH = 8.72$$

Since the conjugate base of a weak acid undergoes hydrolysis (reacts with water), the pH of the solution at the equivalence point is greater than 7.00.

In the case of acetic-NaOH titration, the pH at the equivalence point is about 8.72.

For weak acids with a larger K_a, the pH at the equivalence point is closer to neutral pH; for a smaller K_a, the pH at the equivalence point is higher than neutral pH.

Indicators

Titration endpoint

The *titrant* is a solution of known concentration that is added (titrated) to another solution to determine the concentration of a second chemical species (i.e., analyte).

The *analyte* is the substance whose quantity or concentration is to be determined.

An *indicator* is a substance that changes color to mark a titration's endpoint. Indicators exhibit one color in acid (or protonated form, HIn) and another in base (or deprotonated form, In⁻). Most indicators used in acid-base titration are weak organic acids.

Each indicator has a range of $pH = pK_a \pm 1$, where the change of colors occurs.

A suitable indicator gives the *endpoint* corresponding to the titration's equivalence point. A pH range falls within the sharp increase (or decrease) of pH changes in the titration curves. The endpoint approximates the equivalence point with the known concentration of the titrant to calculate the amount or concentration of the analyte (i.e., unknown quantity or concentration).

Indicator equilibrium

Like a weak acid, an indicator has the following equilibrium in aqueous solution:

$$HIn\ (aq) \leftrightharpoons H^+\ (aq) + In^-\ (aq)$$

$$K_a = \frac{[H^+]\cdot[In^-]}{[HIn]}$$

Rearranging

$$[H^+] = K_a \times ([HIn] / [In^-])$$

$$pH = pK_a + \log([In^-] / [HIn])$$

when

$$[HIn] = 10 \times [In^-]$$

$$pH = pK_a + \log([In^-] / 10[In^-])$$

$$pH = pK_a - 1.0;\ \text{the indicator assumes the color of the acid form.}$$

when

$$[In^-] = 10 \times [HIn]$$

$$pH = pK_a + \log(10[HIn] / [HIn])$$

$$pH = pKa + 1.0;\ \text{the indicator assumes the color of base form.}$$

Common acid–base indicator

Phenolphthalein, the most common acid-base indicator, has $K_a \sim 10^{-9}$. Its acid form (HIn) is colorless, and the conjugate base form (In⁻) is pink.

It is colorless when the solution's pH ≤ 8 when 90% or more of the species are in the acid form (HIn) and pink at pH ≥ 10 when 90% or more of the species are in the conjugate base form (In⁻).

The pH range at which an indicator changes depends on the K_a.

For phenolphthalein, which has $K_a \sim 10^{-9}$, its color changes in the pH range 8–10.

It is a suitable indicator for strong acid-strong base titrations and weak acid-strong base titrations.

pH of common indicators

Indicators	Acid color	Base color	pH Range	Type of Titrations
Methyl orange	orange	yellow	3.2–4.5	strong acid-strong base strong acid-weak base
Bromocresol green	yellow	blue	3.8–5.4	strong acid-strong strong acid-weak base
Methyl red	red	yellow	4.5–6.0	strong acid-strong base strong acid-weak base
Bromothymol blue	yellow	blue	6.0–7.6	strong acid-strong base
Phenol Red	orange	red	6.8–8.2	strong acid-strong base weak acid-strong base

Titration Curves

Interpreting titration curves

A *pH curve* is a graph of the pH of the solution *vs.* the volume of titrant in an acid-base titration.

The *equivalence point* is the point in a titration at which the amount of titrant added is just enough to neutralize the analyte solution completely.

At the equivalence point in an acid-base titration, moles of base = moles of acid, and the solution only contains salt and water.

The *buffering region* pH is usually $pK_a \pm 1$ (or $14 - pK_b \pm 1$), which resists pH changes before the curve's smooth (horizontal) portion is reached.

The data used to plot a pH curve may be obtained by computation *or* measuring a pH directly with the pH meter during titration.

The acid is incrementally added to the alkali (how titrations usually are performed).

Strong acid–strong base titration

The following titration curve is for the reaction:

$$NaOH\ (aq) + HCl\ (aq) \rightarrow NaCl\ (aq) + H_2O\ (l)$$

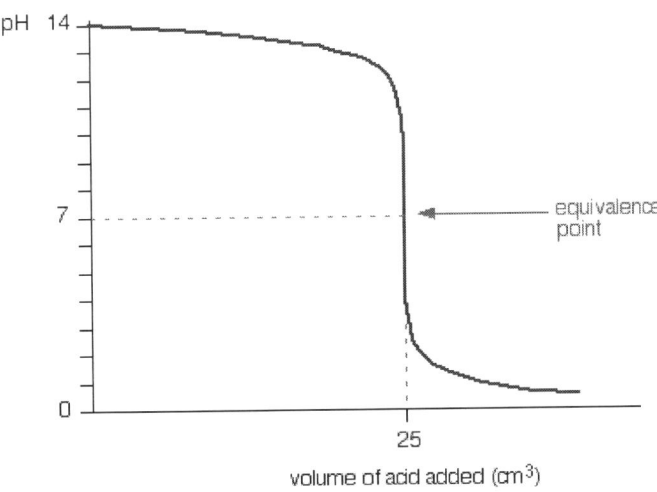

The pH falls a minimal amount until near the equivalence point, then a steep drop.

After the *equivalence point*, it is like the end of the strong acid-strong base reaction.

The buffering region's pH is usually $pK_a \pm 1$ (or $14 - pK_b \pm 1$), which resists pH changes before the curve's smooth (horizontal) portion is reached.

Strong acid–weak base titration

The following titration curve is for the reaction:

$$NH_3\,(aq) + HCl\,(aq) \rightarrow NH_4Cl\,(aq)$$

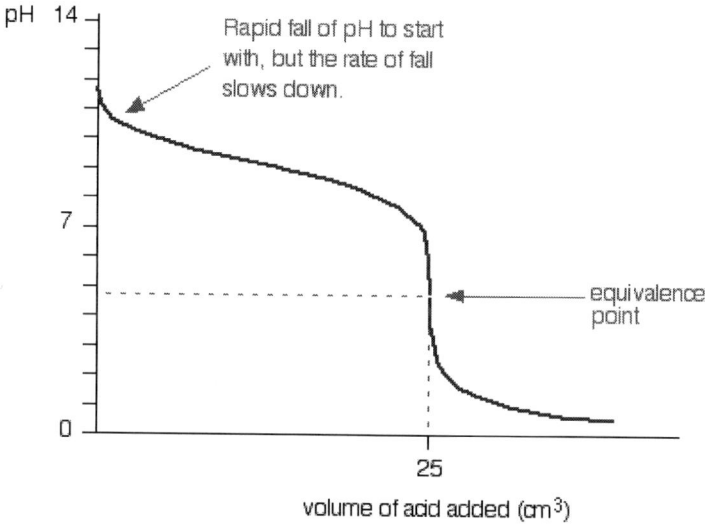

Since it is a weak base rather than a strong base, the titration curve's beginning is vastly different.

The pH decreases steeply as the acid is added, but the curve soon becomes less steep because a buffer solution is created composed of excess ammonia and formed ammonium chloride.

The equivalence point is now acidic (pH ~ 5) but is on the curve's steepest part.

The *equivalence point* is the point in a titration at which the amount of titrant added is just enough to neutralize the analyte solution completely.

At the equivalence point in an acid-base titration, moles of base = moles of acid, and the solution only contains salt and water.

After the *equivalence point*, it is like the end of the strong acid-weak base reaction.

The buffering region's pH is usually $pK_a \pm 1$ (or $14 - pK_b \pm 1$), which resists pH changes before the curve's smooth (horizontal) portion is reached.

Weak acid–strong base titration

The following titration curve is for the reaction:

$$CH_3COOH\ (aq) + NaOH\ (aq) \rightarrow CH_3COONa\ (aq) + H_2O\ (aq)$$

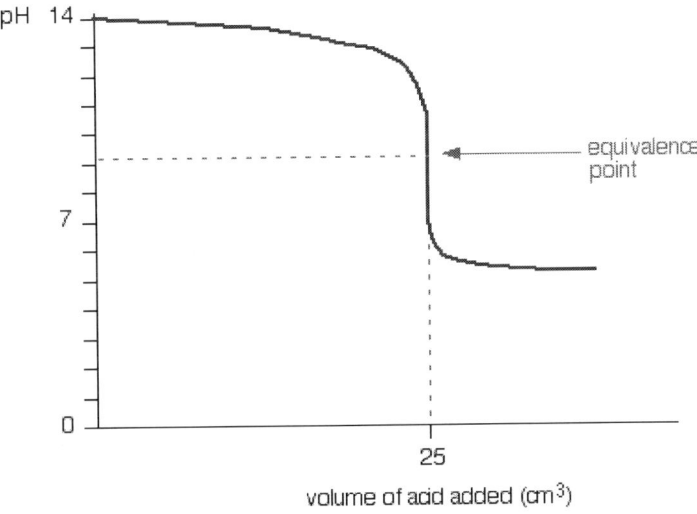

In the first part of the graph, there is an excess of sodium hydroxide, and this part of the curve is like the strong acid-strong base titration.

The *equivalence point* is the point in a titration at which the amount of titrant added is just enough to neutralize the analyte solution completely.

At the equivalence point in an acid-base titration, moles of base = moles of acid, and the solution only contains salt and water.

After the *equivalence point*, it is like the end of the weak acid-strong base reaction.

Once the acid is in excess, there is a difference between forming a buffer solution containing sodium ethanoate and ethanoic acid.

The buffering region's pH is usually $pK_a \pm 1$ (or $14 - pK_b \pm 1$), which resists pH changes before the curve's smooth (horizontal) portion is reached.

Weak acid–weak base titration

The following titration curve is for the reaction:

$$CH_3COOH\ (aq) + NH_3\ (aq) \rightarrow CH_3COONH_4\ (aq)$$

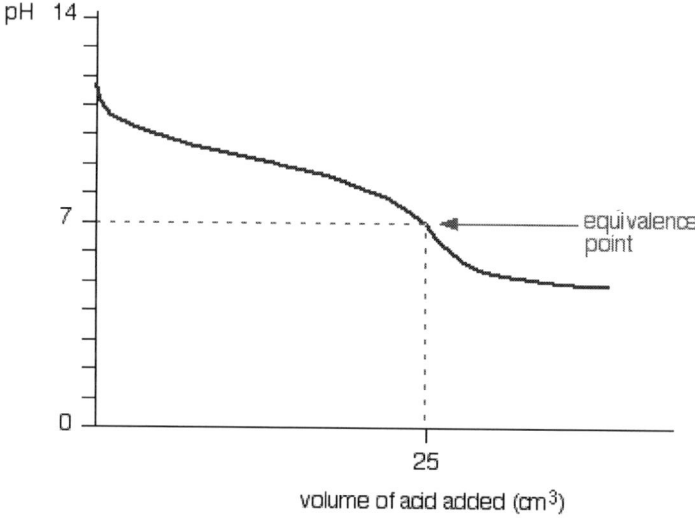

The acid and the base are equally weak, so the equivalence point is around pH 7.

This titration curve is essentially a combination of the two previous graphs.

The *equivalence point* is the point in a titration at which the amount of titrant added is just enough to neutralize the analyte solution completely.

At the equivalence point in an acid-base titration, moles of base = moles of acid, and the solution only contains salt and water.

After the *equivalence point*, it is like the end of the weak acid-weak base reaction.

There is no steep portion of this graph; the lack of a steep portion is an important identifying factor of a weak acid-weak base titration curve.

The buffering region's pH is usually $pK_a \pm 1$ (or $14 - pK_b \pm 1$), which resists pH changes before the curve's smooth (horizontal) portion is reached.

Four titration curves comparison

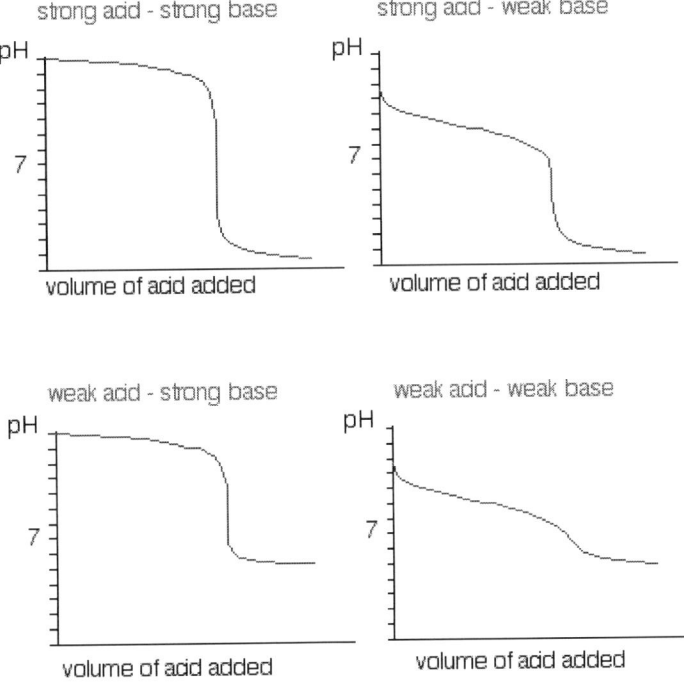

Redox titration transfers electrons

A *redox titration* is a type of titration based on the transfer of electrons.

In redox titration, standard solutions of oxidizing agents are used because solutions of reducing agents may react with oxygen in the air.

The analyzed species must be in a single oxidation state before titration.

For example, when iron ore is dissolved in hydrochloric acid, Fe^{2+} and Fe^{3+}, as iron(II) and iron(III), are present.

However, since it must be in a single oxidation state before titration with a standard $KMnO_4$ solution, iron(III) is reduced to iron(II) by reaction with excess zinc:

$$2\ Fe^{3+}\ (aq) + Zn\ (s) \rightarrow 2\ Fe^{2+}\ (aq) + Zn^{2+}\ (aq)$$

$$MnO_4^-\ (aq) + 5\ Fe^{2+}\ (aq) + 8\ H^+\ (aq) \rightarrow Mn^{2+}\ (aq) + 5\ Fe^{3+}\ (aq) + 4\ H_2O\ (l)$$

$KMnO_4$ and $K_2Cr_2O_7$ are the most used oxidizing agents in redox titration because the solutions change color during reduction; this color change serves as the titration indicator.

$KMnO_4$ is a purple solution but becomes colorless when MnO_4^- is reduced to Mn^{2+}.

The bright orange of $K_2Cr_2O_7$ changes to purplish-blue when $Cr_2O_7^{2-}$ is reduced to Cr^{3+}.

Sample redox titration

For example, a solution contains iron(II) and iron(III) ions. A 50.0 mL solution sample is titrated with 35.0 mL of 0.00280 M $KMnO_4$, which oxidizes Fe^{2+} to Fe^{3+}. The permanganate ion is reduced to manganese(II) ion. Another 50.0 mL solution sample is treated with zinc metal, which reduces Fe^{3+} to Fe^{2+}. The resulting solution is again titrated with 0.00280 M $KMnO_4$ and 48.0 mL is required. What are the concentrations of Fe^{2+} and Fe^{3+} in the solution?

For this problem, several steps are required.

1) The stoichiometric relationship of Fe(II) to permanganate is five to one:

$$MnO_4^- \, (aq) + 5 \, Fe^{2+} \, (aq) + 8 \, H^+ \, (aq) \rightarrow Mn^{2+} \, (aq) + 5 \, Fe^{3+} \, (aq) + 4 \, H_2O \, (l)$$

2) Calculate the moles of Fe(II) reacted:

(0.00280 mol/L) (0.0350 L) = 0.000098 mol of MnO_4^-

(0.0000980 mol Mn)·(5 mol Fe / 1 mol Mn) = 0.000490 mol Fe(II)

3) Determine the total iron content:

(0.00280 mol/L)·(0.0480 L) = 0.0001344 mol of MnO_4^-

(0.0000980 mol Mn)·(5 mol Fe / 1 mol Mn) = 0.000672 mol total Fe

4) Determine Fe(III) in solution and its molarity:

0.000672 mol – 0.000490 mol = 0.000182 mol

(0.000182 mol) / (0.050 L) = 0.00364 M Fe(III)

5) Determine the molarity of Fe(II):

(0.000490 mol) / (0.050 L) = 0.0098 M Fe(II)

Practice Questions

1. Which of the following indicators is green at pH 7?

      I. phenolphthalein   II. bromothymol blue   III. methyl red

  A. I only
  B. II only

  C. III only
  D. I and II only
  E. I, II and III

2. Which of the following is NOT a strong acid?

  A. $HClO_3$
  B. HF

  C. HBr
  D. HCl
  E. HI

3. Which of the following compounds is the strongest base?

  A. ClO_3^-
  B. NH_3

  C. ClO^-
  D. ClO_2^-
  E. ClO_4^-

4. A Brønsted-Lowry acid is defined as a substance that:

  A. acts as a proton donor in any system
  B. acts as a proton acceptor in any system

  C. increases $[H^+]$ when placed in water
  D. decreases $[H^+]$ when placed in water
  E. increases $[H^+]$ in any system

5. Which of the following acts as the best buffer solution?

  A. Strong acids or bases
  B. Strong acids and their salts

  C. Salts
  D. Weak acids or bases and their salts
  E. Strong bases and their salts

6. Which of the following is true if a drain cleaner solution is a strong electrolyte?

      I. Slightly reactive   II. Highly reactive   III. Highly ionized

  A. I only
  B. II only

  C. III only
  D. I and III only
  E. II and III only

7. Which of the following statements describes a basic solution?

A. $[H_3O^+] \times [^-OH] \neq 1 \times 10^{-14}$

B. $[H_3O^+]/[^-OH] = 1 \times 10^{-14}$

C. $[H_3O^+] > [^-OH]$

D. $[H_3O^+] < [^-OH]$

E. $[H_3O^+]/[^-OH] = 1$

8. Which of the following is a general property of an acidic solution?

    I. Neutralizes bases   II. Tastes sour   III. Turns litmus paper red

A. I only

B. II only

C. III only

D. I and III only

E. I, II and III

9. Compared to a solution with a higher pH, a solution with a lower pH has a(n):

A. decreased K_a

B. increased $[^-OH]$

C. increased pK_a

D. increased $[H^+]$

E. decreased $[H^+]$

10. Which of the following statements correctly define an Arrhenius acid?

A. Decreases $[H^+]$ when placed in aqueous solutions

B. Increases $[H^+]$ when placed in aqueous solutions

C. Acts as a proton acceptor in any system

D. Acts as a proton donor in any system

E. Acts as a proton acceptor in aqueous solutions

11. A Brønsted-Lowry base is a(n):

A. electron acceptor

B. proton acceptor

C. electron donor

D. proton donor

E. proton donor and electron acceptor

12. Which of the following is the conjugate base of water?

A. ^-OH (*aq*)

B. H^+ (*aq*)

C. H_2O (*l*)

D. H_3O^+ (*aq*)

E. O^{2-} (*aq*)

13. Which of the following explains why distilled H_2O is neutral?

A. $[H^+] = [OH^-]$

B. Distilled H_2O has no OH^-

C. Distilled H_2O has no H^+

D. Distilled H_2O has no ions

E. None of the above

14. In the following reaction, what does the symbol $\rightleftharpoons$ indicate?

$$^-OH + NH_4^+ \rightleftharpoons H_2O + NH_3$$

A. The rate of the reverse reaction is the same as the forward reaction

B. The forward reaction does not proceed

C. The reaction does not produce an equilibrium

D. The reverse reaction does not proceed

E. The equilibrium depends on the concentrations of the reactants

15. What is the $[H_3O^+]$ of a solution with a pH = 2.34?

A. 1.3×10^1 M

B. 2.3×10^{-10} M

C. 4.6×10^{-3} M

D. 2.4×10^{-3} M

E. 3.6×10^{-8} M

Notes for active learning

Detailed Explanations

1. B is correct.

Bromothymol blue is a pH indicator often used for solutions with neutral pH near 7 (e.g., managing the pH of pools and fish tanks).

Bromothymol blue acts as a weak acid in a solution that can be protonated or deprotonated.

Bromothymol blue appears yellow when protonated (lower pH), blue when deprotonated (higher pH), and bluish-green in neutral solution.

Phenolphthalein is an indicator for acid-base titrations.

A weak acid dissociates protons (H^+ ions) in the solution.

The phenolphthalein molecule is colorless, and the phenolphthalein ion is pink.

It turns colorless in acidic solutions and pink in basic solutions.

With basic conditions, the phenolphthalein (neutral) $\rightleftharpoons$ ions (pink) equilibrium shifts to the right, leading to more ionization as H^+ ions are removed.

Methyl red has a pK_a of 5.1 and is a pH indicator dye that changes color in acidic solutions: it turns red in pH under 4.4, orange in pH 4.4-6.2, and yellow in pH over 6.2.

2. B is correct.

Strong acids dissociate a proton to produce the weakest conjugate base (i.e., most stable anion).

Weak acids dissociate a proton to produce the strongest conjugate base (i.e., least stable anion).

Hydrofluoric (HF) acid has a pK_a of about 3.8 and is weak because it does not dissociate completely.

The F^- anion is the least stable of the halogen anions (due to its small valence shell).

3. C is correct.

Perchloric acid ($HClO_4$) is the strongest acid listed and the weakest conjugate base (i.e., most stable anion).

Hypochlorite (ClO^-) is the strongest base (i.e., least stable anion).

In *oxyacids* (or *oxoacids*), more oxygens increase the acid strength.

The weakest conjugate bases of oxyacids (containing more than one oxygen) are stabilized by resonance.

4. A is correct.

The Brønsted-Lowry acid-base theory focuses on the ability to accept and donate protons (H^+).

A Brønsted-Lowry acid is a substance that donates a proton in an acid-base reaction, while a Brønsted-Lowry base is a substance that accepts the proton.

5. D is correct.

A *buffer* is an aqueous solution that consists of a weak acid and its conjugate base, or vice versa.

Buffered solutions resist changes in pH and are often used to keep the pH at a nearly constant value in many chemical applications. It does this by readily absorbing or releasing protons (H^+) and ^-OH.

When acid is added to the solution, the buffer releases ^-OH and accepts H^+ ions from the acid.

Weak acids or bases and their salts are the best buffer systems.

6. C is correct.

An *electrolyte* is a substance that dissociates into cations (i.e., positive ions) and anions (i.e., negative ions) when placed in a solution.

7. D is correct.

In a basic solution, the concentration of ^-OH is higher than H^+ or H_3O^+.

8. E is correct.

An acid can neutralize a base (i.e., a substance with a pH above 7) to form a salt.

Acids have a sour taste (e.g., lemon juice has a pH of about 2). Acids are known to have a sour taste (e.g., lemon juice) because the sour taste receptors on the tongue detect the dissolved hydrogen (H^+) ions.

An acid is a chemical substance with a pH of less than 7.

Litmus paper is red under acidic conditions and blue under basic conditions.

A pH greater than 7 and feels slippery are qualities of bases, not acids. An acid is a chemical substance with a pH of less than 7, producing H^+ ions in water.

An acid can be neutralized by a base (i.e., a substance with a pH above 7) to form a salt.

Litmus paper is red under acidic conditions and blue under basic conditions.

However, acids are not known to have a slippery feel; this is a characteristic of bases.

Bases feel slippery because they dissolve the fatty acids and oils from the skin and reduce the friction between the skin cells.

9. D is correct.

Acidic solutions contain hydronium ions (H_3O^+).

These ions are in aqueous form because they are dissolved in water.

Although chemists often write H^+ (*aq*), referring to a single hydrogen nucleus (a proton), it exists as the hydrogen atom.

10. B is correct.

An Arrhenius acid increases the concentration of H^+ ions in an aqueous solution.

Since pH is the negative log of the activity of H^+ ions in an aqueous solution, acids have a low pH.

An acid can act as a proton donor, but this is the Brønsted-Lowry definition of an acid, not Arrhenius.

11. B is correct.

Brønsted-Lowry acids are proton donors (e.g., HCl, H_2SO_4).

Brønsted-Lowry bases are proton acceptors (e.g., HSO_4^-, NO_3^-).

Lewis acids are electron-pair acceptors, whereas Lewis bases are electron-pair donors.

12. A is correct.

By the Brønsted-Lowry acid-base theory:

An acid (reactant) dissociates a proton to become the conjugate base (product).

A base (reactant) gains a proton to become the conjugate acid (product).

The definition is expressed as an equilibrium expression

$$acid + base \leftrightarrow conjugate\ base + conjugate\ acid.$$

13. A is correct.

If the $[H_3O^+] = [^-OH]$, it has a pH of 7, and the solution is neutral.

14. A is correct.

The symbol $\rightleftharpoons$ indicates that the reaction is in equilibrium, whereby the *forward reaction rate* equals the *reverse reaction rate*.

Equilibrium refers to the rate and not the relative magnitude of the remaining reactants or the formed products during the reaction.

15. C is correct.

The formula for pH is:

$$pH = -\log[H_3O^+]$$

Rearrange to solve for $[H_3O^+]$:

$$[H_3O^+] = 10^{-pH}$$

$$[H_3O^+] = 10^{-2.34}$$

$$[H_3O^+] = 4.6 \times 10^{-3} \text{ M}$$

Notes for active learning

Notes for active learning

CHAPTER 8

Thermochemistry

- Thermochemistry
- Heat and Temperature
- Temperature Scales
- Heat Capacity and Specific Heat
- Calorimetry
- Heat Transfer
- Thermodynamic Systems
- Endothermic and Exothermic Reactions
- Thermodynamics
- Zeroth Law of Thermodynamics
- First Law of Thermodynamics
- Second Law of Thermodynamics
- Third Law of Thermodynamics
- Enthalpy H
- Bond Energy and Heat of Formation Relationship
- Gibbs Free Energy G
- Spontaneous Reactions and ΔG
- Thermal Expansion
- Kinetic Theory of Gases
- Phase Diagrams
- Colligative Properties
- Heat Transformation
- Pressure $vs.$ Volume Phase Diagrams
- Practice Questions & Detailed Explanations

Thermochemistry

Applied thermochemistry

Thermochemistry:

- explains energy changes, particularly for energy exchange between the system and its surroundings;

- predicts reactant and product quantities during the reaction;

- predicts if a reaction is spontaneous or non-spontaneous, favorable, or unfavorable;

- combines concepts of thermodynamics and chemical bond energy.

Thermochemistry quantities

Thermochemistry commonly includes calculations of such *quantities* as:

heat capacity (c)

heat of combustion

heat of formation

enthalpy (H)

entropy (S)

Gibbs free energy (G)

calories

Energy changes in chemical reactions

Thermochemistry studies heat energy associated with chemical reactions and physical transformations.

Reactions may *release energy* (i.e., exothermic reactions) or *absorb energy* (i.e., endothermic reactions).

Phase change may *release or absorb energy* (e.g., boiling, melting).

Thermochemistry rests on *two* generalizations:

1. Energy change accompanying any transformation is *equal and opposite* to the reverse process's energy change. (Lavoisier and Laplace law proposed in 1780).

2. Hess' Law proposed in 1840 states that the *energy change of any transformation* is the same if the process occurs in one or many steps.

Systems conserve energy

German physicist Julius Robert Mayer (1814-1878) first proposed the *law of the conservation of energy* in 1842, now called the First Law of Thermodynamics.

Conservation of energy law states that *energy is neither created nor destroyed.*

These generalizations were the foundations for *specific heat, latent heat,* and *heat capacity*:

$$\Delta C_p = d\Delta H / dT$$

Integration of the expression $\Delta C_p = d\Delta H / dT$ evaluates the heat of reactions at different temperatures.

Heat and Temperature

Heat as energy

Heat is energy and exists independently from any medium; *heat is a way to transfer energy.*

Heat and temperature are often used interchangeably in everyday vernacular.

However, this usage of interchanging heat and temperature is incorrect.

Temperature as kinetic motion

Temperature measures *kinetic energy* (KE) in a specific medium and depends on the material.

Temperature and heat are connected principles, but heat is the *total energy* due to *molecular motion*. In contrast, temperature is the average measure of kinetic energy in a substance.

Some substances respond drastically to a change in thermal energy. This response depends on the material and heat capacity.

Endothermic *vs.* exothermic processes

> *Endothermic* processes experience heat gained by the system from the surroundings, which equals heat lost by the surroundings.

> *Exothermic* processes cause the system's heat loss to equal the heat gain of the surroundings.

Thermal energy

The quantity of thermal energy transmitted from one body to another measures heat.

An object does *not* "contain" heat but can have a certain amount of thermal energy.

Heat can be measured in joules (J) and calories (cal).

On a microscopic level, thermal energy is related to atomic and molecular *vibrations*.

Matter is made of billions and trillions of small molecules; even in solids, these molecules and atoms are never entirely still.

The movement of molecules determines how much thermal energy they hold.

Calories

Calories relate heat to changes in temperature, which is often useful when solving problems in thermochemistry.

calorie (lowercase *c*) is the heat needed to raise the temperature of one gram of water by one degree Celsius (g/°C).

$$1\ cal = 1\ \frac{g}{°C} = 4.19\ J$$

Food Calories (uppercase *C*) are not equivalent to heat calories. This resembles the relationship between kilograms and grams (1 Calorie = 1,000 calories).

Another difference is that while nutrition Calories measure stored energy just as calories do, they also measure the energy stored in the chemical bonds of food that are broken down and stored when digested.

Calories measure energy in a substance and can be converted to equivalent conventional energy.

For example, the calories contained in 5 lbs. of spaghetti have enough energy to brew a pot of coffee; calories in one piece of cheesecake can power a 60 W incandescent light bulb for 1.5 hours; 222 Big Macs contain enough energy to drive a vehicle for almost 90 miles.

Energy conversion factors

Some useful conversion relationships:

      1 calorie = 4.2 J

      1 Calorie (with capital C) = 1,000 calorie

      1 Calorie (with capital C) = 1 kilocalorie

      1 Calorie = 4,200 J

Conversions for water

      1 gram = 1 cubic centimeter

      1 cubic centimeter = 1 mL

Temperature Scales

Fahrenheit and Celsius

Three commonly used temperature scales are *Fahrenheit, Celsius* and *Kelvin.*

Daniel Fahrenheit (1686-1736) proposed the classic English system of measuring temperature (i.e., °F). in 1724 by dividing the difference between the boiling and freezing point of water into 180 equal degrees (°F).

Anders Celsius (1701-1744) developed the modern metric measuring temperature by dividing the boiling and freezing point of water into 100 equal degrees (°C).

Celsius scale was originally the *centigrade scale.*

In the United States, the temperature unit is *Fahrenheit* (°F).

However, the SI unit for temperature is *Celsius* (°C) with water freezing at 0 °C and boiling at 100 °C (instead of the 32 °F and 212 °F).

Kelvin scale

William Thomas, 1st Baron Kelvin (1824-1907), developed the Kelvin (K) as the third temperature scale. The Kelvin scale is based on the concept of absolute zero (i.e., 0 K, –273.15 °C, or –459.67 °F) and is the most important science scale for temperature.

The size of units on the Kelvin scale is the *same* as for degrees on the Celsius scale.

The conversion between the two scales is:

$$°F = \frac{9}{5}°C + 32$$

Kelvin is another unit of temperature often used in scientific calculations. This scale is often used for gases, as usually, low temperatures are in question.

Absolute zero is the coldest theoretical temperature any substance can have, equal to 0 K.

0 K is equivalent to –273 °C!

Relative temperature scales

Three thermometers display equal temperatures and three standard temperature units are shown.

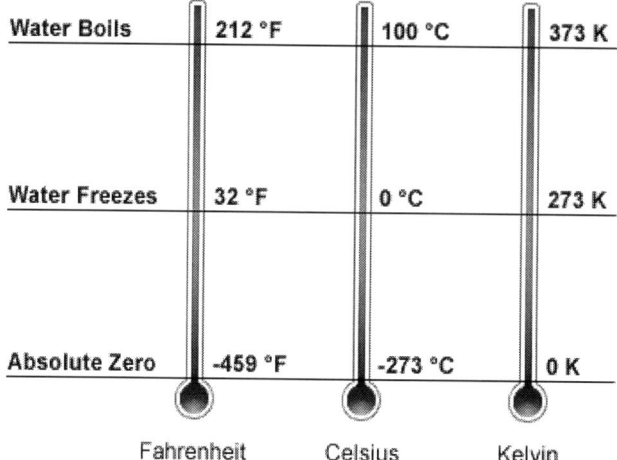

Water Boils	212 °F	100 °C	373 K
Water Freezes	32 °F	0 °C	273 K
Absolute Zero	-459 °F	-273 °C	0 K
	Fahrenheit	Celsius	Kelvin

Comparison between the three temperature scales

Converting between scales

Temperature is the average amount of *kinetic energy* of the particles (i.e., molecules and atoms) of a substance.

Absolute zero when these particles theoretically have no KE and no movement.

Kelvin and Celsius use the same step in their degree measurements, making calculations easier.

For example, a temperature difference is the same in Kelvin as in Celsius, so when calculating a temperature difference in either of these units, the values do not need to be converted.

If required to convert between Kelvin and Celsius, the conversion can be performed by adding 273 to the Celsius temperature to get Kelvin:

$$K = °C + 273$$

The temperature of absolute zero may never be attained. Scientists have achieved temperatures close to absolute zero but have never fully reached it.

In 2003, the lowest temperature recorded was by a group of scientists from MIT (Massachusetts Institute of Technology). Researchers successfully cooled sodium (Na) gas to a temperature of half a billionth of a degree above absolute zero but not at absolute zero.

Fahrenheit and Celsius temperature conversions

Convert between Fahrenheit and Celsius by:

$$°F = (1.8 × °C) + 32$$

For example, what is the temperature in °F for 28 °C?

$$°F = (1.8 × °C) + 32$$

$$°F = (1.8 × 28 °C) + 32$$

$$°F = (50.4) + 32$$

$$°F = 82.4$$

Therefore, 28 °C equals 82.4 °F

For example, what is the temperature in °C for 70 °F?

$$°F = (1.8 × °C) + 32$$

Rearrange:

$$(1.8 × °C) = °F − 32$$

$$(1.8 × °C) = °F − 32$$

$$°C = (°F − 32) / 1.8$$

$$°C = (70 − 32) / 1.8$$

$$°C = 21$$

Therefore, 70 °F equals 21 °C

Kelvin and Celsius temperature conversions

Convert between Kelvin and Celsius by:

$$K = °C + 273$$

For example, what is the temperature in °C for 295 K?

$$K = °C + 273$$

Rearrange:

$$°C = K − 273$$

$$°C = 295 − 273$$

$$°C = 22$$

Therefore, 295 K = 22 °C.

Reference values among temperature scales

	K	°C	°F
Absolute zero	0	−273	−460
Freezing point of water / melting point of ice	273	0	32
Room temperature	298	25	77
Body temperature	310	37	99
Boiling point of water / condensation of steam	373	100	212

Heat Capacity and Specific Heat

Heat capacity, specific heat and molar heat capacity

	Heat capacity	Specific heat capacity	Molar heat capacity
Definition	Amount of heat required to raise temperature of a substance by 1 K (1 °C)	Amount of heat required to raise the temperature of 1 g of a substance by 1 K (1 °C)	Amount of heat required to raise temperature of 1 mole of substance by 1 K (1 °C)
	Ability of a substance to absorb heat	Amount of heat added (or removed) to change the temperature	At constant pressure unless noted
Equation	$C_H = q / \Delta T$	$C_S = q / m\Delta T$ C_S (specific) equals C_P (constant pressure)	$C_M = q / n\Delta T$
Intrinsicality	Intrinsic property	Non-intrinsic property	
Unit	J / °C	J / g °C or J / Kg K	J / mol °C or J / mol K
Property	Extensive property	Intensive property	Intensive property
Substance quantity	Depends on amount	Independent of amount	
Specificity	Non-specific	Highly specific	
Examples	1 L H_2O = 4182 J/kg °C 100 g Cu = 38.5 J/ °C	H_2O = 1 calorie 1 calorie = 4.18 J/g °C Cu = 0.385 J/g °C	

Heat capacity

Heat capacity (or *thermal capacity*) is a physical property of a substance's ability to hold heat.

Heat capacity (C_H) of a substance is the energy required to raise the temperature of a substance by *one degree Celsius*. It is the absorbed heat energy ratio to the resulting temperature change.

Heat capacity is a product of an *object's mass* (*m*) and a *substance's specific heat capacity (c)*.

When measuring the heat capacity of a substance, a variable must be held constant since the heat capacity of a system depends on the temperature, the pressure, and the volume of the system. These are noted with either a c_p or a c_v, respectively.

SI unit for heat capacity is the joule per Kelvin (J/K).

Isobaric and isochoric processes

Gases and liquids are typically measured at constant volume and subjected to pressure for a constant reading.

> *Isobaric process* (c_p) is when *pressure* is held constant, defining the *heat capacity*.

> *Isochoric process* (c_v) is when *volume* is held constant.

Heat capacity *vs.* specific heat capacity

Heat capacity and specific heat of substances are similar.

> *Specific heat* (C_S) is the ratio of energy absorbed to the temperature rise per unit of mass of the substance.

> *Heat capacity* is the ratio of energy absorbed to the rise in temperature.

Specific heat *c* is a characteristic of the substance and thus does not change.

Water has a relatively *high* specific heat capacity; 4.2 J of heat energy is needed to raise one gram of water 1 °C.

High specific heat of the water *helps modulate global temperatures*. Temperature fluctuations during day and night would be much larger if the water's heat capacity were low. This is evident in a desert where water is scarce. Daytime temperatures are high, yet nighttime temperatures can drop to near freezing.

Specific heat

Specific heat capacity (C_S) is the amount of heat required to raise the temperature of 1 gram of a substance by 1 °C.

Specific heat *c* is a characteristic of the substance and thus does not change. Metals tend to have *low* specific heat capacities.

Heat capacity and the specific heat of a substance are similar. *Specific heat capacity* differs from heat capacity because it relates to *the heat energy required to raise one gram of a substance by one degree Celsius.*

> *Specific heat capacity* (or *specific heat*) is the heat capacity of 1 gram of a substance.

> *Molar heat capacity* is the heat capacity of 1 mole of a substance.

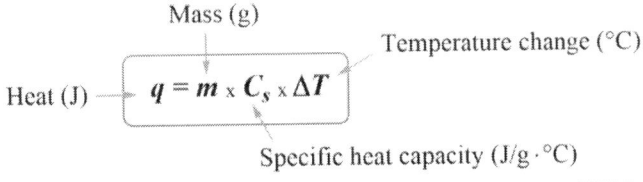

Specific heat capacity

Specific heat relates to heat and temperature

Units for specific heat are calories or joules per gram per °C. For example, the specific heat of water is 1 calorie (or 4.18 joules) per gram per Celsius degree.

The specific heat capacity of water is 4,180 joules per kilogram per degree Celsius (J/kg°C); 4,180 J are needed to raise the temperature of 1 kg of water by 1 °C.

Specific heat is related to heat and temperature by:

$$q = mC_S\Delta T$$

where *q* is the heat flow, *m* is mass measured in grams, C_S is the specific heat capacity of the substance, and ΔT is the temperature change.

Specific heat (C_S) of a substance, heat (Q), and temperature are related by:

$$Q = mC_S\Delta T$$

where Q is the heat transferred to the material (J), *m* is the mass of the object being heated (kg), and ΔT is the temperature change (K).

Calculating specific heat

For example, 3,200 J of heat is added to 1.0 kg of water at 10 °C. (Use specific heat of water C_S = 4,190 J/kg·°C)

Rearrange the specific heat equation to solve ΔT:

$$Q = mC_S\Delta T$$

$$\Delta T = \frac{Q}{mc}$$

$$\Delta T = \frac{3,200\ J}{1.0\ kg \cdot 4,190\ \frac{J}{kg} \cdot °C}$$

$$\Delta T = 0.76\ °C$$

where Q is the heat transferred to the material (J), m is the mass of the object being heated (kg), and ΔT is the change in temperature (K).

ΔT represents the temperature change. This value for the change (not the absolute value) must be either added to the initial temperature to determine the final temperature or subtracted from the final to find the initial.

A reaction must be contained within a system to observe all the products.

Specific heats of selected materials

Material	c(J/kg·K)
Aluminum	879
Concrete	850
Diamond	509
Glass	840
Helium	5,193
Water	4,181

Low *vs.* high specific heat

Substances such as metals (e.g., copper) have a *low specific heat* because it does not take much energy to transfer the heat and excite the molecules.

Raising the value of their average kinetic energy (raising their temperature).

Materials that are difficult to heat (e.g., rubber) have a much *higher specific heat* because more energy is required to raise the temperature by the same amount. The previous table shows that water has a much higher heat capacity than other common materials found on land.

Specific heat modulates environmental temperatures

Raising an ocean by 1 °C takes more energy than raising city sidewalks by the same amount. Thus, land warms faster during the day than water when subjected to the same amount of thermal energy from the Sun. This is the reason for the sea breeze effect.

Hot land heats the air above it, causing it to rise and create a low-pressure system above it. The cooler water creates a high-pressure system due to air cooling down and descending. The difference in pressures creates airflow from the sea to the land during the day as a sea breeze.

At night, the land cools faster than the water (i.e., lower specific heat capacity), and the effect is reversed as air flows from land to the sea.

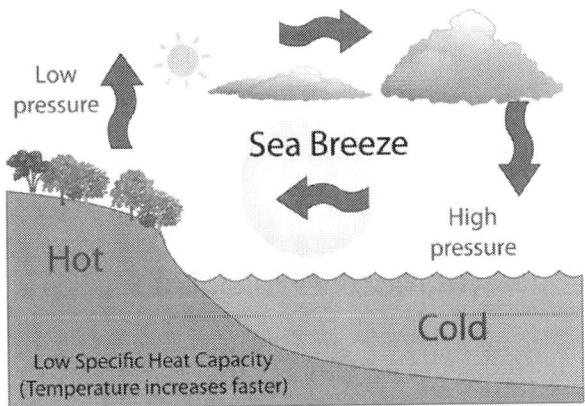

High specific heat of water helps modulate global temperatures

Notes for active learning

Calorimetry

Food samples

Calorimetry is an experimental method to calculate enthalpy changes involving heat flow.

Calorimeters measure the amount of energy in food samples experimentally. Food is placed inside a closed container surrounded by water and burned in the presence of oxygen. Water temperature change is related to the energy released, commonly reported in nutritional Calories (or kilojoules units). Note the term *Calories* in uppercase.

Even though cells within the body combust the molecules differently than a calorimeter, the calorimeter provides an accurate *caloric value* because of the same end products.

Calibration is an experiment in which hot water of known mass and temperature is poured into an insulated cup containing icy water of known mass and temperature. The composite sample equilibrates to a maximum temperature, and a calorimeter constant is determined below.

Nutritional Calories

Nutritional Calories (Cal) are energy units equal to 1,000 calories (cal) and a kilocalorie (kcal).

1 Calorie = 1,000 calories

Nutrient Molecule in Food	Example	Cal/g	kJ/g
Carbohydrate	Table sugar, potatoes, flour	4	17
Protein	Meats, fish, beans	4	17
Fat	Oil, butter	9	38

Foods contain varying amounts of energy, as shown

Calorimeters

Calorimeters are ideally considered an insulated system. The heat liberated by the reaction process is transferred to the other substances within the calorimeter or the calorimeter itself. No heat is lost to the surroundings.

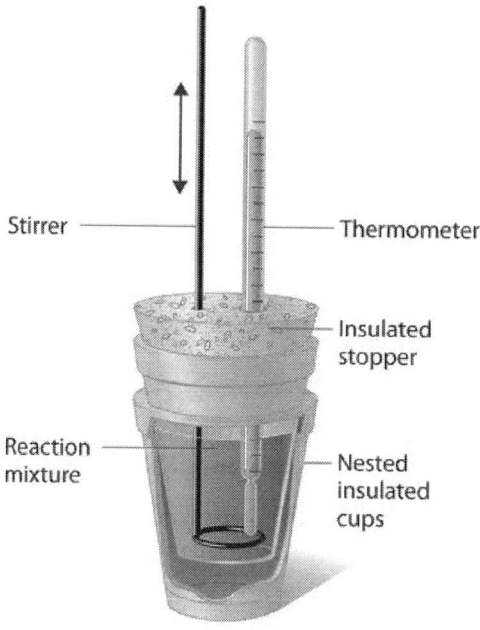

Components of a simple calorimeter

Since no thermal energy is lost in a calorimeter, the heat of a reaction can be measured by inserting a thermometer into a calorimeter and recording the temperature readings before and after the reaction. A stirrer ensures that all calorimeter contents are mixed and react uniformly.

Constant-pressure calorimeters

Constant-pressure and *constant-volume calorimeters* are two types of calorimeters.

Constant-pressure calorimeter is where heat flow is measured at constant atmospheric pressure.

A perfect calorimeter does not exist; no calorimeter completely insulates heat flow. The calorimeter itself, as part of the surroundings, absorbs some heat. Therefore, the calorimeter must be calibrated to obtain accurate results. Styrofoam cups are simple, constant-pressure calorimeters often used in introductory chemistry classes. Although a Styrofoam cup's temperature-insulating properties are reasonably good, the cup absorbs heat.

Calorimeter constant K_{cal}

Calorimeter constant K_{cal} is determined by a composite sample equilibrating at max. temp.

Basic principle of calorimetry:

heat lost by hot water = heat gained by cool water + heat gained by calorimeter

Determine the mathematical equations:

$$-m_{hot}c_{water}(T_{final} - T_{intial, hot}) = m_{cool} \, c_{water}(T_{final} - T_{initial, cool}) + K_{cal} \, (T_{final} - T_{initial, cool})$$

Sign is negative for heat lost by the hot water because heat loss is negative. Substitute known masses and temperatures into the expression. Since the cool water was in the cup, the calorimeter's initial and cool temperatures were the same.

For example, suppose 60.1 g of water at 97.6 °C is poured into a cup calorimeter containing 50.3 g of water at 24.7 °C. The final temperature of the combined water samples reaches 62.8 °C. What is the calorimeter constant? (Use c_{water} = 4.184 J/g°C)

$$-(60.1 \text{ g}) \cdot (4.184 \text{ J/g°C}) \cdot (62.8 \text{ °C} - 97.6 \text{ °C})$$
$$= [(50.3 \text{ g}) \cdot (4.184 \text{ J/°C}) \cdot (62.8 \text{ °C} - 24.7 \text{ °C})] + [K_{cal} \, (62.8 \text{ °C} - 24.7 \text{ °C})]$$

$$-(60.1 \text{ g}) \cdot (4.184 \text{ J/g°C}) \cdot (-34.8 \text{ °C})$$
$$= [(50.3 \text{ g}) \cdot (4.184 \text{ J/°C}) \cdot (38.1 \text{ °C})] + [K_{cal} \, (38.1 \text{ °C})]$$

$$[8.75 \times 10^3 \text{ J}] - [8.02 \times 10^3 \text{ J}] = [K_{cal} \, (38.1 \text{ °C})]$$

$$0.73 \times 10^3 \text{ J} = [K_{cal} \, (38.1 \text{ °C})]$$

$$[0.73 \times 10^3 \text{ J}] / (38.1 \text{ °C}) = K_{cal}$$

$$K_{cal} = 19.2 \text{ J/°C}$$

Heat flow

After calibrating the calorimeter, the specific heat capacity will be determined.

A known quantity of metal is heated to a known temperature and placed in an insulated cup calorimeter containing a known quantity and temperature until the temperature equilibrates.

For example, suppose 28.2 g of an unknown metal is heated to 99.8 °C and placed into a cup calorimeter containing 150.0 g of water at 23.5 °C. The temperature of the water equilibrates at a final temperature of 25.0 °C. What is the specific heat capacity of the metal? (Use the calorimeter constant of 19.2 J/°C).

Consider heat flow:

heat lost $_{hot\ metal}$ = heat gained $_{water}$ + heat gained $_{calorimeter}$

$-m_{hot}C_{water}(T_{final} - T_{intial,\ hot}) = m_{cool}\ c_{water}(T_{final} - T_{initial,\ cool}) + K_{cal}\ (T_{final} - T_{initial,\ cool})$

$-(28.2\ g)·(C_{metal})·(25.0\ °C - 99.8\ °C) =$
  $= [(150.0\ g\ )·(4.184\ J/g°C)·(25.0\ °C - 23.5\ °C)] + [19.2\ J/°C(25.0\ °C - 23.5\ °C)]$
  $-(28.2\ g)·(C_{metal})·(-\ 74.8\ °C)$

$= [(150.0\ g\ )·(4.184\ J/g°C)·(1.5\ °C)] + [19.2\ J/°C(1.5\ °C)]$

$2.11 \times 10^3\ g°C\ (C_{metal}) = (9.4 \times 10^2\ J) + (29\ J)$

$2.11 \times 10^3\ J\ (C_{metal}) = 969\ J$

Solving for specific heat capacity:

$C_{metal} = 969\ J\ /\ 2.11 \times 10^3\ g°C$

$C_{metal} = 0.459\ J/\ g°C$

Constant-volume calorimeters

Constant-volume calorimeter is where heat flow is measured at constant volume.

Bomb calorimeter (decomposition vessel) is the most used constant-volume calorimeter.

Combustion is when oxygen reacts with a substance to produce carbon dioxide (CO_2), water (H_2O), and energy.

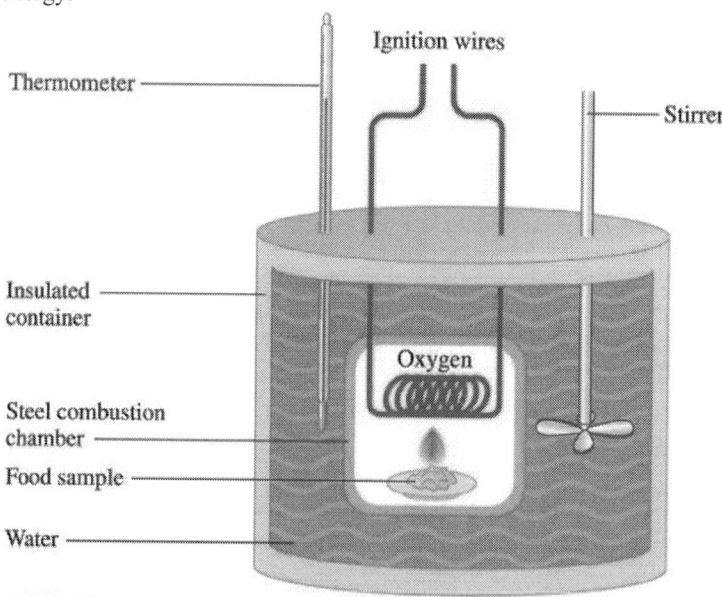

Bomb calorimeter: constant-volume calorimeter

Heat Transfer

Rate of heat transfer

When heat transfers by conduction, the rate of heat transfer (H) can be calculated in joules per second (J/s) or watts (W).

Amount of energy (heat) conducted per unit:

$$H = \frac{kA\Delta T}{t}$$

where ΔT is the temperature difference across the object (K), A is the cross-sectional area (m^2), t is the thickness of the material (m), and k is the thermal conductivity of the material (J/s·m·k).

Heat transfer (H) rate is a material characteristic usually provided in the problem.

Heat transfers between two objects by *conduction, convection,* or *radiation.*

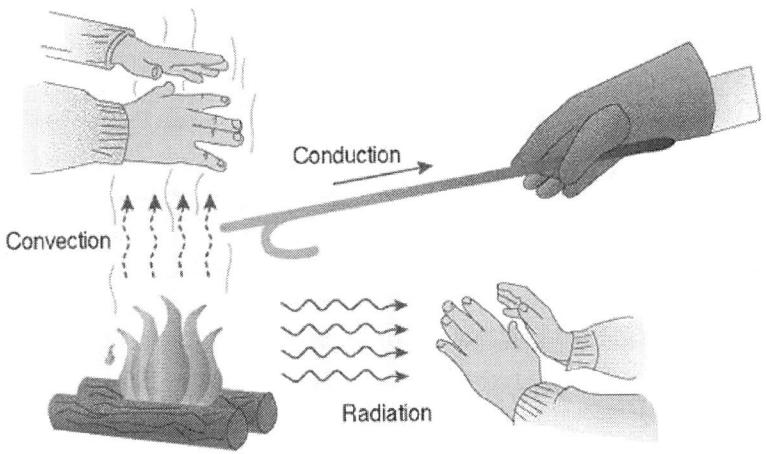

Heat transfer by convection, conduction, and radiation

Heat flows from hotter object

According to the First Law of thermodynamics, *heat always flows from a hotter to a colder system until the two systems are in thermal equilibrium.*

> *Conduction* is the flow of heat between two materials in *direct contact.*

> *Convection* is the flow of energized (e.g., air) molecules that settle and transfer their energy by *direct contact* using conduction.

> *Radiation* is the heat flow (i.e., energy) emitted by matter through *electromagnetic radiation* (e.g., photons or electromagnetic waves).

Thermal conductivities of selected materials

Material	k (J/s·m·K)
Aluminum	237
Concrete	1
Copper	386
Glass	0.9
Stainless Steel	16.5
Water	0.6

Thermal conductivities of common materials

Conduction transfers heat by contact

Conduction is heat transfer between substances in *direct contact* (i.e., must be touching).

When particles of a hotter substance vibrate, these molecules bump into nearby particles and transfer some energy.

Conduction is an essential means of heat transfer for solids, and metals are conductive. It occurs when heat is transferred through *direct contact*.

For example, on a frigid winter night, when a person wraps their hands around a cup of hot cocoa, their fingers become warm because they are in direct contact with the hot mug, which is in direct contact with the hot liquid. The same process is happening at a molecular level.

The liquid in the mug contains heat energy, and the molecules of the hot liquid are in constant motion (high kinetic energy). These molecules continuously collide with the molecules of the mug, increasing their motion. The collisions gradually transfer energy from the hot liquid molecules to the mug molecules.

The mug transfers heat to a person's hands through the same conduction mechanism. The person perceives energy transfer as increasing temperature; thus, their hands feel warm.

Calculating heat conduction

When calculating heat conduction through an object, it is essential to note that the cross-sectional area refers to where the heat is being conducted. For the mug example, the heat is being conducted through its outer curved surface.

When the mug is placed on a table, and the heat conduction to the table is calculated, the cross-sectional area is the area of the bottom of the mug instead of the area of the sidewalls. Therefore, the cross-sectional area is the circumference of the mug times its height. The thickness, in this case, is the thickness of the mug itself.

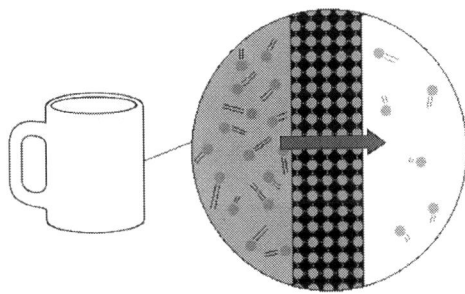

Conduction (heat transfer by direct contact) from the hot liquid to the walls of a mug

Convection transfers heat by motion of particles

Convection refers to the *physical flow of matter*. It is when heat flows energized molecules from one place to another through the movement of fluids.

Convective heat transfer occurs by diffusion (i.e., random Brownian motion of particles) and advection (i.e., larger-scale flowing currents transport matter or heat).

For example, convection occurs when warmer areas of a liquid or gas rise to colder areas, and the cooler liquid or gas replaces the warmer areas.

Gas is fluid, as it can flow and change shape depending on its container.

A pot of water is placed on a hot stove. The water at the bottom of the pot warms first. As the water warms, it rises and is replaced by cooler water. This circulation of the heated water transfers thermal energy from the bottom of the pot to the top.

This movement of molecules is a closed fluid flow pattern of *convection currents*.

Convection ovens work similarly with gas instead of liquid molecules. Air is blown past a heating element to the oven chamber. The hot element energizes the air molecules when passing through the heating element. Energized air molecules are circulated throughout the oven by airflow provided by the fan and continue until all molecules are energized to the same amount.

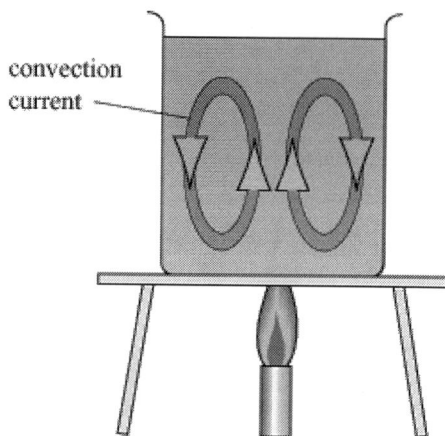

Convention current represented by the circulating water

Radiation transfers heat through electromagnetic waves

Radiation transfers heat *without* contact between the heat source and the heated object. It consists of electromagnetic waves traveling at the speed of light and can occur in a vacuum.

It has *no mass exchange*, and *no medium is required* to transfer heat.

Radiation transfers energy through electromagnetic waves emitted from all objects or surfaces with heat energy and does not require a medium; thus, radiation can occur in a vacuum.

Objects with more energy emit more radiation energy and do not require a medium; thus, radiation can occur in a vacuum.

For example, a hot piece of metal radiates heat as infrared electromagnetic waves. As the metal gets hot, it also radiates heat in the visual spectrum, thus making the metal appear "red hot."

Sun is a radiation source, releasing electromagnetic waves through space and the atmosphere to heat Earth. Similarly, holding hands next to a fire heats hands through radiation.

Radiation can be important even in situations with an intervening medium. An example is the heat transfer between a living entity and its surroundings.

For example, the sun's thermal energy travels through the 93 million miles of space to warm the Earth, even passing through the mesosphere, the coldest part of Earth's atmosphere, which can reach temperatures of -173° C.

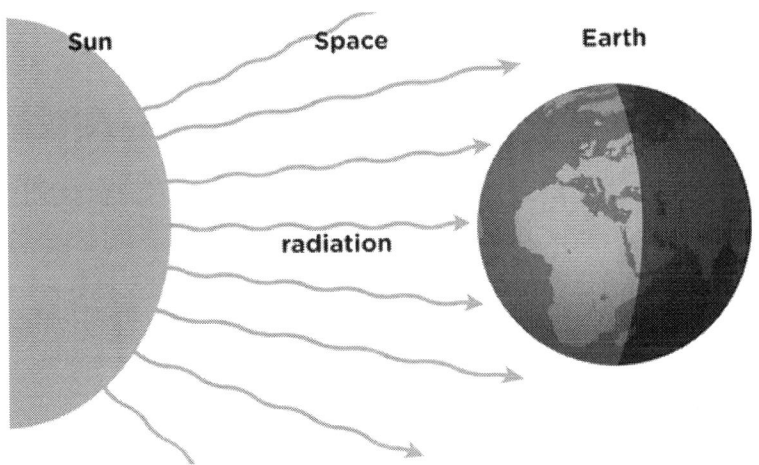

Radiation emitted from the Sun travels to the Earth's surface

Power radiated by an object is related to its temperature:

$$P = e\sigma A(T^4 - T_C^4)$$

where P is the power radiated (W), e is the emissivity of the radiator, σ is Stefan's constant equal to 5.67×10^{-8} W/m^2·K, A is the area of the radiating surface (m^2), T is the temperature of the radiator (K), and T_C is the temperature of the surroundings (K).

Notes for active learning

Thermodynamic Systems

Thermodynamic variables

Thermodynamics defines variables related to energy of physical and chemical changes.

Thermodynamic variables:

ΔE or ΔU measures the *total internal energy* of a system

q measures the *heat* of a system

w measures the *work* of a system

ΔH measures the *enthalpy change* of a system – whether a process is exothermic or endothermic

ΔG measures the *Gibbs free energy change* of a system – whether a process is spontaneous or nonspontaneous

ΔS measures the *entropy change* of the system or surroundings – whether the randomness of the system or surroundings is increasing or decreasing

Steady-state processes occur without a change in the internal energy.

System and surroundings

Thermodynamic system is a body of matter or radiation confined by partitions, with defined permeabilities separating it from its surroundings.

System defines the parameters of the universe being studied, such as the substances involved in a chemical reaction or a phase change.

Surroundings are everything other than the system being studied.

Isolated, closed, and open systems

Thermodynamic system describes processes and matter within a boundary.

> *Isolated system* has no heat transfer, work, or matter with its surroundings.

> *Closed system* may transfer heat and work to its surroundings but *not* matter between the system and surroundings.

> *Open system* can transfer heat, work, or matter between the system and surroundings.

Note: *heat* and *temperature* are different quantities.

Heat is a form of *thermal energy that can leave or enter a system.*

Temperature measures the *average kinetic energy* (KE) of particles in a system.

Characteristics of thermodynamic systems

System	Exchanges with surrounding	Total amount of energy	Example	Illustration
Open	Energy & Matter	Not remain constant	Solution in an open flask	
Closed	Only Energy	Not remain constant	Solution in a sealed flask	
Isolated	Neither energy nor matter	Remains constant	Sealed flask in a thermos flask	

State functions

State function describes the energy required to change the system into another equilibrium state.

State functions (i.e., state quantities) are a group of equations representative of the properties of a system, which depend *only* on the current state of the system, not the path used to achieve the current state.

State functions are quantities with values *not* dependent on the path used to measure the value.

State functions ($\Delta E°$, $\Delta H°$, $\Delta S°$, $\Delta G°$) depend only on the *initial and final states* of the system.

State function variables (e.g., pressure, volume, temp) are a function *of other state variables*.

Thermodynamic state functions include *enthalpy* (*H*), *entropy* (*S*), and *Gibbs free energy* (*G*).

Path functions

Process functions quantitatively describe the transition between equilibrium states of a thermodynamic system.

Path functions depend on the *path taken* to reach one state from another. Therefore, different routes give different quantities (e.g., work, heat).

Quantities, such as work (*w*) and heat (*q*), are not state functions, but are *path functions*, represented by *lowercase letters*.

Notes for active learning

Endothermic and Exothermic Reactions

Bonds absorb and release heat

Some chemical reactions produce energy, while others require heat energy.

Energy is used to *form or break chemical bonds.*

Enthalpy change (ΔH) describes energy changes associated with forming or cleaving a chemical bond.

Unit for enthalpy (H) is the Joules (J).

Joule is the same unit used for energy; it can be expressed as energy per mol (J/mol).

Exothermic reactions release heat

Exothermic reactions, the energy needed to initiate a reaction is less than the released energy, *liberating heat.*

Exothermic reactions have a change in *enthalpy of less than zero.*

$\Delta H < 0$

Energy is *released* into the surroundings because stronger, more stable bonds form in the products than reactants.

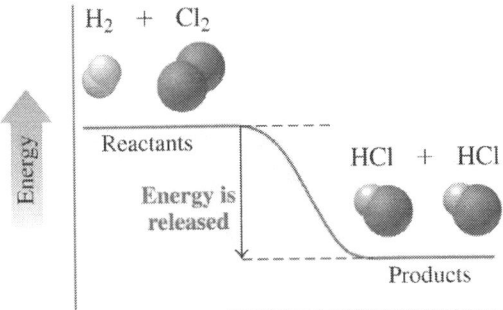

$$H_2 (g) + Cl_2 (g) \rightarrow 2 \, HCl (g) + 185 \, kJ$$

In exothermic reactions, heat is released, and the sign of ΔH is negative ($-\Delta H$); the energy of the products is *less* than the energy of the reactants using bond-dissociation energies.

energy is released $\rightarrow \Delta H < 0$

$\Delta H < 0 \rightarrow$ enthalpy decreases

Exothermic reactions can be represented in a chemical reaction with *heat* on the product side:

$$A + B \rightarrow C + heat$$

For example, the chemical reaction for hydrogen and chlorine gas is:

$$H_2 (g) + Cl_2 (g) \rightarrow 2 HCl (g) + 185 \text{ kJ}$$

$$\Delta H = -185 \text{ kJ/mol (heat released)}$$

Heat is released (i.e., exothermic reaction), and ΔH is negative $(-\Delta H)$.

Another example of an exothermic reaction is fuel burning (i.e., combustion).

Endothermic reactions absorb heat

Endothermic reactions have a change in *enthalpy greater than zero.*

$$\Delta H > 0$$

Endothermic reactions require (net) energy to break bonds.

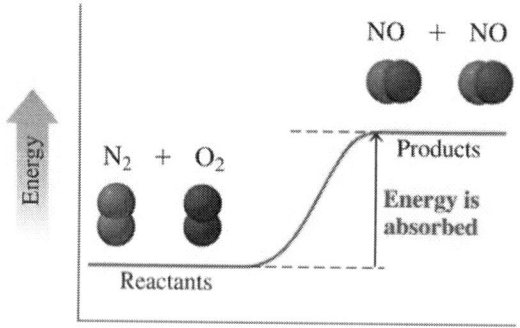

$$N_2 (g) + O_2 (g) + 181 \text{ kJ} \rightarrow 2 NO (g)$$

In endothermic reactions, weaker bonds form in the products than in reactants.

For example, an endothermic chemical reaction represented with the heat as a reactant:

$$heat + NH_4NO_3 (s) \rightarrow NH_4^+ (aq) + NO_3^- (aq)$$

The above reaction is the ionic compound ammonium nitrate dissolving in water (H_2O), dissociating into constituent ions, NH_4^+ and NO_3^-.

Endothermic reactions include melting ice and cooking an egg (heat is required for processes).

In an endothermic reaction, heat is absorbed, and the sign of ΔH is $(+\Delta H)$; the energy of the products is greater than the energy of the reactants.

Amount of energy absorbed is calculated using bond-dissociation energies.

energy is absorbed $\rightarrow \Delta H > 0$

$\Delta H > 0 \rightarrow$ enthalpy increases

Endothermic reactions can be represented in a chemical reaction with *heat* on the reactant side:

A + B + *heat* $\rightarrow$ C

For example, nitrogen gas (N_2) and oxygen gas (O_2) form nitrogen monoxide (NO):

N_2 (g) + O_2 (g) + 181 kJ $\rightarrow$ 2 NO (g)

$\Delta H = +181$ kJ (heat added)

Heat is absorbed (i.e., endothermic reaction), and ΔH is positive.

Exothermic and endothermic reactions comparison

Figure below depicts an endothermic (absorbing) and exothermic reaction (producing heat).

Reaction is exothermic on the left because heat is *produced and released* to the surroundings.

Reaction is *endothermic* on the right because heat is *absorbed* from the surroundings.

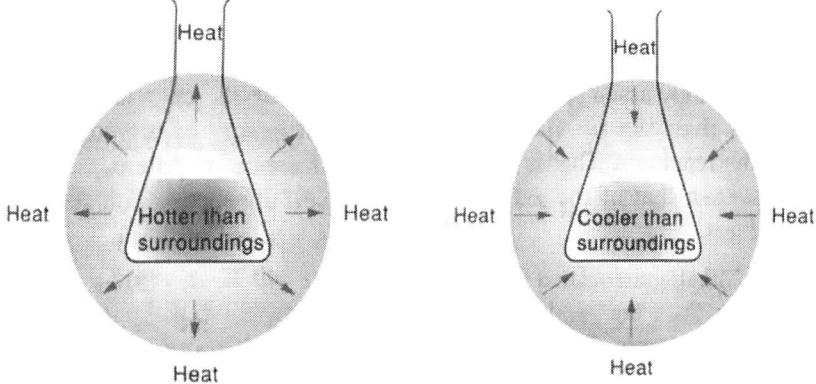

Exothermic process releases heat (left), and the endothermic process absorbs heat (right)

Reaction	Energy Change	Heat in the Equation	Sign of ΔH
Endothermic	Heat absorbed	Reactant side	Positive (+)
Exothermic	Heat released	Product side	Negative (−)

For example, is the following reaction endothermic or exothermic?

$$O_2N–NO_2 \rightarrow O_2N + NO_2$$

Reaction is endothermic (i.e., it absorbs heat) because it forms two products from a single reactant.

Breaking chemical bonds *requires energy*.

Thermodynamics

Heat, work and temperature

Thermodynamics relates heat, work, and temperature to energy, entropy, and properties of matter and radiation.

Thermodynamics answers several fundamental questions regarding *whether* a reaction occurs, to what *extent*, and whether it *releases or absorbs heat*.

An important concept in thermodynamics is the thermodynamic system, the region under study. Everything else is the surroundings.

Systems are separated from the surroundings by a boundary. Transfers of energy as work, heat, or matter between the system and surroundings occur through the boundary.

Four laws of thermodynamics

Thermodynamics states four laws characterizing heat and energy relationships by describing the behavior of physical quantities such as temperature, heat, work, energy, and entropy.

Zeroth Law of thermodynamics (or *concept of temperature*) states that **heat flows from the hotter body to the colder** *when placed in contact until thermal equilibrium.*

First Law of thermodynamics (or *conservation of energy*) states that *the total energy in a system remains constant as* **energy can neither be created nor destroyed**.

Second Law of thermodynamics (or *concept of entropy*) states that *the universe's natural order* **favors a net increase** *in the overall entropy (i.e., disorder).*

Third Law of thermodynamics (or *entropy at absolute zero*) states that *the entropy of a perfectly ordered, pure crystalline solid is* **zero at the temperature of absolute zero** *(equal to 0 K, −273.15 °C or −459.67 °F).*

Notes for active learning

Zeroth Law of Thermodynamics

Thermal equilibrium

Zeroth Law of Thermodynamics is the logical predecessor of the First and Second Laws of Thermodynamics.

Zeroth Law states that *if two systems are in thermal equilibrium with a third system, the two initial systems are also in thermal equilibrium.*

Three systems in thermal equilibrium contain the same amount of heat energy (same temperature) and do *not* exchange heat, fundamental principle underlying thermodynamics.

For example, if system A is in equilibrium with B, and system C is in equilibrium with B, then system A and C must be in equilibrium, and all three systems must be at the same temperature.

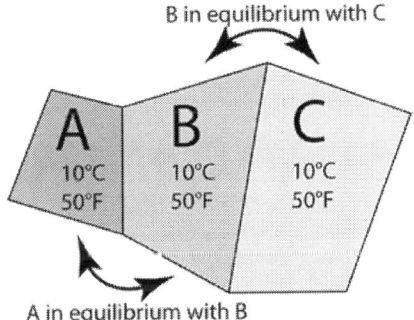

Zeroth Law of Thermodynamics with three systems in thermal equilibrium

However, it is vital to notice that although no net heat transfer occurs between systems in equilibrium, thermal energy is technically transferred between the related systems.

The exchange of thermal energy requires that every unit of energy passed from any system has the same value of energy passed back into the system. This holds even if the two systems contain materials with different specific heats.

This observation of thermal energy transfer means that there must be a property that can be considered the same upon which heat transfer depends and that property is temperature.

Concept of temperature

Zeroth Law of thermodynamics states that *heat flows from the hotter to the colder body when in contact until thermal equilibrium.*

It establishes that *temperature is a core, measurable property of matter.*

Separated by a permeable barrier, two systems are in thermal equilibrium, and *no heat* transfers.

After thermal equilibrium is achieved, no heat transfers, defining the *concept of temperature.*

Expressing thermal equilibrium

If two systems are in thermal equilibrium with a third system, each is in thermal equilibrium.

Thermal equilibrium is expressed by:

$$T_A = T_B$$

and

$$T_C = T_B$$

then

$$T_A = T_C$$

Zeroth Law is crucial for calculations of thermodynamics (i.e., heat and energy relationships).

First Law of Thermodynamics

Conservation of energy

Energy, like matter, is always conserved; it can neither be created nor destroyed.

First Law of Thermodynamics states that *energy can neither be created nor destroyed, and the total energy in a system remains constant. Energy is conserved in a chemical reaction.*

First Law of thermodynamics relates to the *internal energy* of a substance.

Internal energy increases if heat is added to the system or if work is done on the system.

Disordered motion of particles

Change in total *internal energy* of a system equals contributions by heat (q) and work (w):

$$\Delta U = q - w$$

where ΔU is the change in internal energy, q is the contribution from heat and w from work.

Internal energy refers to the microscopic energy of the disordered motion of particles and is represented by the symbol (U).

Internal energy cannot be seen and has no apparent effect on the object's motion, but it does play a part in the thermodynamic properties of the substance. This principle is the foundation for the law of *conservation of energy*.

For example, a glass of water does not look like it contains kinetic or potential energy. Water molecules are whizzing around on the atomic scale, moving hundreds of m/s. If the water is dumped from the glass, its internal energy does not affect its motion.

Water

Thermodynamic properties of the substance include potential and kinetic energy

Sign conventions

$\Delta U = q - w$ uses the following sign conventions:

q is *positive* if heat is *absorbed into* the system (i.e., heating)

q is *negative* if heat is *released from* the system (i.e., cooling)

w is *positive* if work is done *by* the system (i.e., compression)

w is *negative* if work is done *on* the system (i.e., expansion)

First Law of thermodynamics:

$\Delta U = q + w$

where q is *positive* if heat is *absorbed into* the system and *negative* if heat is *released from* the system; work (w) is positive if done *on the* system or negative if done *by* the system.

Heat (q) and work (w) absorbed or released

Use the sign conventions in the chart below:

q	+	Heat **absorbed** by the system	E_{system} increases
q	–	Heat **released** by the system	E_{system} decreases
w	+	Work done **on** the system	E_{system} increases
w	–	Work done **by** the system	E_{system} decreases

Heat or energy *flowing out* of the system is *negative*.

Heat or energy *flowing into* the system is *positive*.

Other natural sciences use the opposite sign convention; note the sign convention used.

Internal energy

The difference between the amount of heat added to the system (Q) and the work done by the system (W) is equal to the total change in internal energy (ΔU):

$$\Delta U = \Delta Q - \Delta W$$

As with other energy, the unit of internal energy is the joule (J), although it is also expressed using calories (cal).

By convention, work done *on a system* is positive, and work done *by a system* is negative.

Occasionally, the above equation is written as $\Delta U = \Delta Q + \Delta W$, and ΔW is defined as the work done on the system. Be sure to note which form is being used.

Calculating internal energy

The quantities in $\Delta U = \Delta Q + \Delta W$ can determine what a thermodynamic system is experiencing.

For example, in a chemical reaction, if ΔQ is a quantity less than zero ($\Delta Q < 0$), the reaction is losing heat. This reaction is *exothermic,* and heat is a product.

If ΔQ is a quantity greater than zero ($\Delta Q > 0$), the reaction absorbs heat from its surroundings. This is an *endothermic* reaction, and heat is a reactant.

Internal energy is expressed:

$$\Delta U = q + w$$

where ΔU is the change in internal energy, q is the contribution from heat and w from work.

For example, consider internal energy, heat, and work. For a process in which 3.4 kJ of heat flows out of the system while the system in the surroundings does 4.8 kJ of work, what is the internal energy?

Calculate internal energy:

$$\Delta U = q + w$$

$$\Delta U = -3.4 \text{ kJ} + (-4.8 \text{ kJ})$$

$$\Delta U = -8.2 \text{ kJ}$$

where ΔU is the change in internal energy, q is *negative* as heat is *released from* the system, and work (w) is negative when done *by* the system.

Use the appropriate heat (q) and work (w) signs.

Energy has flowed from the system to the surroundings by -8.2 kJ.

Notes for active learning

Second Law of Thermodynamics

Concept of entropy (*S*)

Second Law of thermodynamics is that *the universe favors a net increase in the overall entropy.*

    Isolated systems increase in entropy and *never decrease.*

    Open systems decrease entropy as entropy increases for the surroundings.

Second Law of Thermodynamics states *that in an isolated system (no transfer of heat, work, or matter), a process increases entropy or stays constant, but entropy never decreases.*

Entropy is a measure of the randomness or disorder of a system. Entropy measures the system's thermal energy per unit temperature, which is unavailable for useful work.

Work requires ordered molecular motion.

Open systems transfer energy

An exception to the Second Law of Thermodynamics (i.e., entropy cannot decrease) is an *open system* where heat, work, or matter can be transferred to surroundings; the system's entropy decrease is related to an increase in the entropy of the surroundings.

Second Law states that *heat flows spontaneously* from a hot object to a cold object but never spontaneously flows in the opposite direction. A bowl of ice never gets colder in a warm room; the ice always gets warmer.

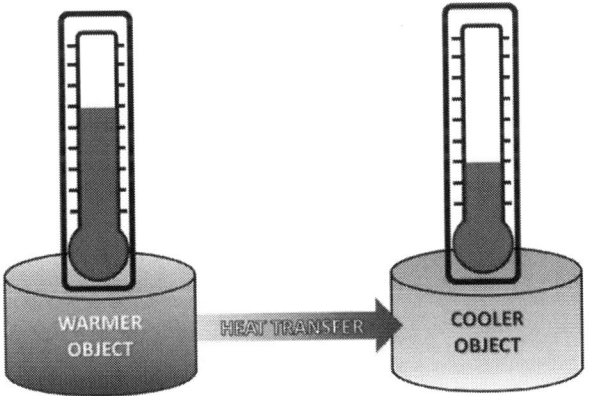

Heat flows spontaneously from a hot object to a cold object

Entropy measures disorder

Entropy measures disorder within a system and likelihood—how likely a system resembles the same state twice or how likely a reaction is to happen.

Entropy of a natural system never decreases, as it is the natural tendency of all things to move towards maximum disorder—maximum entropy.

For example, gas expands to fill space with more degrees of freedom to arrange particles—more *multiplicities*.

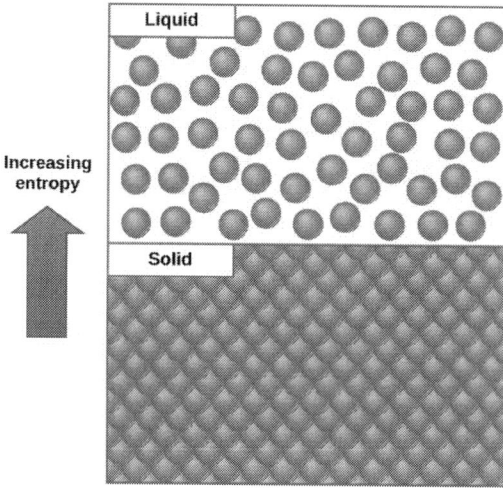

A solid has less entropy than a liquid, which has much less entropy than a gas

According to the Second Law of Thermodynamics (i.e., entropy can stay constant or increase but never decrease), entropy is given by:

$$\Delta S \geq \frac{Q}{T}$$

where ΔS is the change in entropy of a process (J/K), Q is heat transferred (J), and T is the temperature (K).

If a system strives to increase entropy (i.e., disorder), it does not mean that the system cannot become more orderly. This means that to do so, outside energy must be transferred into the system, and it cannot be done on its own.

Nonspontaneous and spontaneous processes

Nonspontaneous processes require energy (or work) for the system to decrease entropy.

Spontaneous processes move towards randomness without energy input to increase entropy.

For example, a valve connects a container with air and another container in a vacuum. When the valve is opened, the air flows from the container to the container in a vacuum. No work or energy is needed for this to happen. The process is *spontaneous*, and entropy *increases*.

For the reverse to occur, energy must be *added to the system* to re-pressurize the original container with air. This process is an example of a *nonspontaneous* reaction.

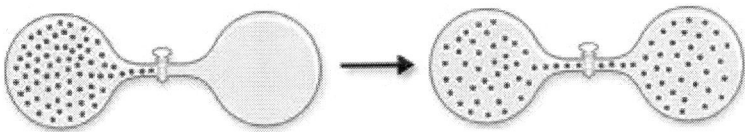

Air flows from the container to the container in a vacuum

Nonspontaneous reactions do not occur without the input of energy

The universe's entropy always *increases* when a *spontaneous process* occurs.

Entropy of the universe is expressed as:

$$\Delta S_{universe} > 0$$

Reversible and irreversible processes

Thermodynamic processes are *reversible* or *irreversible*.

Reversible processes are idealized, while the irreversible process frequently occurs in nature.

> *Reversible processes*
>
> $$\Delta S = q / T$$
>
> Irreversible processes
>
> $$\Delta S > q / T$$

where ΔS is the change in entropy, q is the contribution from heat, and T is the temperature (K).

Natural processes are *irreversible*, so entropy change is greater than the heat transfer over temperature. Due to the irreversible nature of processes, the universe's entropy is increasing.

Reactions at equilibrium are *reversible* with no net change to the system or surroundings and proceed in both directions.

Reversible processes reverse direction if some change is made to some system property:

$$\Delta S_{total} = \Delta S_{system} + \Delta S_{surroundings}$$

ΔS of the system equals in magnitude, but *opposite in sign*, to the ΔS of the surroundings.

Entropy related to heat and enthalpy

Entropy change of surroundings is related to heat (q) flow, enthalpy (ΔH), and temperature:

$$\Delta S_{surroundings} = -q / T$$

or (at constant pressure)

$$\Delta S_{surroundings} = -\Delta H / T$$

Combining these two equations:

$$\Delta S_{surroundings} = \Delta S_{system} - (\Delta H_{system} / T)$$

where ΔS is the change in entropy, ΔH for enthalpy, q is the contribution from heat, and T is the temperature (K).

ΔS from tabulated values

ΔS is calculated from standard molar entropies as for ΔH and ΔG by heats of formation.

Equation to calculate ΔS from tabulated values:

$$\Delta S°_{rxn} = \Sigma \, n(\Delta S°_f \text{ of products}) - \Sigma \, n(\Delta S°_f \text{ of reactants})$$

where n represents moles of each reactant (or product) as a balanced chemical equation, and ΔS is the change in entropy.

For example, determine the entropy change (ΔS) for the above chemical equation as (ΔH):

$$16 \, H_2S \, (g) + 8 \, SO_2 \, (g) \rightarrow 16 \, H_2O \, (l) + 3 \, S_8 \, (s)$$

$\Delta S°_{rxn}$ as a measure of the entropy change of a system:

$$\Delta S°_{rxn} = [(16 \text{ mol}) \cdot (S°_f \text{ of } H_2O \, (l)) + (3 \text{ mol}) \cdot (S°_f \text{ of } S_8 \, (s))]$$
$$- [(16 \text{ mol}) \cdot (S°_f \text{ of } H_2S \, (g)) + (8 \text{ mol}) \cdot (S°_f \text{ of } SO_2 \, (g))]$$

From the table of standard *entropies of formation*, substitute:

$$\Delta S°_{rxn} = \Sigma \, n(\Delta S°_f \text{ of products}) - \Sigma \, n(\Delta S°_f \text{ of reactants})$$

$$\Delta S°_{rxn} = [(16 \text{ mol}) \cdot (69.91 \text{ J/mol·K}) + (3 \text{ mol}) \cdot (31.80 \text{ J/mol·K})]$$
$$- [(16 \text{ mol}) \cdot (205.79 \text{ J/mol·K}) + (8 \text{ mol}) \cdot (161.92 \text{ J/mol·K})]$$

$$\Delta S^{\circ}{}_{rxn} = [1{,}118 \text{ J/K} + 95.40 \text{ J/K}] - [3{,}292.6 \text{ J/K} - 1{,}295.4 \text{ J/K}]$$

$$\Delta S^{\circ}{}_{rxn} = -3{,}375 \text{ J/K}$$

where $\Delta S^{\circ}{}_{rxn}$ is a measure of the entropy change of a system, and the negative sign accounts for the heat of the surroundings opposite in sign to the heat of the system.

Entropy calculations

For example, the reaction enthalpy (ΔS_{surr}) was –1,876 kJ at a standard temperature of 298 K. The entropy change was calculated as –3,375 J/K (or –3.375 kJ/K). What is the entropy change of the surroundings?

Using the equation:

$$\Delta S_{surroundings} = - \Delta H / T$$

$$\Delta S_{surroundings} = -(- 1{,}876 \text{ kJ}) / 298 \text{ K}$$

$$\Delta S_{surroundings} = 6.30 \text{ kJ/K or } \Delta S_{surroundings} = 6.30 \times 10^3 \text{ J/K}$$

where $\Delta S_{surroundings}$ is change in entropy, ΔH is change in enthalpy, and T is temperature (K).

As a *positive* value, the entropy of the surroundings *increased.*

Total entropy change is the sum of the entropy change of the system and the surroundings:

$$\Delta S_{total} = \Delta S_{system} + \Delta S_{surroundings}$$

$$\Delta S_{total} = -3.375 \times 10^3 \text{ J/K} + 6.30 \times 10^3 \text{ J/K}$$

$$\Delta S_{total} = 2.92 \times 10^3 \text{ J/K}$$

where ΔS_{total} is the total entropy change, ΔS_{system} is the entropy change of the system, and $\Delta S_{surroundings}$ is the entropy change of the surroundings.

Entropy measures degrees of freedom

Entropy (S) measures *randomness* (or *disorder*) from the number of ways (i.e., *degrees of freedom*) a thermodynamic system can be arranged.

Entropy (or disorder) of the universe is *increasing.*

Entropy uses energy divided by temperature; the unit is Joules per Kelvin (J/K).

Entropy is expressed as *entropy per unit mass* (J/K·kg) or *unit amount of substance* (J/K·mol).

For example, imagine taking a piece of paper, tearing it into small pieces, and throwing them in the air. How much energy is needed to restore it to its original form?

Consider tearing paper as *spontaneous* and putting it together as a *nonspontaneous* process.

Temperature and entropy relationship

The higher the entropy and lower the energy (i.e., temperature), the more stable the system.

Higher entropy + lower energy (lower temperature) → *more stability*

Lower entropy + higher energy (higher temperature) → *less stability*

As *temperature increases, entropy increases*:

$T = 0$ K

particles are in equilibrium lattice positions, and S is relatively low

$T > 0$ K

molecules vibrate, S increases

As T increases, more violent vibrations occur, and the entropy (S) increases.

Entropy for gasses, liquids, and crystals

Entropy differences between phases of matter (i.e., solid, liquid, gas) are related to the *degrees of freedom* of kinetic energy (KE).

Solids have low entropy because particles forming a crystal solid are packed tightly.

Particles of a solid are confined to vibrating in fixed positions and limited in *degrees of freedom*.

Liquids are intermediate in entropy as intermolecular forces hold particles as a liquid but have increased *degrees of freedom* to move.

Gases are high in entropy because the gas particles have complete *freedom* to move with a few particles arranged in an orderly fashion.

Low-pressure gas experiences little to no intermolecular forces of attraction or repulsion.

Entropy of phase changes

Entropy changes accompanying phase changes can be estimated, at least qualitatively.

Freezing decreases entropy as the relatively disordered liquid becomes a well-ordered solid.

Boiling results in a substantial increase in entropy as the liquid becomes a more highly disordered gas.

Sublimation (i.e., from a solid to a gas) is the phase transition with the greatest entropy change for a substance.

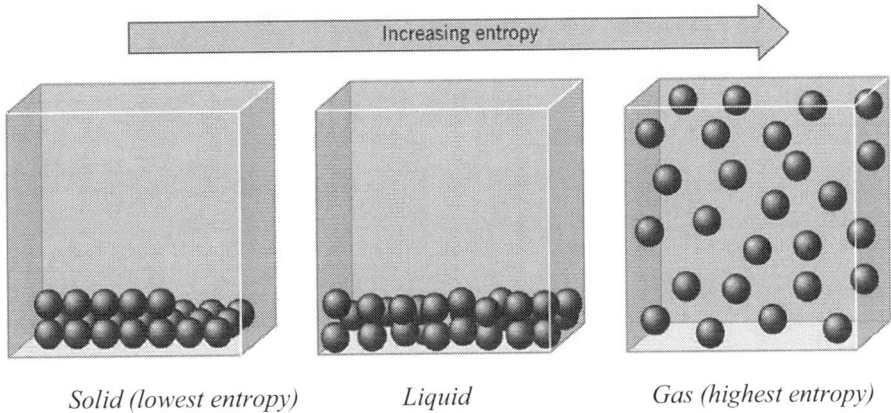

Solid (lowest entropy)   *Liquid*   *Gas (highest entropy)*

The figure displays a crystalline solid, liquid, and gas with entropy (i.e., randomness) increasing to the right.

Entropy for gases

Entropy *increases* as gas volume *increases* (i.e., decreasing pressure).

Consider a gas separated from a vacuum by an impermeable partition (figure on the left).

Removing the partition (center image) allows more energy distribution.

Gas expands to achieve higher particle distribution (figure on the right).

Gas molecules' configuration is *random* and *more positive S* (more disorder).

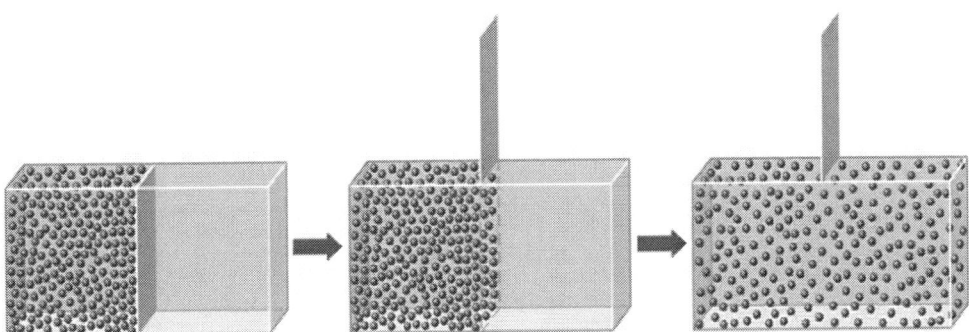

Gas particles are highly ordered on the left (low S) and random on the right (high S)

Number of particles affects entropy reactions, producing more particles as a *positive ΔS*.

Adding particles to a system (below on the right) increases the number of ways energy can be distributed.

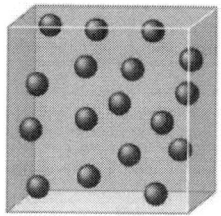 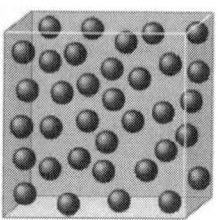

Low entropy   *High entropy: more disorder*

Third Law of Thermodynamics

Absolute zero

Third Law of thermodynamics states that *as the temperature of a system approaches absolute zero, the entropy approaches a constant value*.

Typically, this constant value is zero. In this definition, the entropy of a system is related to the number of microstates possible.

Microstates refer to the atomic configurations possible with the equation:

$$S = k \cdot ln \, (\Omega)$$

where S is the entropy (J/K), k is Boltzmann's constant (1.381×10^{-23} J/K), and Ω is the number of microstates.

Only one microstate is possible at absolute zero, and so Ω equals 1:

$$S = k \cdot ln(1) = k \cdot 0 = 0$$

The entropy of a system at absolute zero is zero.

Minimum energy state

However, some systems have more than one minimum energy state as the temperature approaches absolute zero: the entropy levels off at a value other than zero.

Cooling any process to absolute zero in a finite number of steps is impossible.

In left image, $T = 0$ can be reached following infinite steps (step lines between X_1 and X_2).

In the image on the right, $T = 0$ is reached by a finite number of steps, each getting closer to zero but never to zero.

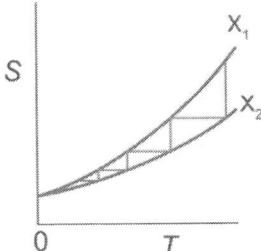 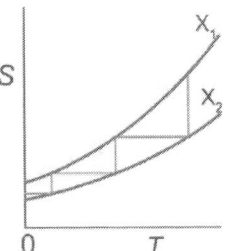

Entropy vs. temperature illustrating the step to approach absolute zero

Entropy at absolute zero

Third Law of Thermodynamics states that *the entropy of a perfectly ordered, pure crystalline solid is zero at absolute zero.*

At *absolute zero* (the coldest temperature possible on the Kelvin scale equals 0 K, –273.15 °C, or –459.67 °F), the particles' average kinetic energy (KE) equals zero.

Particles' average kinetic energy (KE) increases as entropy increases.

With increasing temperature (i.e., KE), substances transition to liquid and then gaseous state.

For a substance, the gas state has the highest entropy.

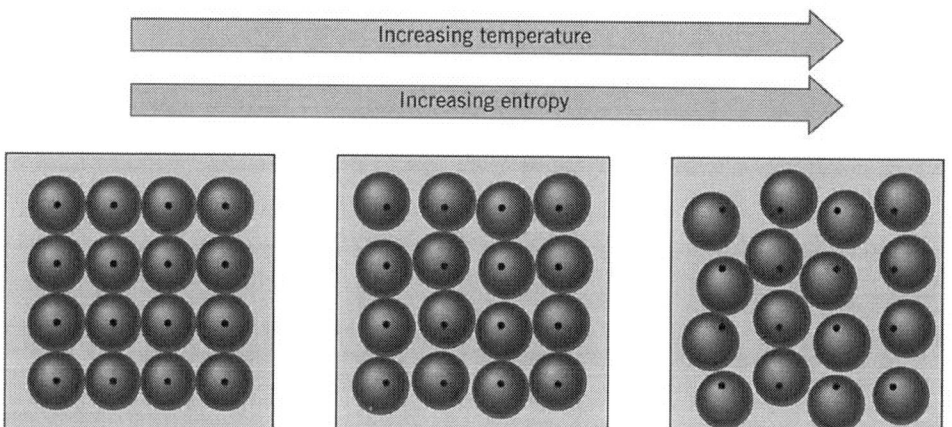

Crystalline solids have low KE with particles arranged orderly.

Liquid particles vibrate from higher KE, and entropy (*S*) is moderate.

Gas particles have large KE, and entropy (*S*) is large.

Determining ΔS for reactions

For example, determine the sign of ΔS for the reaction:

$$CaCO_3\,(s) + 2\,H^+\,(aq) \rightarrow Ca^{2+}\,(aq) + H_2O + CO_2\,(g)$$

$\Delta n_{gas} = 1\,\text{mol} - 0\,\text{mol}$

$\Delta n_{gas} = 1\,\text{mol}$

Since Δn_{gas} is positive, ΔS is positive.

For example, determine the sign of ΔS for the reaction:

$$2\,N_2O_5\,(g) \longrightarrow 4\,NO_2\,(g) \;+\; O_2\,(g)$$

$\Delta n_{gas} = 4\,\text{mol} + 1\,\text{mol} - 2\,\text{mol}$

$\Delta n_{gas} = 3\,\text{mol}$

Since Δn_{gas} is positive, ΔS is positive.

For example, determine the sign of ΔS for the reaction:

$$OH^-\,(aq) \;+\; H^+\,(aq) \longrightarrow H_2O$$

$\Delta n_{gas} = 0\,\text{mol}$

$\Delta n_{gas} = 1\,\text{mol} - 2\,\text{mol}$

$\Delta n_{gas} = -1\,\text{mol}$

Since Δn is negative, ΔS is negative.

Notes for active learning

Enthalpy

Energy changes

Enthalpy (ΔH) of a system is the amount of heat used at constant pressure in a system or reaction and is given in SI units of Joules.

Energy changes in the system can be written as:

$$\Delta U = \Delta H - P\Delta V$$

where ΔH is change in system enthalpy (J), P is pressure (Pa), and ΔV is volume change (m³).

During an isobaric process (pressure is held constant), such as that conducted in an open container, the change in enthalpy equals the amount of heat transferred during the process. Most reactions do not generate many gaseous products, so little work is associated with the reaction.

Energy change of the system equals the change in enthalpy:

$$\Delta E \approx \Delta H$$

Enthalpy is equal to the transfer of thermal energy (heat); it is only valid at *constant pressure*.

Enthalpy changes can occur at standard temperatures and pressures.

Standard state refers to substances in their most stable form (i.e., the lowest-energy state).

Standard state in thermodynamics refers to one-atmosphere pressure (1 atm) and a temperature of 25 °C (298 K).

For example, oxygen is O_2 (diatomic gas) at the standard state, and carbon is C (solid graphite).

Standard heats of reaction (ΔH)

Standard heat of reaction (*standard enthalpy of reaction* ΔH°_{rxn}) is the change in heat content when one mole of the matter is transformed by a chemical reaction under standard conditions (1 atm and 298 K).

Theoretical and experimental methods calculate the *enthalpy change* of a chemical reaction.

General equation to calculate the heat of a reaction (ΔH) from the difference (delta or Δ) in the energy of the reactants and products is:

$$\Delta H = H_{products} - H_{reactants}$$

Standard heats of formation ($\Delta H°_f$)

Standard heat of formation (*standard enthalpy of formation* $\Delta H°_f$) is the change in heat content when one mole of a compound or molecule forms from constituent elements in standard states.

Standard enthalpy of formation has been determined for many molecules, and the $\Delta H°_f$ for the most common molecules are reported and tabulated data.

When choosing $\Delta H°_f$ from a table, choose the value corresponding to the appropriate state of matter (solid, liquid, aqueous, gas).

Standard heat of formation for an element in its standard state equals 0 kJ/mol.

For example, calculate the heat of reaction from the heat of formation and determine if the process is exothermic or endothermic for the balanced chemical equation:

$$16 \; H_2S \; (g) + 8 \; SO_2 \; (g) \rightarrow 16 \; H_2O \; (l) + 3 \; S_8 \; (s)$$

From thermodynamic quantities, use values for the heat of formation of each component in the reaction; set the equation to calculate the heat of the reaction:

$$\Delta H = H_{\text{products}} - H_{\text{reactants}}$$

$$\Delta H°_{rxn} = [16 \; \text{mol}(\Delta H°_f \; \text{of} \; H_2O \; (l)) + 3 \; \text{mol}(\Delta H°_f \; \text{of} \; S_8 \; (s))]$$
$$- [16 \; \text{mol}(\Delta H°_f \; \text{of} \; H_2S \; (g)) + 8 \; \text{mol}(\Delta H°_f \; \text{of} \; SO_2 \; (g))]$$

$$\Delta H°_{rxn} = [16 \; \text{mol}(-285.8 \; \text{kJ/mol}) + 3 \; \text{mol}(0 \; \text{kJ/mol})]$$
$$- [16 \; \text{mol}(-20.2 \; \text{kJ/mol}) + 8 \; \text{mol}(-296.8 \; \text{kJ/mol})]$$

$$\Delta H°_{rxn} = [-4{,}573 \; \text{kJ} + 0 \; \text{kJ}] - [-323 \; \text{kJ} + (-2{,}374 \; \text{kJ})]$$

$$\Delta H°_{rxn} = -4{,}573 \; \text{kJ} + 323 \; \text{kJ} + 2374 \; \text{kJ}$$

$$\Delta H°_{rxn} = -1{,}876 \; \text{kJ}$$

$\Delta H°_{rxn}$ is negative, and the reaction is exothermic; the system releases heat to the surroundings.

Enthalpy for exothermic and endothermic reactions

Exothermic reactions have a change in enthalpy of less than zero.

$$\Delta H < 0$$

Endothermic reactions have a change in enthalpy greater than zero.

$$\Delta H > 0$$

Since enthalpy (H) is a state function, it represents a system property; it can be used to describe a reaction or an equation for a reaction.

Calculating the heat of reaction (ΔH)

Calculate the heat of reaction (ΔH) using these steps:

1) Identify the given and needed quantities.

2) Write an expression using the heat of reaction and molar mass needed.

3) Write the conversion factors, including the heat of the reaction.

4) Set the expression to calculate heat.

Hess' Law for heat of summation

In 1840, Swiss-born Russian chemist Germain Hess (1802-1850) published Hess' Law of constant heat summation.

Hess' Law describes an essential relationship in physical chemistry that *total enthalpy change (ΔH) during the complete course of a chemical reaction is independent of the number of steps.*

Hess' Law states that the change of enthalpy in a reaction is equal to the sum of the products' enthalpy minus the sum of the reactants' enthalpy:

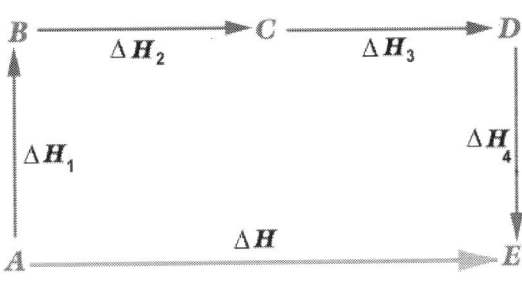

$$\Delta H_{rxn} = \Sigma H_{products} - \Sigma H_{reactants}$$

Enthalpy is a *state function* (i.e., depends only on initial and final states, *not* the reaction's path).

Hess' Law of Heat Summation is expressed

$$\Delta H°_{rxn} = \Sigma\, n(\Delta H°_f \text{ of products}) - \Sigma\, n(\Delta H°_f \text{ of reactants})$$

where n represents the moles of each reactant or product in the balanced chemical equation, and the Greek letter Σ means the sum of the variables.

Hess' Law states that the total energy change throughout a reaction is the same, regardless of whether the reaction occurs in one step or several steps.

Calculating heat of summation

For example, how much heat (kJ) is absorbed when 1.65 grams of NO gas is produced in the following reaction?

$$N_2 (g) + O_2 (g) \rightarrow 2\ NO\ (g)$$

$$\Delta H = +181\ kJ/mol$$

Step 1: State the given and needed quantities.

Given:

1.65 grams of NO

$\Delta H = +181\ kJ/mol$

Solve:

Heat absorbed in kJ

$$N_2 (g) + O_2 (g) \rightarrow 2\ NO\ (g) \qquad \Delta H = +181\ kJ/mol$$

1.65 g ? kJ

Step 2: Write an expression for the heat of the reaction and the molar mass needed.

grams of NO → moles of NO → kilojoules of energy

molar mass *heat of reaction*

Step 3: Write the conversion factors, including the heat of the reaction.

1 mole of NO = 30.01 g of NO:

30.01 g NO / 1 mol NO or 1 mol NO / 30.01 g NO

ΔH for 2 moles of NO = +181 kJ/mol

therefore:

2 mol NO / +181 kJ/mol or +181 kJ/mol / 2 mol NO

Step 4: Set the expression to calculate the heat.

1.65 g NO × (1 mol NO/30.01 g NO) × (+181 kJ/mol/2 mol NO)

1.65 g̶ ̶N̶O̶ × (1 m̶o̶l̶ ̶N̶O̶/30.01 g̶ ̶N̶O̶) × (+181 kJ/mol/2 m̶o̶l̶ ̶N̶O̶)

1.65 × (1/30.01) × (+181 kJ/mol/2) = 4.98 kJ/mol

4.98 kJ/mol of nitrogen gas is absorbed when 1.65 g of nitrogen monoxide is produced from nitrogen and oxygen.

Quantity relationships in chemical reactions

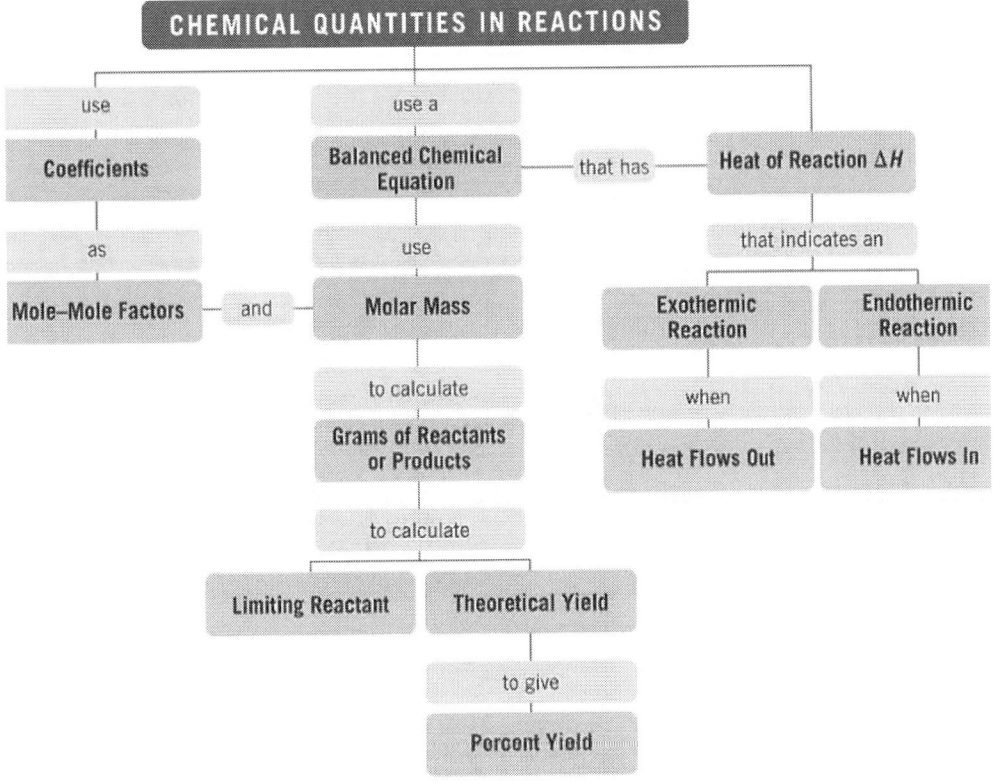

Interrelationship of chemical processes and quantitative outcomes

Thermochemical systems to calculate enthalpy

In addition to Hess' Law of Heat Summation, alternative methods can calculate enthalpy.

For example, if a series of reactions whose enthalpies are known to create an overall reaction that gives the reaction sought, the sum of those enthalpies is the enthalpy of the entire reaction.

Calculating enthalpy change by *thermochemical systems of equations* is below.

For example, from Hess' Law, determine the enthalpy for the reaction:

$$2\,N_2\,(g) + 5\,O_2\,(g) \longrightarrow 2\,N_2O_5\,(g) \qquad\qquad \Delta H = ?$$

Use the following heats of reaction:

Equation (1) $2\,H_2\,(g) + O_2\,(g) \longrightarrow 2\,H_2O\,(l)$ $\qquad\qquad \Delta H = -571.7\ \text{kJ}$

Equation (2) $N_2O_5\,(g) + H_2O\,(l) \longrightarrow 2\,HNO_3\,(l)$ $\qquad\qquad \Delta H = -92\ \text{kJ}$

Equation (3) $N_2\,(g) + 3\,O_2\,(g) \longrightarrow 2\,HNO_3\,(l)$ $\qquad\qquad \Delta H = -348.2\ \text{kJ}$

Manipulate the above equations, so sum the desired equation. Inspect the compounds in each equation and choose an equation with a compound appearing once in the equations.

Oxygen (O_2) is in equations (1) and (3), so it is *not* suitable to begin solving the problem.

Dinitrogen pentoxide (N_2O_5) is in equation (2), so equation (2) could begin solving the problem.

Nitrogen is in equation (3), so equation (3) could begin solving the problem.

For example, solve the problem using nitrogen (N_2) and equation (3). Equation (3) contains one mole of nitrogen as a reactant, and two moles of nitrogen are needed.

Since N_2 is a reactant in both, the equation is not reversed, but equation (3) must be multiplied by two for the required two moles of nitrogen (N_2).

What is done to manipulate the chemical equation is applied to the heat of the reaction.

Heat of reaction is multiplied by two:

2 × Equation (3):

$$2\,N_2\,(g) + 6\,O_2\,(g) + 2\,H_2O\,(g) \rightarrow 4\,HNO_3\,(l)$$

$$\Delta H = \mathbf{2}(-348.7 \text{ kJ})$$

$$\Delta H = -696.4 \text{ kJ}$$

Dinitrogen pentoxide (N_2O_5) is introduced since only equation (2) contains (N_2O_5).

In equation (2), N_2O_5 appears as one mole of reactant.

Reverse the equation and multiply by two.

The sign of the heat of reaction is reversed and doubled:

−2 × Equation (2):

$$4\,HNO_3\,(l) \longrightarrow 2\,N_2O_5\,(g) + 2\,H_2O\,(l)$$

$$\Delta H = -\mathbf{2}(-92 \text{ kJ})$$

$$\Delta H = +184 \text{ kJ}$$

Four moles of nitric acid cancel, which is convenient since it is not in the equation sought.

2 × Equation (3):

$$2\,N_2\,(g) + 6\,O_2\,(g) + 2\,H_2O\,(g) \longrightarrow 4\,\cancel{HNO_3\,(l)}$$

$$\Delta H = \mathbf{2}(-348.7 \text{ kJ})$$

$$\Delta H = -696.4 \text{ kJ}$$

$-2 \times$ Equation (2):

$$4 \, \cancel{HNO_3 \, (l)} \longrightarrow 2 \, N_2O_5 \, (g) + 2 \, H_2O \, (l)$$

$$\Delta H = -\mathbf{2}(-92 \text{ kJ})$$

$$\Delta H = +184 \text{ kJ}$$

Cancel two moles of H_2 and H_2O and one mole of O_2 by reversing equation (1).

Reverse the sign for the heat of reaction:

$2 \times$ Equation (3):

$$2 \, N_2 \, (g) + \cancel{6} \, O_2 \, (g) + 2 \, H_2O \, (g) \longrightarrow 4 \, \cancel{HNO_3 \, (l)}$$

$$\Delta H = \mathbf{2}(-348.7 \text{ kJ})$$

$$\Delta H = -696.4 \text{ kJ}$$

$-2 \times$ Equation (2):

$$4 \, \cancel{HNO_3 \, (l)} \longrightarrow 2 \, N_2O_5 \, (g) + 2 \, H_2O \, (l)$$

$$\Delta H = -\mathbf{2}(-92 \text{ kJ})$$

$$\Delta H = +184 \text{ kJ}$$

$-1 \times$ Equation (1):

$$2 \, H_2O \, (l) \longrightarrow 2 \, H_2 \, (g) + O_2 \, (g)$$

$$\Delta H = -\mathbf{1}(571.7 \text{ kJ})$$

$$\Delta H = +571.1 \text{ kJ}$$

Summing the chemical equations gives the desired reaction.

Summing the heat of reactions gives the heat of the reaction for the *overall process*:

$$2 \, N_2 \, (g) + 5 \, O_2 \, (g) \longrightarrow 2 \, N_2O_5 \, (g)$$

$$\Delta H = +59 \text{ kJ}$$

ΔH values used in calculations need not be for actual, measurable processes.

Values from the table of heats of formation have been calculated *via* Hess' Law and can be used for further calculations.

Notes for active learning

Bond Energy and Heat of Formation Relationship

Bond-dissociation energy

Bond-dissociation energy is the standard enthalpy change when a bond is cleaved, and it measures a bond's strength.

Bond energy refers to the heat required to break the chemical bonds of *one mole of gaseous molecules* to give *separate gaseous atoms*. Bond energy depends on the types of atoms bonding.

Since energy *input* is required to pull two atoms apart, bond-dissociation energy and bond energy are positive quantities associated with an *endothermic* process. (i.e., absorb energy)

Bond dissociation is the reverse process of *bond formation*; the energy is negative and associated with an exothermic process.

For example, the H–H single bond has a bond energy of 436 kJ per mole, but the H–O single bond has a bond energy of 464 kJ per mole.

Common bond energies

Bond	Bond Energy (kJ/mole)	Bond	Bond Energy (kJ/mole)
H–H	436	N–N	159
H–C	414	O–O	138
H–N	389	Cl–Cl	243
H–O	464	C=O	803
H–F	569	N=O	631
H–Cl	431	O=O	498
C–O	351	C≡C	837
C–C	347	N≡N	946

The bond energies in the above chart can be used to estimate $\Delta H°_f$

Bond energy for calculating the heat of formation

For example, calculate $\Delta H°_f$ for CH_3OH (g) in the following reaction.

$$C(g) \quad + \quad 4H(g) \quad + \quad O(g) \xrightarrow{\hspace{3cm}} \boxed{4}$$

$$\boxed{1} \uparrow \qquad\qquad \boxed{2} \uparrow \qquad\qquad \boxed{3} \uparrow \qquad\qquad\qquad \downarrow$$

$$C(s) \quad + \quad 2H_2(g) \quad + \quad \tfrac{1}{2}O_2(g) \longrightarrow CH_3OH(g)$$

$$C\,(g) \quad + \quad 4\,H\,(g) \quad + \quad O\,(g) \qquad\quad \text{Step 4}$$

$$\text{Step 1} \uparrow \qquad\qquad \text{Step 2} \uparrow \qquad\quad \text{Step 3} \uparrow \qquad\qquad\qquad \downarrow$$

$$C\,(s) \quad + \quad 2\,H_2\,(g) \quad + \quad \tfrac{1}{2}\,O_2\,(g) \quad \rightarrow \quad CH_3OH\,(g)$$

Step 1: **break C–C bonds** (positive energy)

Step 2: **break H–H bonds** (positive energy)

Step 3: **break O–O bond** (positive energy)

Step 4: **form four C–H bonds** (negative energy)

$$\Delta H°_f\,CH_3OH\,(g) = \Delta H°_f\,C\,(g) + 4 \times \Delta H°_f\,H\,(g) + \Delta H°_f\,O\,(g) - \Delta H_{atom}\,CH_3OH\,(g)$$

$$\Delta H°_f\,CH_3OH\,(g) = [716.7 + (4 \times 217.9) + 249.2]\,kJ - \Delta H_{atom}\,CH_3OH\,(g)$$

$$\Delta H°_f\,CH_3OH\,(g) = 1{,}837.5\,kJ - \Delta H_{atom}\,CH_3OH\,(g)$$

$$\Delta H°_f\,CH_3OH\,(g) = 1{,}837.5\,kJ - 3D_{C-H} + D_{C-O} + D_{O-H}$$

$$\Delta H°_f\,CH_3OH\,(g) = 1{,}837.5\,kJ - (3 \times 412) + 360 + 463$$

$$\Delta H°_f\,CH_3OH\,(g) = 1{,}837.5\,kJ - 2{,}059\,kJ$$

$$\Delta H°_f\,CH_3OH\,(g) = -222\,kJ$$

$\Delta H°_f$ of CH_3OH (g) is calculated as –222 kJ using bond energies.

According to the chart, $\Delta H°_f$ of CH_3OH (g) was experimentally determined to be –201 kJ/mol; estimated bond energies are within 10% (–222 kJ/mol *vs.* –201 kJ/mol).

Gibbs Free Energy (*G*)

Reaction spontaneity

In 1870, Josiah Willard Gibbs (1839-1903) expressed *Gibbs free energy* (*G*) as the maximum amount of energy present in molecules available *to do non-expansion work* at constant temperature and pressure.

Free energy change (Δ*G*) is the difference of free energy in the product and reactant molecules.

Gibbs free energy Δ*G* equation:

$$\Delta G = \Delta H - T\Delta S$$

where *G* is Gibbs free energy, measured in Joules (J).

Calculating Gibbs free energy

Because Δ*G* depends on signs, it can be calculated from estimated values.

> If Δ*G* is equal to *zero*, the reaction is at *equilibrium*.

> If Δ*G* is *negative*, the reaction is *spontaneous*.

> If Δ*G* is positive, the reaction is *nonspontaneous*.

For example, if Δ*G* is *negative* at a given temperature and pressure, the reaction is *spontaneous*.

Chemical bonds may break in the reactants when molecules react, and others form in products. Breaking bonds requires energy, while making bonds releases energy, free energy exchanges during a reaction. Not all chemical bonds have the same strength or require the same energy.

Gibbs free energy change (Δ*G*) equals the maximum energy produced during a reaction harnessed as work.

Spontaneous *vs.* nonspontaneous reactions

Table approximates Δ*G* and whether the reaction is spontaneous or nonspontaneous.

ΔH	ΔS	Result
-	+	Spontaneous at *all* temperatures
+	+	Spontaneous at *high* temperatures
-	-	Spontaneous at *low* temperatures
+	-	Not spontaneous at *any* temperatures

Reaction energy diagrams

A chemical reaction's thermodynamics can be represented on a *reaction energy diagram* showing the energy of the reactants, products, activation energy, and free energy change ΔG.

Energy appears on the *y*-axis, and the reaction progress (time) from the starting reactants to the final products appears on the *x*-axis.

Each reaction has an *energy of activation* (E_{ac}) or *energy barrier* before the reactants transform into products.

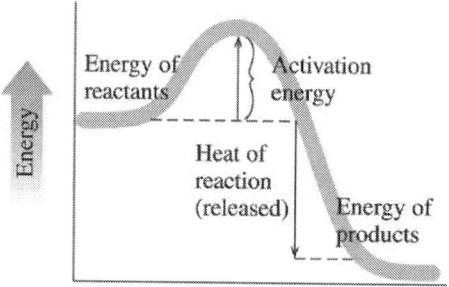

Reaction diagram of energy vs. time (reaction progress)

Gibbs free energy equation

There are several methods to calculate Gibbs free energy ΔG.

Gibbs free energy is a state function, so one method is analogous to the methods for enthalpy and entropy.

Gibbs free energy expression of $\Delta G°_f$ products *minus* $\Delta G°_f$ reactants:

$$\Delta G°_{rxn} = \Sigma\, n(\Delta G°_f \text{ of products}) - \Sigma\, n(\Delta G°_f \text{ of reactants})$$

where *n* represents moles of each product or reactant given by the balanced chemical equation coefficients.

Like enthalpy, the standard Gibbs free energy of formation ($\Delta G°_f$) of elements in their standard states is 0 kJ/mol.

Notice the similarity of Gibbs free energy to *Hess' Law of constant heat summation.*

Spontaneous Reactions and ΔG

ΔG for forward and reverse reactions

Gibbs free energy (ΔG) determines a reaction as spontaneous or nonspontaneous.

$\Delta G < 0$, the reaction is spontaneous in the *forward* reaction

$\Delta G = 0$, the reaction is at *equilibrium* (i.e., rate of forward = rate of reverse reaction)

$\Delta G > 0$, the reaction is spontaneous in the *reverse* direction

ΔG for spontaneous and nonspontaneous reactions

Spontaneous describes a process occurring independently without the input of energy from outside the system, related to the Second Law of Thermodynamics.

ΔG is negative, the reaction is *spontaneous* (or exergonic).

ΔG is positive, the reaction is *nonspontaneous* (or endergonic).

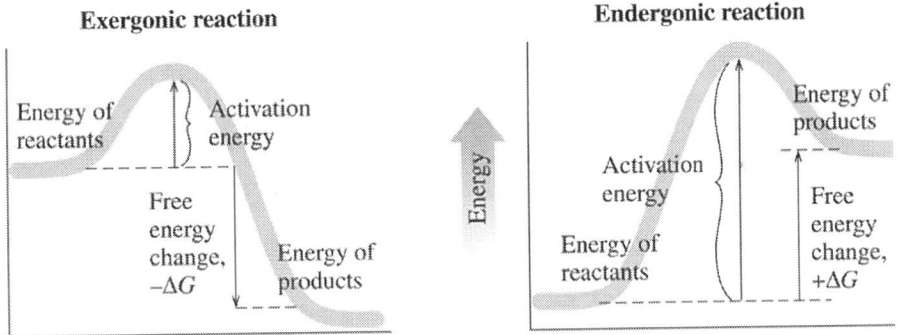

Entropy measures disorder

Entropy (S) is a thermodynamic quantity measuring the spread of a system's energy or disorder.

An increase in entropy of a system corresponds to an increase in the system's randomness or disorder.

$\Delta S° < 0$ represents a decrease in entropy

$\Delta S° > 0$ represents an increase in entropy

For example, forming sugar crystals in a supersaturated sugar water solution *increases entropy*.

Entropy increases because the crystals form spontaneously, resulting in an entropy increase.

Gibbs free energy relates enthalpy, temperature, and disorder

Enthalpy, temperature, and entropy relate free energy by:

$$\Delta G° = \Delta H° - T\Delta S°$$

where $\Delta G°$ is the change in free energy, $\Delta H°$ is the change in enthalpy, T is temperature, and $\Delta S°$ is the change in entropy

An exothermic reaction may *not* be spontaneous because a large, negative ΔS makes it nonspontaneous.

Do *not* assume an endothermic reaction is not spontaneous because a large, *positive* ΔS can make it *spontaneous*.

ΔH	ΔS	Sign of ΔG	Spontaneous
–	+	$\Delta G = (-) - [T\,(+)] = -$	Always, regardless of T
+	–	$\Delta G = (+) - [T\,(-)] = +$	Never, regardless of T
+	+	$\Delta G = (+) - [T\,(+)] = ?$	Depends; spontaneous at high T, $-\Delta G$
–	–	$\Delta G = (-) - [T\,(-)] = ?$	Depends; spontaneous at low T, $-\Delta G$

Table displays how ΔH, ΔS, and T affect spontaneity.

Temperature-controlled reactions

Temperature-controlled reactions are spontaneous at some temperatures but not at others in the pie chart's bottom left and top right quadrants.

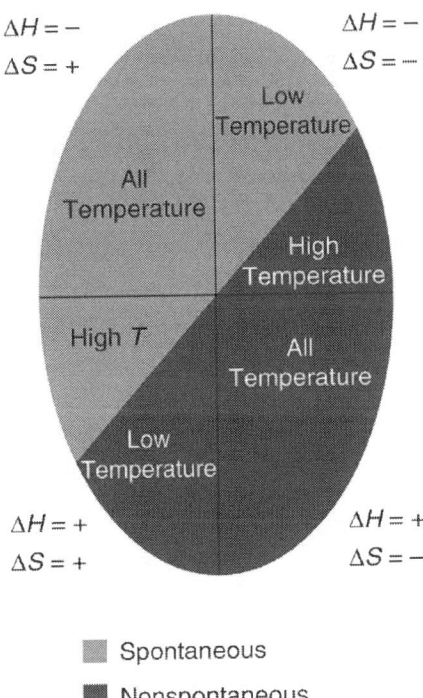

Spontaneous

Nonspontaneous

Temperature-controlled reactions

For example, if the enthalpy change is –1,876 kJ, the entropy change is –3.375 kJ/K, and the temperature is 298 K (standard temperature), is reaction spontaneous? (Units must be consistent).

Entropy is often expressed in units of J/K and converted to kJ/K.

Substituting these quantities into the equation for *Gibbs free energy*:

$$\Delta G° = \Delta H - T\Delta S$$

$$\Delta G° = -1876 \text{ kJ} - [(298 \text{ K}) \cdot (-3.375 \text{ kJ/K})]$$

$$\Delta G° = -870.2 \text{ kJ}$$

Since $\Delta G°$ is *negative*, the reaction is *spontaneous*.

Calculating $\Delta G°$ by chemical equilibrium

Use chemical equilibrium to calculate $\Delta G°$:

$$\Delta G° = -RT \ln K$$

$$\Delta G° = -2.303RT \log K$$

where R is the ideal gas law constant ($R = 8.314$ J/mol·K), T is the temperature in Kelvin, and K is the equilibrium constant

ΔG implies the *direction* and *extent* of a chemical reaction; it does NOT imply a *rate* for the reaction.

Spontaneous reactions may not occur quickly as they may take years, depending on kinetics (i.e., rate).

Calculating $\Delta G°$ by electrochemistry

Using electrochemistry to calculate $\Delta G°$:

$$\Delta G° = -nFE°$$

where n represents the moles of electrons transferred, F is Faraday's constant (96,485 C/mol), and $E°$ is the standard potential for an electrochemical cell

Calculations of Gibbs free energy involve substituting appropriate quantities into the expression.

Thermal Expansion

Thermal expansion of materials

Thermal expansion is the tendency of matter to *change volume* with temperature changes. Most materials expand when the temperature rises and contract when the temperature falls.

Thermal expansion occurs through *heat transfer*.

When material changes temperature, it shrinks or expands, depending upon the temperature difference.

As the temperature increases, the molecules within the material gain energy and vibrate at higher rates. This vibration increases the distance between the molecules and causes the material to expand.

If the material does not experience a phase change (e.g., liquid to gas), this expansion of material can be related to the change in temperature.

Coefficients of thermal expansion (α)

Thermal expansion coefficient is an empirically derived constant that describes the expansion of a given material during a temperature rise and depends on the geometry of the material.

Coefficient of thermal expansion (α) is the degree of expansion divided by the change in temperature.

In solid materials, the magnitude of thermal expansion can be calculated to account for:

linear expansion:	$\Delta L / L_0 = \alpha \Delta T$
area expansion:	$\Delta A / A_0 = 2\alpha \Delta T$
volumetric expansion:	$\Delta V / V_0 = 3\alpha \Delta T$

All three calculations require the thermal coefficient of the material. The value of the coefficient α depends on the type of expansion.

Linear expansion

The *coefficient of linear expansion* is a measure of the expansion (or contraction) per unit of length of a material that occurs when the temperature is increased or decreased by 1 °C. This term is also known as *expansivity*.

The *coefficient of thermal expansion* (α) is given as:

$$\alpha_{linear} = \frac{\Delta L}{l_i \cdot \Delta T}$$

where α_{linear} is the coefficient of thermal expansion (K^{-1}), ΔL is the change in length (m), l_i is the initial length (m), and ΔT is the change in temperature (K)

Thermal expansion equation for linear expansion is:

$$\Delta L = \alpha L_0 \Delta T$$

where ΔL is the change in length (m), L_0 is the initial length (m), ΔT is the change in temperature (K), and α is the coefficient of linear expansion(K^{-1})

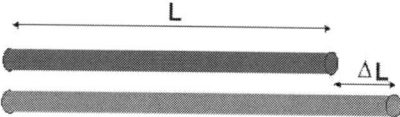

A bar of material is at an initial temperature and has a length denoted as L.

If the bar is heated such that its temperature rises, it expands linearly by some length ΔL.

For example, a metal rod of 7.00 m is heated to 35 °C. If the rod's length expands to 7.12 m after some time, calculate the linear expansion coefficient (Use room temperature = 27 °C).

Calculate the change in length ΔL from the initial $L_0 = 7.00$ m and expanded length $L = 7.12$ m.

$$\Delta L = 7.12 \text{ m} - 7.00 \text{ m}$$

$$\Delta L = 0.12 \text{ m}$$

calculate ΔT from the initial room temperature $T_0 = 27$ °C and the heated $T = 35$ °C.

$$\Delta T = 35 \text{ °C} - 27 \text{ °C}$$

$$\Delta T = 8 \text{ °C}$$

convert ΔT to Kelvin

$$\Delta T = 8 + 273$$

$$\Delta T = 281 \text{ K}$$

linear expansion expression:

$$\Delta L / L_0 = \alpha \Delta T$$

isolate α

$$\alpha = \Delta L / (L_0 \times \Delta T)$$

$$\alpha = 0.12 \text{ m} / (7 \text{ m} \times 281 \text{ K})$$

$$\alpha = 6.1 \times 10^{-5}$$

The coefficient of linear expansion α for the material is 6.1×10^{-5}.

Area expansion

Objects can undergo area expansion due to the same principles.

Thermal expansion equation for area expansion is:

$$\Delta A = \gamma A_0 \Delta T$$

where ΔA is the change in the area (m^2), and γ is the coefficient of area expansion (K^{-1}).

The *coefficient of area expansion* is expressed as:

$$\alpha_{area} = \frac{\Delta A}{A_i \cdot \Delta T}$$

where α_{area} is the coefficient of thermal expansion (K^{-1}), ΔA is the change in the area (m^2), A_i is the initial area (m^2), and ΔT is the change in temperature (K)

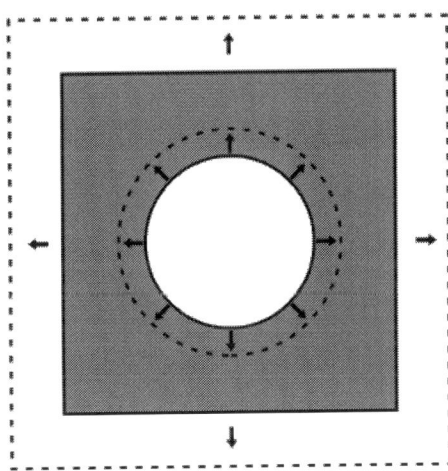

Volume expansion

In general, when a substance is heated, the molecules increase kinetic energy and move more; thus, greater average separation is maintained, increasing volume (expansion).

Volume expansion:

$$\Delta V = \beta V_0 \Delta T$$

where ΔV is the volume change (m^3), and β is the coefficient of volume expansion (K^{-1}).

The *coefficient of volume expansion* is expressed as:

$$\alpha_{volume} = \frac{\Delta V}{V_i \cdot \Delta T}$$

where α_{volume} is the coefficient of thermal expansion (K^{-1}), ΔV is the change in volume (m^3), V_i is the initial volume (m^3), and ΔT is the change in temperature (K).

Notes for active learning

Kinetic Theory of Gases

Macroscopic properties of gas

Thermal energy (or *internal energy*) is the sum of the *kinetic energies* of the particles. Thermal energy of a gas is directly proportional to the molecules' kinetic energy (KE).

The higher the kinetic energy of the molecules, the more thermal energy the system possesses.

The lower the kinetic energy of the molecules, the less thermal energy the system possesses.

The figure below illustrates a closed container of gas where the gas molecules are in random motion throughout the container.

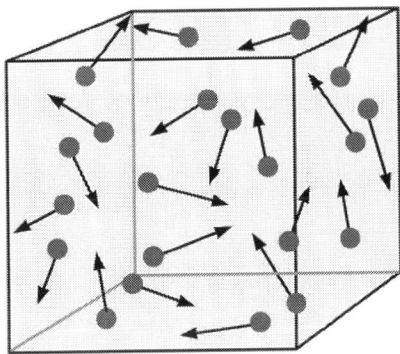

Gas molecules exhibit kinetic energy from random motion in the container

Kinetic Molecular Theory

Kinetic Molecular Theory of Gases relates the microscopic values of atomic kinetic energy to quantities like temperature and pressure.

Kinetic molecular theory (KMT) uses atomic theory to describe the behavior of gases. This theory developed over 100 years, culminating in an 1857 paper by German physicist Rudolf Clausius (1822-1888).

Kinetic molecular theory explains the macroscopic properties of a gas (e.g., pressure and temperature) in terms of its microscopic components (e.g., molecules or atoms). It originated in the ancient idea that matter consists of tiny invisible atoms in rapid motion. The idea was revived and developed between the 17th and 19th centuries to explain gases' properties.

Five assumptions for gases

Five premises for the kinetic theory of gases:

1. Gases are comprised of molecules. Molecules may be treated as perfect spheres of mass (*point masses*), and the space between each of these masses is many times greater than the diameter of the gas molecules.

2. Motion of each gas molecule is *random*, with no order or pattern to either the direction or magnitude of the velocity.

3. Molecules follow *Newton's Laws of Motion*. Each gas molecule moves in a straight line with a constant velocity. If gas molecules collide, the molecules exert an equal but opposite force on the other.

4. Collisions of molecules are *perfectly elastic*; they *lose no kinetic energy*.

    Therefore, since intermolecular attractive and repulsive forces are nonexistent, the gas molecules' total kinetic energy (KE) remains *constant* before and after the collision.

5. Average kinetic energy (*KE*) for gas particles is proportional to the absolute temperature, regardless of the chemical identity or atomic mass.

    At 0 K (absolute zero or –273 °C), the molecules (and orbiting electrons) are not moving and have no volume.

These rules display gases as ideal substances and only approximate their actual behavior. However, their description is remarkably accurate. From these assumptions, laws are derived to describe the behavior of gases.

Boltzmann's constant

Final assumption (#5) for KE is expressed by:

$$KE = \tfrac{1}{2}\, mv^2$$

$$\tfrac{1}{2}\, mv^2 = (3/2)\, k_\mathrm{B} T$$

So,

$$KE = (3/2)\, k_\mathrm{B} T$$

where *KE* is kinetic energy, *m* is mass, *v* is velocity, *T* is the temperature (Kelvin), and k_B is *Boltzmann's constant* ($k_\mathrm{B} = 1.38 \times 10^{-23}$ J/K).

Boltzmann's constant relates macroscopic and microscopic behavior of gases to their particles.

Root-mean-square speed

KE = (3/2) $k_B T$ equation is significant because it states that a gas particle's average kinetic energy is proportional to its absolute temperature. Increasing the temperature of the gas particles increases the total speed and overall energy.

Gas atoms are infinitesimally small; nearly impossible to measure a particle's speed accurately.

Speed of gas particles is defined by the *root-mean-square speed* (u_{rms}) by the equation:

$$u_{rms} = \sqrt{\frac{3RT}{MM}}$$

where R is ideal gas constant, T is absolute temperature (K), and MM is molar mass of gas.

Root-mean-square speed written with the *Boltzmann constant*, k_B:

$$u_{rms} = \sqrt{\frac{3kT}{m}}$$

where k is the Boltzmann constant, T is the absolute temperature (K), and m is the mass of one gas molecule.

Boltzmann distribution plot

Plotting each gas particle's speed distribution at a given temperature shows a slightly asymmetric curve. This speed-distribution curve is the *Maxwell-Boltzmann distribution curve*.

Boltzmann distribution plot shows that the curve's peak corresponds to the most probable speed.

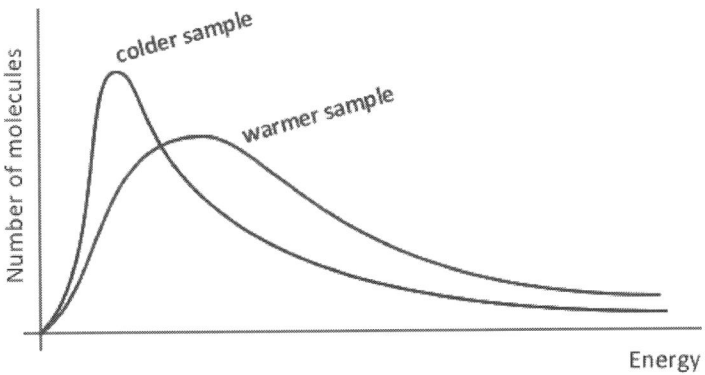

Maxwell-Boltzmann Distribution curves of molecular speeds at two temperatures

If the temperature (K) remains constant, the total kinetic energy remains unchanged.

However, the energy is distributed in several ways, and the gas particles travel at many speeds. This distribution changes continually as the gas atoms collide with each other and with the container walls.

The curve flattens (i.e., the net area under the curve is greater) and shifts to the right at higher temperatures, indicating that more gas molecules move at higher speeds and possess more kinetic energy.

Diffusion and effusion

Diffusion is when a substance (solute or particle) spreads from a high-concentration region to a lower concentration. It is explained using the kinetic theory of motion.

According to the kinetic theory, heavier gases diffuse more slowly than lighter gases because of the difference in their travel speed.

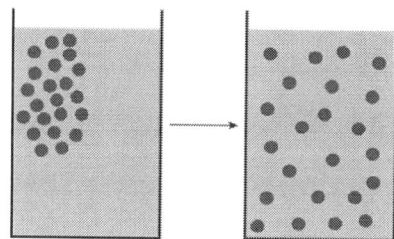

Diffusion of solutes in a solvent

Effusion is the flow of gas particles under pressure from one compartment to another through a small opening, as shown below.

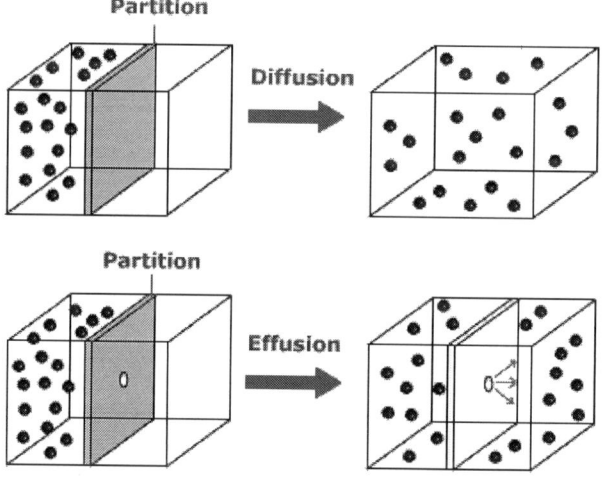

Diffusion and effusion of gas particles

Graham's Law

Scottish chemist Thomas Graham (1805-1869) developed *Graham's Law*, which states that *at constant temperature and pressure conditions, the rates at which two gases diffuse are inversely proportional to the square root of their molar masses.*

Graham's Law is expressed by:

$$\frac{r_1}{r_2} = \sqrt{\frac{MM_2}{MM_1}}$$

where r_1 and r_2 are the diffusion rates of gas 1 and gas 2, respectively, and MM_1 and MM_2 are the molar masses of the gases.

Graham's Law applies to the effusion of gas particles, and the equation for effusion is the same as the equation for diffusion.

Ideal Gas Law

Ideal Gas Law explains the relationship between pressure (P), volume (V), and temperature (T).

Gas law is expressed as:

$$PV = nRT$$

where n is number of moles, and R is universal gas constant with a value of 8.314 J/ mol·K.

One mole is equal to 6.023×10^{23} molecules. Technically, a mole is the number of hydrogen atoms in one gram of hydrogen. Atoms are tiny, so counting them in moles is more practical than counting them individually. Ideal Gas Law applies to the pressure a gas exerts on a cylinder with a moving wall.

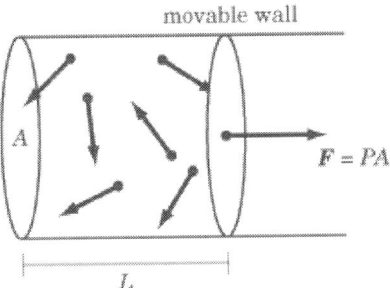

Force that the gas exerts on the wall equals F = PA

Since pressure equals $P = F/A$, the force the gas exerts on the wall equals $F = PA$.

If this force moves the wall back a length of L, the cylinder volume increases by $\Delta V = LA$.

By solving for A and substituting this into the equation for force, the result is:

$$F = P\Delta VL \text{ or } P\Delta V = FL$$

Work is *force multiplied by distance traveled*. By pushing the wall at a distance of L with a force of F, the gas has done work equal to FL.

When gas does work, it symbolizes a change in energy.

If a change in PV is equal to a change in energy, then PV is the total energy of the gas.

For the ideal gas law, nRT is the expression for the total kinetic energy of the gas molecules.

Ideal gas law (PV = nRT) can be related to the number of molecules (N) and Boltzmann's constant (k), which has a value of 1.381×10^{-23} J/K:

$$PV = NkT$$

Other gas laws derive from the Ideal Gas Law by holding specific variables constant.

The number of moles (n) and the gas constant (R) are constant, so four other gas laws result from the Ideal Gas Law:

> Boyle's Law
>
> Charles' Law
>
> Combined Gas Law
>
> Closed Container Law

Ideal gases follow KMT (isothermal process)

As described previously, *kinetic molecular theory (KMT)* is based on a theoretical ideal gas.

Ideal gases follow *five* assumptions of KMT in which gas molecules:

> move in random motion
>
> are point masses with no volume,
>
> there are no intermolecular forces,
>
> collisions are elastic and
>
> kinetic energy of molecules is proportional to temperature.

Compared to an ideal gas, the behavior of a real gas is complicated. By conceptualizing ideal gas behavior, real gas behavior is easier to comprehend.

In the 17th and 18th centuries, scientists realized several relationships between gases' properties. Boyle's, Charles', and Avogadro's Law relate pressure, volume, and temperature properties.

These laws were developed empirically and are individual cases of the ideal gas equation:

$$PV = nRT$$

Boyle's Law (isothermal process)

In 1661, Robert Boyle (1627-1691) studied the compressibility of gases. He observed that *the volume of a fixed amount of gas at a given temperature is inversely proportional to the pressure exerted on the gas.*

Boyle's Law applies when the temperature of a gas is held constant. It states that an increase in pressure causes a decrease in volume or that a decrease in pressure causes an increase in volume.

Boyle's Law is expressed as:

$$P_1V_1 = P_2V_2$$

where subscripts 1 and 2 are same gas sample under two sets of pressure and volume conditions.

A plot of volume *vs.* pressure for a gas illustrates that Boyle's Law is a particular case of the ideal gas law, where n (moles of gas) and T (temperature) are constant (figure below).

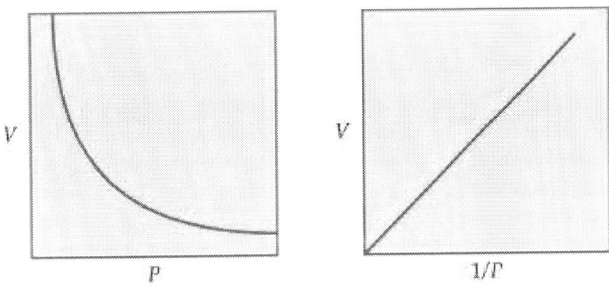

Boyle's Law: as pressure increases, volume decreases

For example, what is the new pressure in a container of gas in a 15.0 L container at a pressure of 5.00 atm when the volume decreases to 0.500 L?

Substitute the known quantities (P_1, V_1, and V_2) into Boyle's Law equation to solve P_2. Note: the volume units need to be consistent. In this example, volumes are expressed in units of liters.

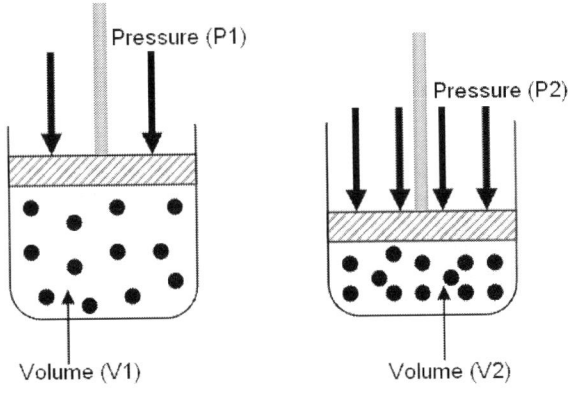

Boyle's Law is where n (moles of gas) and T (temperature) are constant

Boyle's Law:

$$P_1V_1 = P_2V_2$$

Solve for P_2:

$$P_1V_1 / V_2 = P_2$$

$$[(5.00 \text{ atm}) \cdot (15.0 \text{ L})] / 0.500 \text{ L} = P_2$$

$$P_2 = 150 \text{ atm}$$

Charles' Law (isobaric process)

French scientist Jacques Charles (1746-1823) developed the relationship between temperature and gas volume.

Charles' Law states that *at constant pressure, the volume of a gas is directly proportional to its absolute temperature (Kelvin).* It describes the behavior of a gas under constant pressure.

In this relationship, volume and temperature are *directly proportional*; when temperature increases, volume increases; when the temperature decreases, the volume decreases.

Charles' Law is expressed as:

$$\frac{V_1}{T_2} = \frac{V_2}{T_2}$$

where subscripts 1 and 2 refer to the same gas sample at different temperature and volume.

Temperature vs. volume plot for a gas illustrates that Charles' Law is another application of the ideal gas law, where n and P are constant, as shown below.

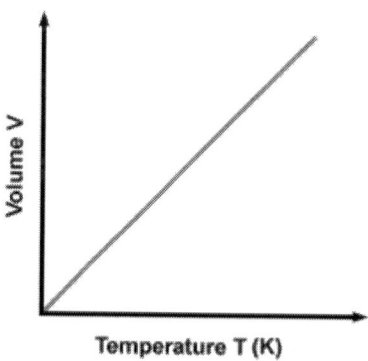

Charles' Law states that as temperature increases, volume increases

For example, if there is 25.0 L of gas at 0 °C, and the temperature is raised to 100 °C, what is the volume of the gas?

First, the temperature is converted to Kelvin.

$$T_1 = 0\ °C + 273 = 273\ K$$

$$T_2 = 100\ °C + 273 = 373\ K$$

Substitute the known quantities into the equation for Charles' Law.

$$V_1 / T_1 = V_2 / T_2$$

Rearrange and solve for V_2:

$$V_1\ T_2 / T_1 = V_2$$

$$(25.0\ L)·(373\ K) / 273\ K = V_2$$

$$V_2 = 34.2\ L$$

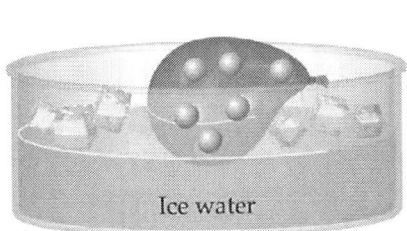
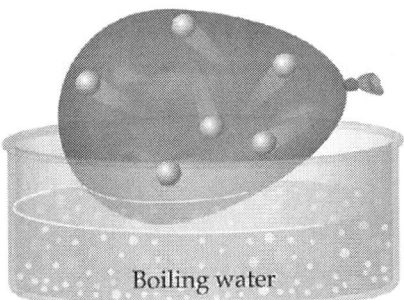

Volume and temperature are directly proportional, when temperature increases, volume increases

Gay-Lussac's Law

In 1802, French chemist Louis Gay-Lussac (1778-1850) proposed *Gay-Lussac's Law*, which states that *the pressure of a given sample of gas is directly proportional at a constant volume to its absolute temperature (Kelvin)*.

Gay-Lussac's Law is expressed by:

$$\frac{P_1}{T_1} = \frac{P_2}{T_2}$$

where subscripts 1 and 2 refer to the same gas sample with different temperatures and pressure.

Temperature vs. pressure plot for a gas illustrates that Gay-Lussac's Law is another particular case of the ideal gas law, where n and V are constant.

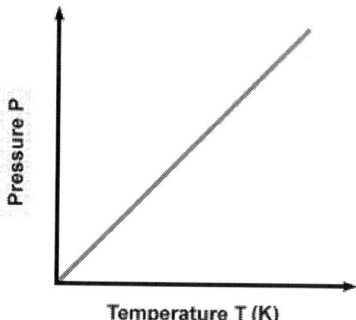

Gay-Lussac's Law states that as temperature increases, pressure increases

For example, suppose a gas is at 30.0 atm pressure and 100 °C, and the temperature changes to 400 °C. What is the new pressure of the gas?

Convert the temperature to units of Kelvin.

$T_1 = 100 \ °C + 273 = 373 \ K$

$T_2 = 400 \ °C + 273 = 673 \ K$

Gay-Lussac's Law solves P_2 by substituting the quantities into the equation.

$$\frac{P_1}{T_1} = \frac{P_2}{T_2}$$

$P_2 = [(30.0 \ atm) \cdot (673 \ K)] / 373 \ K$

$P_2 = 54.1 \ atm$

Closed Container Law (constant volume)

Closed Container Law describes the behavior of a gas under constant volume. In such cases, pressure and temperature are directly proportional:

$$\frac{P_i}{T_i} = \frac{P_f}{T_f}$$

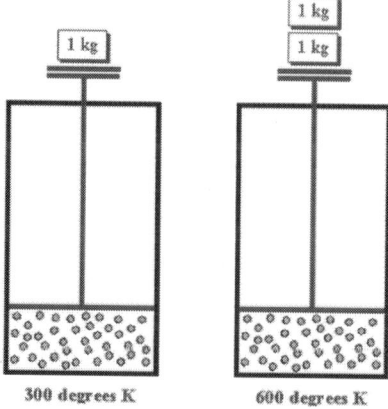

Combined Gas Law

Combined Gas Law amalgamates Boyle's and Charles' Laws. The combined gas law is *not* the ideal gas law (PV = nRT); it is expressed as an equation to relate changes in temperature, volume, and gas pressure.

Combined Gas Law relates the pressure, volume, and temperature of a gas and is expressed as:

$$\frac{P_i V_i}{T_i} = \frac{P_f V_f}{T_f}$$

where subscripts 1 and 2 refer to the same gas sample but under different temperatures, pressure, and volume.

For example, suppose a gas is at 15.0 atm pressure, with a volume of 25.0 L and a temperature of 300.0 K. What would the volume of the gas be at standard temperature and pressure? (Use standard pressure = 1.00 atm and standard temperature = 273.0 K)

Use the *combined gas law* at standard temperature (T_2) and pressure (P_2):

$(P_1 V_1) / T_1 = (P_2 V_2) / T_2$ solve for V_2:

$(P_1 \times V_1 \times T_2) / (T_1 \times P_2) = V_2$

$V_2 = [(15.0 \text{ atm}) \cdot (25.0 \text{ L}) \cdot (273.0 \text{ k})] / [(300.0 \text{ K}) \cdot (1.00 \text{ atm})]$

$V_2 = 341 \text{ L}$

Avogadro's Law

The volume of a gas in a container is affected by pressure, temperature, and moles of gas.

The relationship between the quantity of gas and its volume was derived by Italian scientist Amadeo Avogadro (1776-1856).

Avogadro's Law states that *gases at a given temperature and pressure occupying a volume are directly proportional to the number of moles of gas present.*

Avogadro's Law is expressed as:

$$\frac{n_1}{V_1} = \frac{n_2}{V_2}$$

where n_1 and n_2 are moles of gas 1 and gas 2, and V_1 and V_2 are the volumes of gas 1 and gas 2.

A plot of *volume vs. moles* of gas (shown below) illustrates that Avogadro's Law is a particular case of the ideal gas law where T and P are held constant.

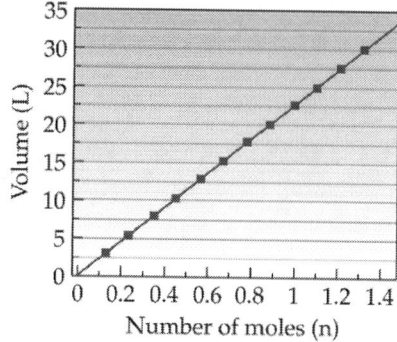

Avogadro's Law states that as moles of a gas increase, volume increases

For example, if 8.00 moles of a gas occupy a volume of 4.00 L at constant pressure and temperature, what volume of gas would 16.0 moles of this gas occupy at the same temperature and pressure?

Use *Avogadro's Law*

$$n_1 / V_1 = n_2 / V_2$$

where n_1 and n_2 are moles of gas 1 and gas 2, and V_1 and V_2 are the volumes of gas 1 and gas 2.

Solve for the V_2

$$V_2 = (V_1 \times n_2) / n_1$$

$$V_2 = [(4.00 \text{ L}) \cdot (16.0 \text{ mol})] / 8.00 \text{ mol}$$

$$V_2 = 8 \text{ L}$$

Phase Diagrams

Phases of substances

Most substances exist in solid, liquid, or gas form.

Solids have a definite shape and volume (e.g., ice). The molecules in a solid vibrate at a fixed position and a solid cannot be compressed.

Liquids have a definite volume but take their container shape (e.g., water). The molecules move about but are close and bound by intermolecular forces.

Gases have neither a definite volume nor a definite shape; they take the container's volume and shape (e.g., steam or water vapor). The molecules move independently and are not held together by intermolecular forces, making gas easily compressible.

Phases at equilibrium

Phase diagrams indicate the relationship between temperature, pressure, and associated phases, plotting pressure, temperature, or volume for thermodynamically distinct phases (solid, liquid, or gas) at equilibrium.

Phase of a substance (i.e., solid, liquid, or gas) mostly depends on temperature and pressure.

If the pressure is exceptionally high, a substance is less likely to be in a gas phase because the pressure brings the molecules so close that the substance either liquefies or sublimes.

Phase change diagram

When an ice block is placed in the Sun, it slowly absorbs thermal energy through radiation and undergoes phase changes. The ice cube changes from solid to liquid to gas.

> *Melting point* of a substance is the temperature at which it changes from a solid to a liquid, and it has the same temperature range as the *freezing point*; the reverse reaction occurs at the same temperature.

> *Boiling point* (or vaporization) of a substance is the temperature at which it changes phase from a liquid to a gas, the same temperature range as the *condensation point*.

From the graph below, once the ice cube absorbs enough heat to melt (i.e., reaches 0 °C), its temperature remains constant.

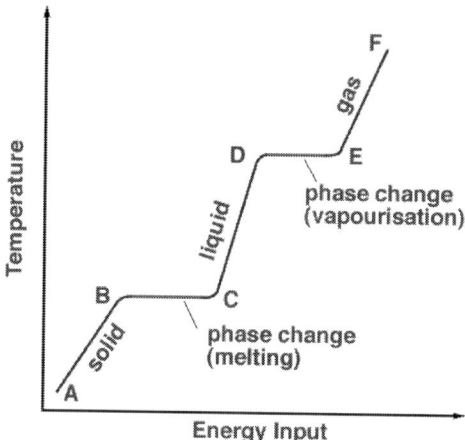

Phase diagram relates temperature changes and states

Melting is points B to C on the graph where all molecules have reached a temperature of 0 °C.

Ice melts only when this occurs, as illustrated by the graph's horizontal plateau from B to C.

When liquid water changes to a gas, the liquid absorbs enough heat to *vaporize* (i.e., at 100 °C).

Liquid temperature remains constant from point D to E until all molecules reach 100 °C.

As illustrated by the *horizontal plateau* from D to E, the *liquid vaporizes* only when this occurs.

Heat during phase changes

Although the energy (heat) input increases with temperature, it does not remain constant as the temperature does during a phase change. Technically, heat converts the potential energy stored within each atom to kinetic energy before changing phase.

As energy is continuously added to the process, the substance uses the added heat to transition each molecule *before increasing* the temperature.

Pressure *vs.* temperature relationship

Phase diagram graphically displays the phases of a substance of pressure *vs.* temperature.

The graph is split into three disproportionate parts, representing the substance's three phases.

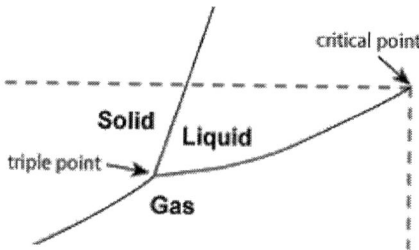

PV phase diagram with triple point and critical point labeled. Dashed horizontal line shows the critical pressure, and vertical dashed line shows the critical temperature.

Triple point, critical point, and critical temperature

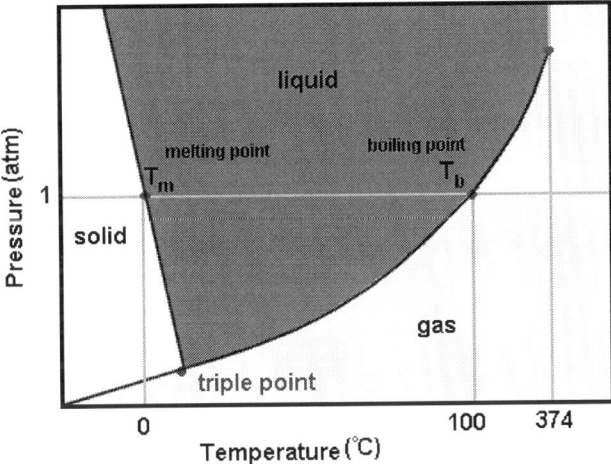

PV phase diagram with triple point and critical point labeled.

Melting point and *boiling point* are found by drawing a horizontal line at 1 atm of atmospheric pressure. The horizontal line intersects with a line on the phase diagram; the substance changes phases at that temperature.

Triple point is temperature and pressure at which three phases of matter coexist in equilibrium.

Critical point is temperature and pressure at which liquids and gases become indistinguishable.

Critical temperature is when a liquid cannot form, no matter the pressure.

Phase line boundaries

Solid-liquid boundary is where the *solid and liquid* exist in equilibrium.

Solid-gas boundary is where *solid and gas* exist in equilibrium.

Liquid-gas boundary is where *liquid and gas* exist in equilibrium.

Positive slope graphs

Solid-liquid phase boundary (or fusion curve) for most substances, has a positive slope (i.e., rise over run), so the *melting point increases with pressure*. This is true when the solid phase is denser than the liquid phase.

The greater the pressure, the closer the molecules of the substance are brought to each other, increasing the effect of intermolecular forces. Thus, the substance requires a higher temperature for its molecules to have enough energy to break the fixed pattern of the solid phase and enter the liquid phase.

A similar concept applies to liquid–gas phase changes.

Negative slope graphs

Water is an exception, which has a solid-liquid boundary with a negative slope so that the melting point decreases with pressure.

Ice (solid water) is less dense than liquid water, as ice floats on water.

At a molecular level, ice is less dense because it has a more extensive network of hydrogen bonding, which requires a greater separation of water molecules.

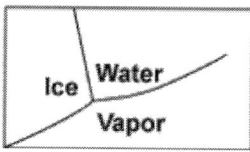

Phase diagram of pressure vs. pressure for water differs because the solid-liquid boundary is slanted to the left. Water (liquid) is denser than ice (solid), if the pressure increases, ice melts into water.

Colligative Properties

Solute *vs.* solvent ratio

Colligative properties depend on the ratio of solute particles to solvent but *not* on the type of chemical species.

Colligative properties include:

1) the relative lowering of vapor pressure,

2) the elevation of boiling point,

3) the depression of freezing point, and

4) osmotic pressure.

Colligative properties calculations incorporate the van 't Hoff factor (*i*), which measures a solute's effect on the solution's colligative properties.

van 't Hoff factor is:

$$\frac{[\text{concentration of particles produced when a substance dissolves}]}{[\text{concentration of a substance calculated from its mass}]}$$

Concentration is converted to reflect the total number of particles in the solution. (e.g., 5 g of NaCl dissociates into Na^+ and Cl^-).

For example, glucose ($C_6H_{12}O_6$) has an *i* value of 1 because it does not dissociate in solution. Nonelectrolytes do not dissociate in solution (e.g., CH_4 and O_2). The salt NaCl has an *i* value of 2 because it dissociates into 1 Na^+ and 1 Cl^- (2 resulting ions per one reactant molecule).

Boiling point elevation ($\Delta T_b = K_b m$)

Adding a solute to a solvent stabilizes the solvent in the liquid phase, thus lowering the solvent molecules' tendency to move to the gas or solid phases. Therefore, the boiling point increases.

The *boiling point elevation* is proportional to the decrease of vapor pressure in a dilute solution:

$$\Delta T_b = k_b \cdot m \cdot i$$

where ΔT_b is the increase in boiling point, k_b is the molal boiling point constant (a given value determined experimentally), m is molality (mol solute/kg solvent), and i is the van 't Hoff factor.

For example, what is the boiling point elevation if 6.4 g of ammonia is dissolved in 0.3 kg of water? (Use the k_b for water = 0.52 °C/m and the molar mass of ammonia = 17.031 g/mol)

Convert the mass of ammonia (NH_3) into moles:

> molecular mass = mass / moles

> moles = mass / molecular mass

> moles = (6.4 g) / (17.031 g/mol)

> moles = 0.38 mol NH_3

Calculate the molality of the solution:

> molality = (moles of solute) / (mass (in kg) of solvent)

> m = (0.38 mol NH_3) / (0.3 kg of water)

> m = 1.27 m

Calculate the boiling point elevation:

> $\Delta T_b = k_b \cdot m \cdot i$

> $\Delta T_b = (0.52 °C/m) \cdot (1.27 m) \cdot (1)$

> $\Delta T_b = 0.66 °C$

The boiling point increased by 0.66 °C due to adding 6.4 g of NH_3.

Freezing point depression ($\Delta T_f = K_f m$)

Solute particles in mixtures increase the strength of intermolecular (e.g., dipole-dipole) bonds. It is more difficult to boil the solution (i.e., boiling point elevation) or freeze it (i.e., freezing point depression).

Freezing point depression (i.e., lowering the freezing point) is calculated by the following equation: (note the negative sign indicates that the change is a decrease).

$$\Delta T_f = -k_f m \cdot i$$

where ΔT_f is the decrease in freezing point, k_f is the molal freezing point constant (a given value determined experimentally), m is the molality (mol solute/kg solvent), and i is the van 't Hoff factor (given since it is determined experimentally).

For example, a 48.0 g sample of a nonelectrolyte (does not dissociate in solution) is dissolved in 500.0 g of water to produce a solution with a freezing point of –3.5 °C. What is the molar mass of the compound? (Use k_f of water = 1.86 °C/m, the freezing point of pure water = 0 °C)

The freezing point of water = 0 °C, and the freezing point depression, $\Delta T_f = 3.5$ °C.

Calculate the moles using the freezing point depression expression:

$$\Delta T_f = k_f \cdot m \cdot i$$

$$3.5 \text{ °C} = (1.86 \text{ °C/m}) \cdot (x / 0.5 \text{ kg}) \cdot (1)$$

$$3.5 \text{ °C} = (3.72 \text{ °C/m}) \cdot (x)$$

$$3.5 \text{ °C} / (3.72 \text{ °C/m}) = x$$

$$x = 0.94 \text{ mol nonelectrolyte}$$

Calculate molar mass by dividing the sample's mass by moles in the sample.

Molar mass = mass / moles

Molar mass = 48.0 g / 0.94 mol

Molar mass = 51.1 g/mol

The molar mass of the unknown nonelectrolyte is 51.1 g/mol.

Notes for active learning

Heat Transformation

Fusion and vaporization

Latent heat (*latent energy* or *heat of transformation*) is the energy released or absorbed by a thermodynamic system during a constant-temperature process — usually a first-order phase transition (i.e., constant temperature during "mixed-phase" states).

Latent heat is the energy supplied or extracted to change the state of a substance without changing its temperature (i.e., first-order phase transition).

Examples include *latent heat of fusion* (solid to liquid) and *latent heat of vaporization* (liquid to gas) for phase changes when a substance condenses or vaporizes at a specified temperature and pressure.

Heat and phase changes

Energy is released for:

Condensation: gas changes phase into liquid

Freezing: liquid transforms into a solid

Deposition: gas changes phase directly into solid

Reverse processes use the same amount of energy.

Vaporization (reverse of condensation): liquid changes phase into a gas

Melting (reverse of freezing): solid changes phase into liquid

Sublimation (reverse of deposition): solid changes phase directly into a gas

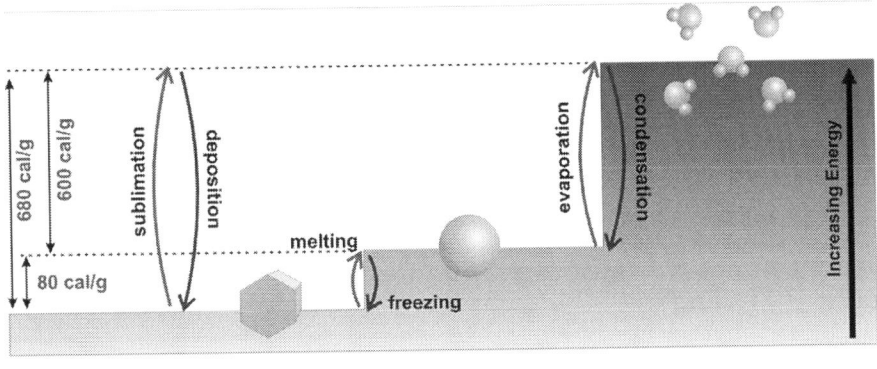

Latent heat

Latent heat is the amount of heat required for a phase change. Latent heat has been reported for most substances. In general, only computing the total heat is required for the reaction.

Heat required to convert a substance from one phase to another is:

$$Q = m \cdot l$$

where l is the latent heat (J/kg), and m is the mass of the substance (kg).

Latent heats are sometimes given in molar heat or joules per mol (J/mol). Divide by the molecular weight of the substance if given the mass in grams.

If any of these processes are reversed, the latent heat is negative (reverse process).

Heat of fusion (ΔH_{fusion})

Heat of fusion (ΔH_{fusion}) is the energy to melt a given quantity of solid to liquid at a constant temperature; the energy required to melt the solid times the number of moles.

Heat of fusion (ΔH_{fusion}) is the energy to melt a given quantity of solid to liquid at a constant temperature; the energy required to melt the solid times the number of moles.

Heat of fusion ΔH_{fusion} is expressed as:

$$q = n \times \Delta H_{fusion}$$

where q is heat energy, n is the number of moles of the solid, and ΔH_{fusion} is the heat of fusion.

Most substances have reported reference values for the heat of fusion (ΔH_{fusion}).

If the particles become excited (solid to a liquid, liquid to a gas), the change in entropy is always positive because heat is added to the system.

For this process, use latent heat of fusion. The total heat required is the heat of fusion.

If the particles become less excited (gas to a liquid, liquid to solid), the change in entropy is always negative because heat is removed from the system.

A negative change in entropy results in a more ordered molecular structure.

The total heat required for this process is the *heat of vaporization*.

Heat of vaporization (ΔH_{vap})

Heat of vaporization (ΔH_{vap}) is the energy to vaporize a given quantity of liquid to gas at a constant temperature.

It is the energy required to vaporize the liquid times the number of moles of that liquid.

Most substances have reported reference values for the heat of vaporization (ΔH_{vap}).

Heat of vaporization ΔH_{vap} is expressed as:

$$q = n \times \Delta H_{vap}$$

where q is heat energy, and n is number of moles of the liquid, and ΔH_{vap} is heat of vaporization.

Sublimation is when a solid bypasses the liquid phase and transitions directly into the gas phase.

The heat required for *sublimation* sums up the *heat of fusion* and the *heat of vaporization*.

Deposition is the reverse reaction of sublimation.

$$Q_{sub} = Q_{fus} + Q_{vap}$$

Total heat of reaction

For a process with multiple phase changes, the total heat of the reaction is the sum of the separate heat needed for each phase change and the addition of heat.

For these problems, consider the First Law of Thermodynamics, which states that heat must be conserved (heat gained = heat loss), and the equation for heat ($Q = mc\Delta T$), where c is the specific heat capacity, and m is the mass.

For example, when calculating the change in heat (change in entropy) when 2 kg of ice at 0 °C is warmed to 120 °C, a few equations are needed using the known values.

- melting point = 0 °C

- boiling point = 100 °C

- latent heat of fusion = 3.33×10^5 J/kg

- latent heat of vaporization = 2.26×10^6 J/kg

- specific heat capacity = 4,186 J/kg °C

Heat required to phase change the *ice into liquid water*:

$$Q = m \, l_{fusion}$$

$$Q = (2\text{kg}) \cdot (3.33 \times 10^5 \text{ J/kg}) = 6.66 \times 10^5 \text{ J}$$

Heat required to heat the *water to its boiling point*:

$$Q = mc\Delta T$$

$$Q = (2 \text{ kg}) \cdot (4186 \text{ J/kg °C}) \cdot (100 \text{ °C} - 0 \text{ °C}) = 837{,}200 \text{ J}$$

Heat required to phase change the *liquid into gas:*

$$Q = m \, l_{vap} = (2 kg) \cdot (2.26 \times 10^6 \, J/kg)$$

$$Q = 4.52 \times 10^6 \, J$$

Heat required to heat the water to 120 °C:

$$Q = mc\Delta T$$

$$Q = (2 \, kg) \cdot (4,186 \, J/kg \, °C) \cdot (120 \, °C - 100 \, °C) = 167,440 \, J$$

Total heat required for the process to occur is the *sum of the separate heats:*

$$Q_{total} = Q_{fusion} + Q_{heat\_1} + Q_{vap} + Q_{heat\_2}$$

$$Q_{total} = 6.66 \times 10^5 + 837,200 + 4.52 \times 10^6 + 167,440$$

$$Q_{total} = 6,190,640 \, J$$

Calculating heat of fusion and heat of vaporization

Heat of fusion and heat of vaporization use units of kJ/mol.

However, they may be expressed as J/g, where energy is obtained by multiplying the latent heat by the substance's mass rather than moles.

For example, how much kJ is required when 31.5 g of H_2O melts at its melting point of 0 °C? (Use molar mass for $H_2O = 18$ g/mol and molar heat of fusion for water $\Delta H_{fusion} = 6.02$ kJ/mol).

Use molar mass relationship:

$$mass \, (g) = moles \, (mol) \times molar \, mass \, (g/mol)$$

Rearrange to find the number of moles:

$$moles \, (g) = mass \, (g) \, / \, molar \, mass \, (g/mol)$$

$$moles = (31.5 \, g) \, / \, (18 \, g/mol)$$

$$moles = 1.75 \, mol$$

Use heat of fusion (ΔH_{fusion}):

$$q = n \times \Delta H_{fusion}$$

$$q = 1.75 \, mol \times 6.02 \, kJ/mol$$

$$q = 10.54 \, kJ$$

10.54 kJ heat (i.e., positive value) is required to melt 31.5 g of water.

Pressure *vs.* Volume Phase Diagrams

PV graphs

Pressure vs. volume diagram (*PV* graph) describes system pressure and volume changes.

PV diagrams show the thermodynamic process by graphing pressure against volume (P *vs.* V).

Work done by the system is equal to the area under the curve.

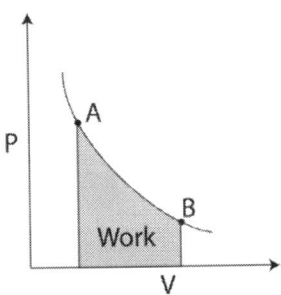

Area under the curve on the above PV diagram equals the work done

Cyclic pressure *vs.* volume reactions

In cyclic reactions, the work done is the *area enclosed* in the graph.

These reactions are often comprised of different processes, as shown below.

work is *done by* the gas when the volume increases and pressure decreases.

work is *done on* the gas when the volume decreases and pressure increases.

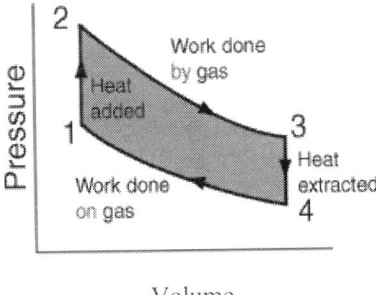

Points 4 to 1 show work done on the gas; points 2 to 3 work done by the gas

Calculating work done by gas

Work can be calculated for the gas phase of a gas and its container.

Gases may *expand* or *compress*; the *pressure* of the gas and *volume changes* determine work:

Figure illustrates:

> *expansion* is work done **by** the system ($\Delta E = q - w$)

> *compression* is work done **on** the system ($\Delta E = q + w$)

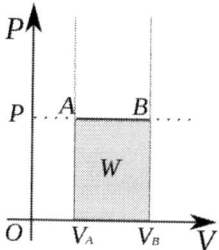

Shaded area (W) represents the work done by the gas

Positive and negative volume changes

Volume change (ΔV) is the final minus initial volume of gas:

$$\Delta V = V_{final} - V_{initial}$$

Volume is increased to *expand* a gas (ΔV is *positive*).

> Gas (of the system) must do work (w) on surroundings; thus, work must be *negative*.

Volume is decreased to compress a gas (ΔV is *negative*).

> Surroundings must do work (w) on the system; thus, work must be *positive*.

Thermodynamic work cycles

Processes may labeled *adiabatic, isothermal, isobaric,* and *isochoric.*

Adiabatic process has no transfer of heat ($Q = 0$); the change in energy equals the work.

Adiabatic processes are often confused with the isothermal process because it is sometimes assumed that no heat transfer means no change in temperature. However, this is not true—in isothermal instances, heat energy must be allowed into (or out of) the reaction to maintain a constant temperature.

Adiabatic processes, unlike an isothermal reaction, occur quickly and have a PV graph with the *same shape* but are *much steeper* than isothermal processes.

Adiabatic processes

Adiabatic process is when energy transfer within the system results in *work only*.

For example, a rapid expansion or contraction of gas is nearly adiabatic.

Adiabatic processes have *no heat* transfer.

Adiabatic processes *cannot decrease entropy*.

temperature is constant

$q = 0$

$\Delta E = w$

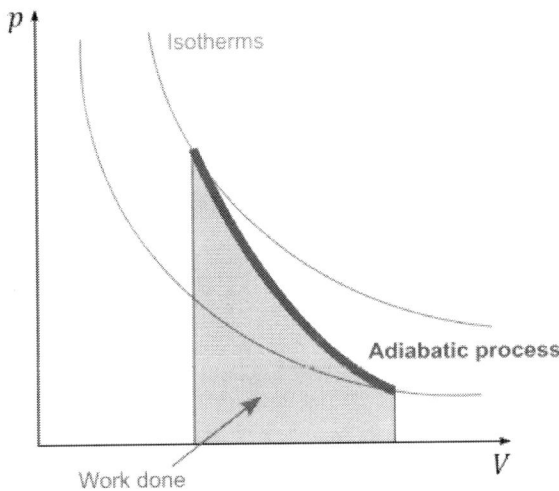

Adiabatic process PV diagrams of work with isothermal lines called isotherms

Adiabatic processes include a tire pump or a sharp expulsion of breath.

Isochoric processes

Isochoric process (or *constant-volume process, isovolumetric process*, and *isometric process*) is when the volume of the closed system remains *constant*; these include those inside a closed and rigid container or a constant volume thermometer.

volume is constant

$w = 0$

$\Delta E = q$

Thermodynamic process is the addition or removal of heat; the isolation of the contents of the container establishes the closed system, and the inability of the container to deform imposes the *constant-volume* condition.

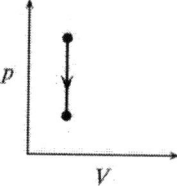

Isochoric process PV diagrams with constant volume show a vertical line

Isochoric processes include heating (or cooling) contents of a sealed, inelastic container.

Isothermal processes

Isothermal reaction keeps temperature constant ($\Delta T = 0$). The change in internal energy must stay the same, as an increase in thermal energy (heat) increases temperature ($\Delta U = 0$).

temperature is constant

$\Delta T = 0$

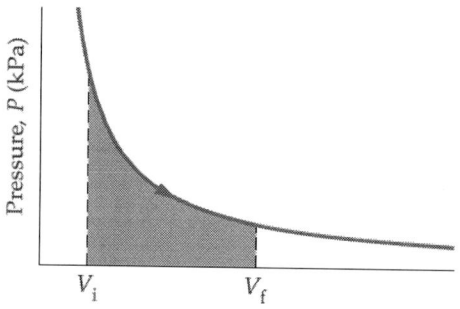

Isothermal process PV diagram at constant temperature

Isothermal process is when the temperature of the system remains constant; typically, when a system is in contact with an outside thermal reservoir. A change in the system allows the system to adjust temperature via the reservoir through heat exchange.

Isobaric processes

Isobaric process is when the pressure of the system remains constant.

Isobaric reactions keep the pressure constant; work equals the pressure times the volume change ($W = P\Delta V$).

pressure is constant

$\Delta P = 0$

$w = P\Delta V$

Examples of isobaric reactions include a piston in an engine or a flexible container open to Earth's atmosphere.

When pressure is constant, the PV graph of an isobaric process is a horizontal line.

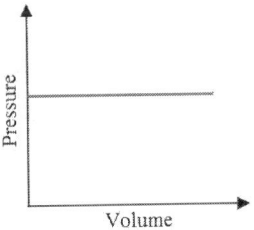

Isochoric process is an isovolumetric process because the volume is constant.

Since there is no movement, the work is zero, and the change in energy is due to the amount of heat added to the system.

$W = 0$

$\Delta E = Q$

Heat transferred to the system does work and *changes* the internal energy (U) of the system.

Thermodynamic work processes summary

Process	Constant	PV Diagram	Ideal Gas Law	First Law of Thermodynamics
Isobaric	Pressure	Horizontal line	$V \propto T$	$\Delta U = Q + W$
Isochoric	Volume	Vertical line	$P \propto T$	$\Delta U = Q$
Isothermal	Temperature	Curved line	$PV \propto T$	$\Delta U = 0$
Adiabatic	No heat exchanged	Curved line (jumps to different isotherm)	$PV = nRT$ (only "nR" are constant)	$\Delta U = W$

Thermodynamic work processes, with relationship to Ideal Gas Law (PV = nRT)

Practice Questions

1. If a stationary gas has a kinetic energy of 500 J at 25 °C, what is its kinetic energy at 50 °C?

A. 125 J

B. 450 J

C. 540 J

D. 1,120 J

E. 270 J

2. What is the purpose of the hollow walls in a closed hollow-walled container that effectively maintains the temperature inside?

A. Traps air trying to escape from the container, which minimizes convection

B. Act as an effective insulator, which minimizes convection

C. Acts as an effective insulator, which minimizes conduction

D. Provides an additional source of heat for the container

E. Reactions occur within the walls that maintain the temperature within the container

3. How much heat energy (in Joules) is required to heat 21.0 g of copper from 21.0 °C to 68.5 °C? (Use specific heat c of Cu = 0.382 J/g·°C)

A. 462 J

B. 188 J

C. 522 J

D. 662 J

E. 381 J

4. The species in the reaction $KClO_3\,(s) \rightarrow KCl\,(s) + 3/2O_2\,(g)$ have the values for standard enthalpies of formation at 25 °C. At constant physical states, assume that the values of $\Delta H°$ and $\Delta S°$ are constant throughout a broad temperature range. Which of the following conditions may apply to the reaction? (Use $KClO_3\,(s)$ with $\Delta H°_f = -391.2$ kJ mol^{-1} and KCl (s) with $\Delta H°_f = -436.8$ kJ mol^{-1})

A. Nonspontaneous at low temperatures, but spontaneous at high temperatures

B. Spontaneous at low temperatures but nonspontaneous at high temperatures

C. Nonspontaneous at all temperatures over a broad temperature range

D. Spontaneous at all temperatures over a broad temperature range

E. No conclusion can be drawn about spontaneity based on the information

5. Determine the value of $\Delta E°_{rxn}$ for this reaction, whereby the standard enthalpy of reaction ($\Delta H°_{rxn}$) is -311.5 kJ mol^{-1}:

$$C_2H_2\,(g) + 2\,H_2\,(g) \rightarrow C_2H_6\,(g)$$

A. -306.5 kJ mol^{-1}

B. -318.0 kJ mol^{-1}

C. $+346.0$ kJ mol^{-1}

D. $+306.5$ kJ mol^{-1}

E. $+466$ kJ mol^{-1}

6. The process of $H_2O\,(g) \rightarrow H_2O\,(l)$ is nonspontaneous under pressure of 760 torr and temperatures of 378 K because:

A. $\Delta H = T\Delta S$

B. $\Delta G < 0$

C. $\Delta H > 0$

D. $\Delta H < T\Delta S$

E. $\Delta H > T\Delta S$

7. A fuel cell contains hydrogen and oxygen gas that react explosively, and the energy converts water to steam, which drives a turbine to turn a generator that produces electricity. The fuel cell and the turbine represent which forms of energy, respectively?

A. Electrical and mechanical energy

B. Electrical and heat energy

C. Chemical and mechanical energy

D. Chemical and heat energy

E. Nuclear and mechanical energy

8. The thermodynamic systems with high stability tend to demonstrate:

A. maximum ΔH and maximum ΔS

B. maximum ΔH and minimum ΔS

C. minimum ΔH and maximum ΔS

D. minimum ΔH and minimum ΔS

E. none of the above

9. Which is true for the thermodynamic functions G, H and S in $\Delta G = \Delta H - T\Delta S$?

A. G refers to the universe, H to the surroundings, and S to the system

B. G, H, and S refer to the system

C. G and H refers to the surroundings and S to the system

D. G and H refer to the system and S to the surroundings

E. G and S refers to the system and H to the surroundings

10. Whether a reaction is endothermic or exothermic is determined by:

A. an energy balance between bond breaking and bond-forming, resulting in a net loss or gain of energy
B. the presence of a catalyst
C. the activation energy
D. the physical state of the reaction system
E. none of the above

11. What is the term for a reaction that proceeds by releasing heat energy?

A. Endothermic reaction
B. Isothermal reaction

C. Exothermic reaction
D. Nonspontaneous
E. Endergonic

12. The bond dissociation energy is:

    I. useful in estimating the enthalpy change in a reaction
    II. the energy required to break a bond between two gaseous atoms
    III. the energy released when a bond between two gaseous atoms is broken

A. I only
B. II only

C. I and II only
D. I and III only
E. I, II and III

13. Which of the following statements is true for the following reaction? (Use the change in enthalpy, $\Delta H° = -113.4$ kJ/mol and the change in entropy, $\Delta S° = -145.7$ J/K mol)

    $2 NO\ (g) + O_2\ (g) \rightarrow 2 NO_2\ (g)$

A. The reaction is at equilibrium at 25 °C under standard conditions
B. The reaction is spontaneous at only high temperatures
C. The reaction is spontaneous only at low temperatures
D. The reaction is spontaneous at all temperatures
E. $\Delta G°$ becomes more favorable as temperature increases

14. Which law explains the observation that the amount of heat transfer accompanying a change in one direction is equal in magnitude but opposite in sign to the amount of heat transfer in the opposite direction?

A. Law of Conservation of Mass

B. Law of Definite Proportions

C. Avogadro's Law

D. Boyle's Law

E. Law of Conservation of Energy

15. Calculate the value of $\Delta H°$ of reaction using provided bond energies.

$H_2C = CH_2\,(g) + H_2\,(g) \rightarrow H_3C–CH_3\,(g)$

C–C: 348 KJ

C=C: 612 kJ

H–H: 436 kJ

C≡C: 960 kJ

C–H: 412 kJ

A. −348 kJ

B. +134 kJ

C. −546 kJ

D. −124 kJ

E. −238 kJ

Detailed Explanations

1. C is correct.

Temperature is a measure of the average kinetic energy of the molecules.

Kinetic energy is proportional to temperature.

In most thermodynamic equations, temperatures are expressed in Kelvin, so convert Celsius to Kelvin:

$$25 \text{ °C} + 273.15 = 298.15 \text{ K}$$

$$50 \text{ °C} + 273.15 = 323.15 \text{ K}$$

Calculate the kinetic energy using simple proportions:

$$KE = (323.15 \text{ K} / 298.15 \text{ K}) \times 500 \text{ J}$$

$$KE = 540 \text{ J}$$

2. C is correct.

An *insulator* reduces conduction (i.e., thermal energy transfer through matter).

Air and vacuum are excellent insulators. Storm windows, which have air wedged between two glass panes, work by utilizing this conduction principle.

3. E is correct.

Heat capacity is the amount of heat needed to increase the whole sample's temperature by 1 °C.

Specific heat is the energy required to increase *1 gram* of the sample by 1 °C.

Heat = mass × specific heat × change in temperature:

$$q = m \times c \times \Delta T$$

$$q = 21.0 \text{ g} \times 0.382 \text{ J/g·°C} \times (68.5 \text{ °C} - 21.0 \text{ °C})$$

$$q = 21.0 \text{ g} \times 0.382 \text{ J/g·°C} \times (47.5 \text{ °C})$$

$$q = 381 \text{ J}$$

4. D is correct.

Calculate the value of:

$$\Delta H_f = \Delta H_{f\,product} - \Delta H_{f\,reactant}$$

$$\Delta H_f = -436.8 \text{ kJ mol}^{-1} - (-391.2 \text{ kJ mol}^{-1})$$

$$\Delta H_f = -45.6 \text{ kJ mol}^{-1}$$

To determine spontaneity, calculate the Gibbs free energy:

$$\Delta G = \Delta H^\circ - T\Delta S$$

The reaction is spontaneous if ΔG is negative:

$$\Delta G = -45.6 \text{ kJ} - T\Delta S$$

Typical ΔS values are around 100–200 J.

If the ΔH value is –45.6 kJ (or –45,600 J), the value of ΔG would be negative unless $T\Delta S$ is less than –45,600.

Therefore, the reaction would be spontaneous over a broad range of temperatures.

5. A is correct.

Relationship between enthalpy (ΔH) and internal energy (ΔE):

Enthalpy = Internal energy + work (for gases, work = PV)

$$\Delta H = \Delta E + \Delta(PV)$$

Solving for ΔE:

$$\Delta E = \Delta H - \Delta(PV)$$

According to ideal gas law:

$$PV = nRT$$

Substitute ideal gas law to the previous equation:

$$\Delta E = \Delta H - \Delta(nRT)$$

R and T are constant, which leaves Δn as the variable.

The reaction is:

$$C_2H_2 (g) + 2 H_2 (g) \rightarrow C_2H_6 (g).$$

There are three gas molecules on the left and one on the right, which means $\Delta n = 1 - 3 = -2$.

continued...

In this problem, the temperature is not provided.

However, the presence of degree symbols ($\Delta G°$, $\Delta H°$, $\Delta S°$) indicates that those are standard values, measured at 25 °C or 298.15 K.

Double-check the units before performing calculations; ΔH is in kilojoules, while the gas constant (R) is 8.314 J/mol K. Convert ΔH to Joules before calculating.

Use the Δn and T values to calculate ΔE:

$$\Delta E = \Delta H - \Delta n RT$$

$$\Delta E = -311{,}500 \text{ J/mol} - (-2 \times 8.314 \text{ J/mol K} \times 298.15 \text{ K})$$

$$\Delta E = -306{,}542 \text{ J}$$

$$\Delta E \approx -306.5 \text{ kJ}$$

6. E is correct.

The spontaneity of a reaction is determined by evaluating the Gibbs free energy, or ΔG.

$$\Delta G = \Delta H - T\Delta S$$

A reaction is spontaneous if $\Delta G < 0$ (or ΔG is negative).

For ΔG to be negative, ΔH must be less than $T\Delta S$.

A reaction is nonspontaneous if $\Delta G > 0$ (or ΔG is positive and ΔH is greater than $T\Delta S$).

7. C is correct.

Potential Energy	Kinetic Energy
Stored energy and the energy of position (gravitational)	**Energy of motion: motion of waves, electrons, atoms, molecules and substances**
Chemical Energy Chemical energy is the energy stored in the bonds of atoms and molecules. Biomass, petroleum, natural gas, propane, and coal are examples.	**Radiant Energy** Radiant energy is electromagnetic energy that travels in transverse waves. It includes visible light, x-rays, gamma rays, and radio waves.
Nuclear Energy Nuclear energy is the energy stored in the nucleus of an atom, and it is the energy that holds the nucleus together. The nucleus of the uranium atom is an example.	**Thermal Energy** Thermal energy (or heat) is the internal energy in substances, vibration, and the movement of atoms and molecules. Geothermal energy is an example.

Stored Mechanical Energy	Motion
Stored mechanical energy is energy stored in objects by applying force. Compressed springs and stretched rubber bands are examples.	The movement of objects or substances from one place to another. Wind and hydropower are examples.
Gravitational Energy	**Sound**
Gravitational Energy is the energy of place or position. Water in a reservoir behind a hydropower dam is an example of gravitational potential energy. When the water is released to spin the turbines, it becomes kinetic energy.	Sound is energy movement through substances in longitudinal (compression/rarefaction) waves.
	Electrical Energy
	Electrical energy is the movement of electrons. Lightning and electricity are examples.

8. C is correct.

Gibbs free energy:

$$\Delta G = \Delta H - T\Delta S$$

Stable molecules are *non-spontaneous* because they have a negative (or relatively low) ΔG.

ΔG is most negative (most stable) when ΔH is small, and ΔS is large.

9. B is correct.

All thermodynamic functions in $\Delta G = \Delta H - T\Delta S$ refer to the system.

10. A is correct.

ΔH refers to enthalpy (or heat).

> *Endothermic* reactions have heat as a reactant.

> *Exothermic* reactions have heat as a product.

Endothermic reactions absorb energy to break strong bonds to form a less stable state (i.e., positive enthalpy).

Exothermic reactions release energy while forming stronger bonds and a more stable state (i.e., negative enthalpy).

continued…

The reaction is *nonspontaneous* (i.e., endergonic) when products are *less stable* than reactants and ΔG is *positive.*

The reaction is *spontaneous* (i.e., exergonic) when products are *more stable* than reactants and ΔG is *negative.*

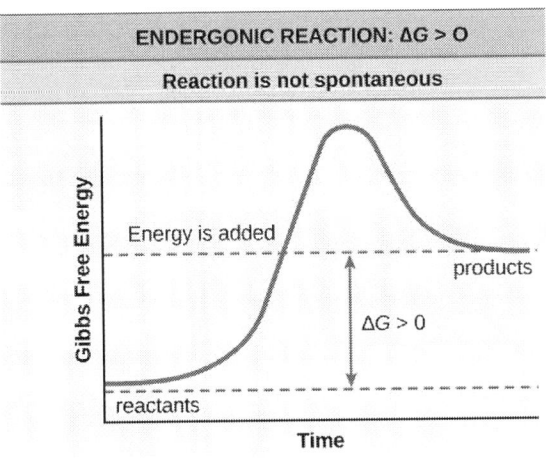

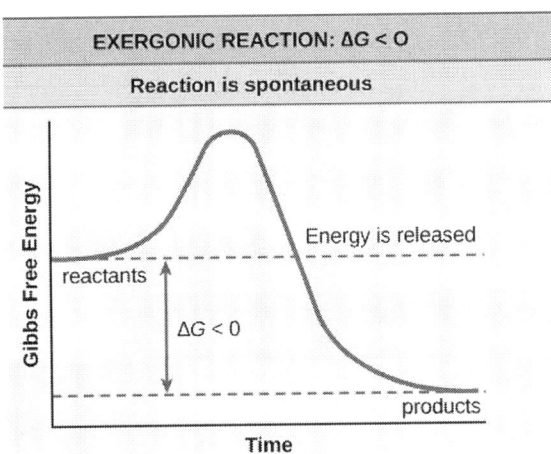

Exothermic reactions *release heat* and cause the surroundings' temperature to rise (i.e., a net loss of energy).

Endothermic reactions *absorb heat* and cool the surroundings (i.e., a net gain of energy).

11. C is correct.

ΔH refers to enthalpy (or heat).

Endothermic reactions have *heat as a reactant*.

Exothermic reactions have *heat as a product*.

Exothermic reactions *release heat* and cause the surroundings' temperature to rise (i.e., a net loss of energy).

Exothermic reactions release energy while forming stronger bonds forming a more stable state (i.e., negative enthalpy).

Endothermic processes *absorb heat* and cool the surroundings (i.e., a net gain of energy).

Endothermic reactions absorb energy to break strong bonds to form a less stable state (i.e., positive enthalpy).

The reaction is nonspontaneous (i.e., endergonic) if products are less stable than reactants and ΔG is positive.

The reaction is spontaneous (i.e., exergonic) if products are more stable than reactants and ΔG is negative.

12. C is correct.

Bond dissociation energy is the energy required to break a bond between two gaseous items and is useful in estimating the enthalpy change in a reaction.

13. C is correct.

To predict spontaneity of reaction, use the Gibbs free energy equation:

$$\Delta G = \Delta H° - T\Delta S$$

The reaction is spontaneous if ΔG is negative.

Substitute the given values to the equation:

$$\Delta G = -113.4 \text{ kJ/mol} - [T \times (-145.7 \text{ J/K mol})]$$

$$\Delta G = -113.4 \text{ kJ/mol} + (T \times 145.7 \text{ J/K mol})$$

Important: note that the units are not identical; $\Delta H°$ is in kJ, and $\Delta S°$ is in J.

Convert kJ to J (1 kJ = 1,000 J):

$$\Delta G = -113,400 \text{ J/mol} + (T \times 145.7 \text{ J/K mol})$$

It can be predicted that the value of ΔG would be negative if the value of T is small.

If T increases, ΔG approaches a positive value, and the reaction is nonspontaneous.

14. E is correct.

The *Law of Conservation of Energy* states that an isolated system's total energy (e.g., potential or kinetic) remains constant and conserved.

Energy can be neither created nor destroyed but is transformed from one form to another.

For instance, chemical energy is converted to kinetic energy in a firecracker's explosion.

The *Law of Conservation of Mass* states that for any system closed to transfers of matter and energy, the mass of the system must remain constant over time.

The *Law of Definite Proportions* (Proust's Law) states that a chemical compound contains the same proportion of elements by mass.

The law of definite proportions forms the basis of stoichiometry (i.e., from the known amounts of separate reactants, the amount of the product can be calculated).

Avogadro's Law is an experimental gas law relating the volume of a gas to the amount of substance (i.e., moles) of gas present.

Avogadro's Law states that equal volumes of gases have the same number of molecules at the same temperature and pressure.

Boyle's Law is an experimental gas law that describes how the pressure of a gas tends to increase as the volume of a gas decreases. It states that the absolute pressure exerted by a given mass of an ideal gas is inversely proportional to the volume it occupies if the temperature and amount of gas remain unchanged within a closed system.

15. D is correct.

Reactant bonds are broken in a chemical reaction, and new bonds form to create products.

Therefore, in bond dissociation problems,

ΔH reaction = sum of bond energy in reactants – sum of bond energy in products

For $H_2C=CH_2 + H_2 \rightarrow CH_3–CH_3$:

$\Delta H_{reaction}$ = sum of bond energy in reactants – sum of bond energy in products

$\Delta H_{reaction} = [(C=C) + 4(C–H) + (H–H)] – [(C–C) + 6(C–H)]$

$\Delta H_{reaction} = [612 \text{ kJ} + (4 \times 412 \text{ kJ}) + 436 \text{ kJ}] – [348 \text{ kJ} + (6 \times 412 \text{ kJ})]$

$\Delta H_{reaction} = –124 \text{ kJ}$

Remember that this is the opposite of ΔH_f problems,

where: $\Delta H_{reaction}$ = (sum of ΔH_f products) – (sum of ΔH_f reactants)

Notes for active learning

CHAPTER 9

Electrochemistry

Electrochemistry is **the study of chemical processes that cause electrons to move**. This movement of electrons is called electricity, which can be generated by movements of electrons from one element to another in a reaction known as an oxidation-reduction ("redox") reaction.

Electrochemistry studies the interconversion of electrical and chemical energy. It utilizes spontaneous oxidation-reduction reactions for electrical energy to drive nonspontaneous reactions. Electrochemistry involves an oxidation-reduction process. Some electrochemistry applications include batteries, corrosion control, metallurgy, and electrolysis.

- Electrochemical Cells
- Electrolysis
- Galvanic (Voltaic) Cells
- Reduction Potential
- Electrical Potential Difference
- Electron Flow
- Batteries
- Practice Questions & Detailed Explanations

Electrochemical Cells

Galvanic and electrolytic cells

The two types of electrochemical cells are *galvanic* (or Voltaic) and *electrolytic*.

Galvanic (or Voltaic) cells convert chemical energy to electrical energy by *spontaneous* redox reactions. A galvanic cell consists of two different metals (electrodes) connected through a conducting solution (an electrolyte) and externally, completing a circuit. A galvanic (voltaic) cell has a resistor or voltmeter in place of the battery.

Galvanic (Voltaic) cells have a positive cell potential; thus, no electrical input is required.

Electrolytic cells use electrical energy to drive a non-spontaneous redox reaction, often used to decompose chemical compounds via electrolysis. Electrolytic cells convert electrical energy to chemical energy. Electrolytic cells involve non-spontaneous reactions and thus require an external electron source (e.g., battery).

For example, the electrolysis of sodium chloride (NaCl, common salt) forms sodium Na^+ metal and chlorine Cl^- gas; an electric current supplies the energy required to make the reaction proceed.

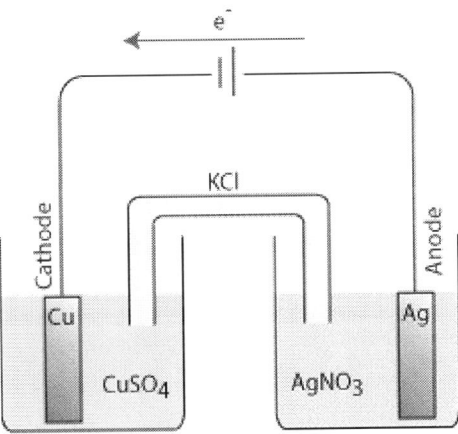

Electrolytic cells use electrical energy to drive non-spontaneous redox reactions

Anode and cathode reactions

Oxidation occurs at the *anode* for an electrochemical cell, and reduction occurs at the *cathode*.

To designate anode and cathode, write a balanced net ionic equation for the spontaneous cell reaction.

- The oxidizing agent (more positive or less negative reduction potential $E°$) is the cathode.

- The oxidizing agent (less positive or more negative reduction potential E°) is the anode.

- Oxidation occurs in the anode half-cell and reduction in the cathode half-cell.

- The anode is negative (-), and the cathode is positive (+).

Mnemonic: **An Ox** = **AN**ode **Ox**idation   **Red Cat** = **RED**uction **CAT**hode

When drawing a galvanic cell, the anode is on the left by convention, with cathode on the right.

Use the mnemonic **ABC** as convention (**A**node / **B**ridge / **C**athode).

Galvanic (or Voltaic) cells	Electrolytic cells
Cell converts chemical energy into electrical energy.	Cell converts electrical energy into chemical energy.
The redox reaction is *spontaneous* and produces electrical energy.	The redox reaction is *not spontaneous*, and electrical energy must be supplied to initiate the reaction.
The electrons are supplied by the species getting oxidized. The electrons migrate from the anode to the cathode in the external circuit.	An *external battery* supplies the electrons which enter through the cathode and come out through the anode
The two half-cells occur in different chambers and are connected through a salt bridge.	Both electrodes are placed in the same vessel in the solution.
The anode is *negative*, and the cathode is the positive electrode.	The anode is *positive*, and the cathode is the negative electrode.
Oxidation occurs at the anode, while reduction occurs at the cathode.	Oxidation occurs at the anode, while reduction occurs at the cathode.

Comparison between galvanic and electrolytic cells

Oxidation and reduction at electrodes

Oxidation occurs at the anode for electrolytic cells, while reduction occurs at the cathode.

Electrons originate from the anode because oxidation (loss of electrons) occurs.

$$M \rightarrow M^+ + e^-$$

Electrons travel to the cathode, where they join cations on the cathode surface, and reduction (gain of electrons) occurs.

$$M^+ + e^- \rightarrow M$$

Mnemonic: OIL RIG: **O**xidation **I**s **L**osing e$^-$, **R**eduction **I**s **G**aining e$^-$.

> *Oxidation* is an *increase in charge* (more positive) due to the *loss* of e$^-$.

> *Reduction* is a *decrease in charge* (more negative) due to the *gain* of e$^-$.

Concentration cells

A *concentration cell* is an electrochemical cell in which the half-cells are of the same type but with different electrolyte concentrations.

The following cell notations are examples of concentration cells:

$$Cu|Cu^{2+} (aq, 0.0010 \text{ M})\|Cu^{2+} (aq, 1.0 \text{ M})|Cu$$

$$Ag|Ag^+ (aq, 0.0010 \text{ M}) \|Ag^+ (aq, 0.10 \text{ M})|Ag$$

In concentration cells, the half-cell with the lower electrolyte concentration serves as an anode half-cell, and one with the higher electrolyte concentration is the cathode half-cell.

At the anode half-cell, an oxidation reaction occurs to increase the electrolyte concentration, and at the cathode half-cell, a reduction reaction occurs to decrease its electrolyte concentration.

Notes for active learning

Electrolysis

Forced current

Electrolysis forces a current through a cell to produce a chemical change, negative cell potential. The electrolysis process extracts metals from molten salts, and a sustained current is applied to the molten salt.

Electrolysis requires potential (voltage) input.

The potential (voltage) input + the cell potential must be > 0 for reactions to occur. Electrolytic cells have negative cell potentials, so a potential input greater than the cell potential's magnitude is needed for electrolysis.

A battery is a contained unit that produces electricity. Commercial batteries are galvanic cells using solids or pastes as reactants to maximize the electrical output per unit mass. A *fuel cell* is a galvanic cell requiring constant reactants to generate electricity.

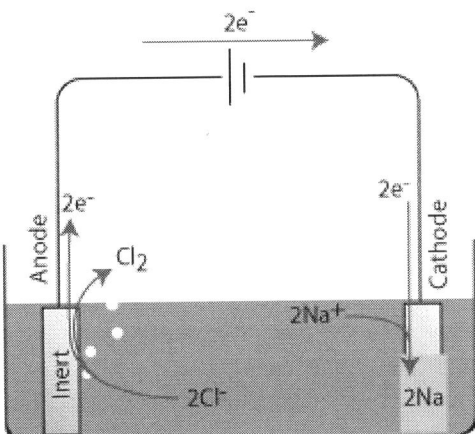

Salt electrolysis charge flow with arrows as how the battery forces electron flow

For example, the following spontaneous reaction occurs in a galvanic (voltaic) cell consisting of $Zn|Zn^{2+}$ (*aq*) and $Cu|Cu^{2+}$ (*aq*) half-cells:

$$Zn\ (s) + Cu^{2+}\ (aq)\ \rightarrow\ Zn^{2+}\ (aq) + Cu\ (s); \qquad E°_{cell} = 1.10\ V$$

In an electrolytic cell, the reverse reaction occurs if a voltage greater than the E° cell is applied, which causes the reverse reaction, and the excess voltage is *overpotential* or *overvoltage*.

Electrolytic cell reactions

Three types of reactions occur in electrolytic cells if sufficient voltage is applied:

- Solute ions or molecules may be oxidized or reduced.

- The solvent can be oxidized or reduced.

- Metal electrode that forms the anode can be oxidized.

Which reactions occur depends on the thermodynamic and kinetic properties of each reaction.

Electric current

An *electric current* flowing through a cell is measured in *ampere* (A), equal to the charge in *coulomb* (C) flowing per second.

1 ampere = 1 coulomb / 1 second

Ampere = Coulomb/second (1 A = 1 C/s)

Coulomb = Ampere × time in sec (1 C = 1 A·s) = total charge

Joule = Coulomb × Volt (1 J = 1 C.V)

Electrolytes are ions that conduct electricity by motion. Without electrolytes, electricity cannot propagate.

Electrolysis of water

The *decomposition of water* is a *nonspontaneous* process with an amount of energy of about 400 kJ/mol of water. The decomposition of water can be effectuated by electrolysis, where O_2 forms at the *anode* and H_2 at the *cathode*. Assumes an electrolytic cell with:

$[H^+] = [OH^-] = 1$ M and $P_{H2} = P_{O2} = 1$ atm

For the electrolysis of pure water, where $[H^+] = [OH^-] = 10^{-7}$ M, the potential for the overall process is -1.23 V.

Anode reaction:	$2\ H_2O \rightarrow O_2 + 4\ H^+ (aq) + 4\ e^-$	$E° = -1.23$ V
Cathode reaction:	$4\ H_2O + 4\ e^- \rightarrow 2\ H_2 + 4\ OH^- (aq)$	$E° = -0.83$ V

Net reaction:	$6\ H_2O \rightarrow 2\ H_2 + O_2 + 4\ (H^+ + OH^-)$	$E° = -2.06$ V
	$2\ H_2O \rightarrow 2\ H_2 + O_2$	

A voltage of 1.23 V is enough for the electrolysis of water. However, an additional voltage (overpotential or overvoltage) of about 1 V is needed to force the electrolysis.

Electrolysis of solution containing mixtures of ions

The metal with the smallest potential is the first to be formed.

For example, consider a solution of Cu^{2+}, Zn^{2+} and Ag^+ electrolyzed with a current with sufficient voltage to reduce the cations.

The reduction potentials of these elements are as follows:

$$Ag^+(aq) + e^- \rightarrow Ag\ (s) \qquad E° = 0.80\ V$$

$$Cu^{2+}(aq) + 2\ e^- \rightarrow Cu\ (s) \qquad E° = 0.34\ V$$

$$Zn^{2+}(aq) + 2\ e^- \rightarrow Zn\ (s) \qquad E° = -0.76\ V$$

Since the reduction of Ag^+ to Ag has the most positive potential, Ag deposits at the cathode before other metals, followed by Cu and Zn, respectively. If the voltage is controlled, starting with the lowest voltage, separating the three metals using electrolysis is possible.

For example, how many grams of Ag is deposited in the cathode of an electrolytic cell if a current of 3.50 A is applied to a solution of $AgNO_3$ for 12 minutes? (Use the molecular mass of Ag = 107.86 g/mol and the conversion of 1 mol $e^- = 9.65 \times 10^4$ C)

Balanced equation:

$$Ag^+ + e^- \rightarrow Ag\ (s)$$

Formula to calculate deposit mass:

mass of deposit = (atomic mass × current × time) / 96,500 C

mass of deposit = [107.86 g × 3.50 A × (12 min × 60 s/min)] / 96,500 C

mass of deposit = 2.82 g

Stoichiometry of electrolysis

The number of substances formed at the anode or cathode is calculated from the current's magnitude (amperes) and time (sec).

For example, if a current of 1.50 A flows through an aqueous solution of $CuSO_4$ for 25.0 minutes, the amount of *charge passing* through the solution is:

charge = (1.50 C/s)·(25.0 minutes)·(60 sec/min)

charge = 2,250 C

The number of *moles of electrons* passing through the solution:

moles of electrons = (2,250 C) × (1 mol e^-)

moles of electrons = 0.0233 mol

Reduction of Cu^{2+} requires 2 mol e^- per mole of Cu:

$$Cu^{2+} + 2\,e^- \rightarrow Cu$$

Cu formed at cathode:

$$(0.0233 \text{ mol } e^-) \times \frac{(1 \text{ mol Cu})}{(2 \text{ mol } e^-)} \times \frac{(63.55 \text{ g Cu})}{(1 \text{ mol Cu})}$$

Cu formed at cathode = 0.741 g Cu

If the process uses a cell with 3.0 V:

energy consumed = 3.0 V × 2250 C

energy consumed = 6,750 C.V

6,750 C.V = 6,800 J

6,800 J = 6.8 kJ

Faraday's Law relates the number of elements

In 1833, Michael Faraday (1791-1867) proposed quantitative laws to express magnitudes of electrolytic effects. The mass of elements deposited at an electrode is directly proportional to the charge (A·sec or Coulombs).

Faraday constant is the electric charge of 1 mole of electrons, a conversion factor between moles and coulombs. Faraday constant (F):

$$F = [6.02 \times 10^{23} \text{ electrons/mol } e^-] / [6.24 \times 10^{18} \text{ } e^-/C]$$

$$F = 96,485 \text{ C/mol } e^-$$

1 mol electrons is 6.022×10^{23} electrons, and a coulomb is the (negative) charge of 6.241×10^{18} electrons. The Faraday constant is the quotient of these.

For electrolysis, divide the charge in coulombs by the Faraday constant to calculate the number of moles that underwent electrolysis.

Current I = coulombs of charge per second

$$I = q / t$$

Faraday's constant:

$$F = \text{coulombs of charge per mol of electron}$$

$$F = \text{total charge / the total moles of electrons}$$

$$F = q / n$$

$$q = It \text{ and } q = nF$$

Thus:

$$It = nF$$

Current × time = moles of e^- × Faraday's constant.

Using this equation, solve for moles of electrons (n). Using the half-equation stoichiometry, determine how many moles of the element are made for every e^- transferred.

For example, 1 mol of Cu is deposited for every 2 moles of electrons for the half-reaction, using Faraday's constant = 96,485 C/mo.

$$Cu^{2+} + 2\ e^- \rightarrow Cu$$

For example, if 1 amp of current passes a cathode for 10 minutes, how much Zn (s) forms in the following reaction? (Use the molecular mass of Zn = 65 g/mole)

$$Zn^{2+} + 2\ e^- \rightarrow Zn\ (s)$$

Calculate the mass of metal deposited at the cathode:

Step 1: *Calculate total charge using current and time*

$$Q = \text{current} \times \text{time}$$

$$Q = 1\ A \times (10\ \text{minutes} \times 60\ \text{s/minute})$$

$$Q = 600\ A \cdot s = 600\ C$$

Step 2: *Calculate moles of electron that has the same amount of charge*

$$\text{moles } e^- = Q\ /\ 96{,}500\ C/mol$$

$$\text{moles } e^- = 600\ C\ /\ 96{,}500\ C/mol$$

$$\text{moles } e^- = 6.22 \times 10^{-3}\ mol$$

Step 3: *Calculate moles of metal deposit*

Half-reaction of zinc ion reduction:

$$Zn^{2+}\ (aq) + 2\ e^- \rightarrow Zn\ (s)$$

$$\text{moles of Zn} = (\text{coefficient Zn}\ /\ \text{coefficient } e^-) \times \text{moles } e^-$$

$$\text{moles of Zn} = (\tfrac{1}{2}) \times 6.22 \times 10^{-3}\ mol$$

$$\text{moles of Zn} = 3.11 \times 10^{-3}\ mol$$

Step 4: ***Calculate mass of metal deposit***

mass Zn = moles Zn × molecular mass of Zn

mass Zn = 3.11×10^{-3} mol × (65 g/mol)

mass Zn = 0.20 g

For example, how long would it take to deposit 4.00 grams of Cu from a $CuSO_4$ solution if a current of 2.5 A is applied? (Use molecular mass of Cu = 63.55 g/mol and conversion of 1 mol e^- = 9.65×10^4 C and 1 C = A·s)

Balanced equation:

$Cu^{2+} + 2\,e^- \rightarrow Cu\,(s)$

Formula to calculate deposit mass:

mass of deposit = atomic mass × moles of electron

Since Cu has 2 electrons per atom, multiply the moles by 2:

mass of deposit = atomic mass × (2 × moles of electron)

mass of deposit = atomic mass × (2 × current × time) / 96,500 C

4.00 g = 63.55 g × (2 × 2.50 A × time) / 96,500 C

time = 4,880 s

time = (4,880 s × 1 min/60s × 1 hr/60 min)

time = 1.36 hr

Galvanic (Voltaic) Cells

Spontaneous oxidation-reduction reactions

A *galvanic cell* is a typical application of electrochemistry as batteries. A galvanic cell (i.e., *voltaic cell*) uses a spontaneous oxidation-reduction reaction to produce an electric current. Galvanic cells produce voltage and were invented between 1780-1799 by Italian physicists Luigi Galvani (1737-1798) and Alessandro Volta (1745-1827).

In galvanic cells, species with more positive reduction potential serve as a cathode, and those with less positive or more negative reduction potential serve as the anode half-cell.

Galvanic cells:

- Oxidation occurs at the anode.

- Reduction occurs at the cathode.

- Salt bridges or porous disks allow ions to flow without extensive mixing of the solutions.

Salt bridge contains a strong electrolyte held in a gel-like matrix.

Porous disk contains tiny passages for the hindered flow of ions.

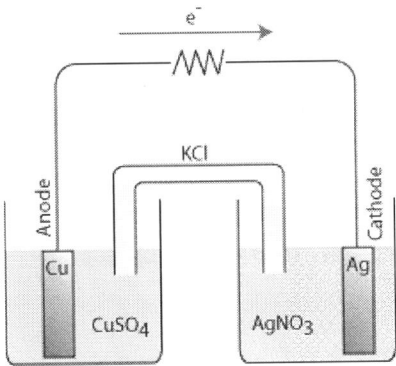

Galvanic (Voltaic) Cell

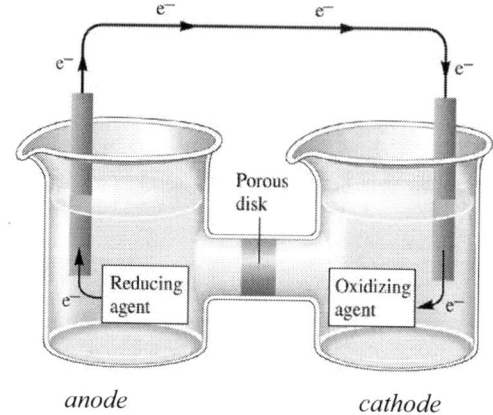

anode   *cathode*

Oxidation-reduction reaction examples

For example, spontaneous oxidation-reduction reactions:

$$Zn\ (s) + CuSO_4\ (aq) \rightarrow Cu\ (s) + ZnSO_4\ (aq)$$

$$Zn\ (s) + Cu\cancel{SO_4}\ (aq) \rightarrow Cu\ (s) + Zn\cancel{SO_4}\ (aq)$$

$$Zn\ (s) + Cu^{2+}\ (aq) \rightarrow Cu\ (s) + Zn^{2+}\ (aq)$$

$$Cr\ (s) + 3\ AgNO_3\ (aq) \rightarrow Cr(NO_3)_3\ (aq) + 3\ Ag\ (s)$$

$$Cr\ (s) + 3\ Ag\cancel{NO_3}\ (aq) \rightarrow Cr\cancel{(NO_3)_3}\ (aq) + 3\ Ag\ (s)$$

$$Cr\ (s) + 3\ Ag^+\ (aq) \rightarrow Cr^{3+}\ (aq) + 3\ Ag\ (s)$$

$$MnO_4^-\ (aq) + 5\ Fe^{2+}(aq) + 8\ H^+(aq) \rightarrow Mn^{2+}(aq) + 5\ Fe^{3+}(aq) + 4\ H_2O\ (l)$$

$$Zn\ (s) + 2\ MnO_2\ (s) \rightarrow ZnO\ (s) + Mn_2O_3\ (s)$$

Redox half-reactions

Oxidation half-reaction is the species losing electrons (increases charge).

Oxidation example,

$$Cu \rightarrow Cu^{2+} + 2\ e^-$$

Reduction half-reaction is the species gaining electrons (decreases charge).

Reduction example,

$$2\ Ag^+ + 2\ e^- \rightarrow 2\ Ag$$

Ten steps for balancing redox half-reactions

1. Split the equation into two half-reactions.

2. Balance atoms other than O or H.

3. Balance O using H_2O. Add water as needed to balance oxygens on the side deficient in O.

4. If in an acidic solution, balance hydrogens using H^+.

    Add H^+ ions as needed to the side deficient in H.

5. In basic solution, add OH^- ions equal to the number of H^+ ions to each side of the chemical equation.

6. The mass is balanced. To balance charge, determine the charge on each side of the reaction and add needed electrons to the positive side, so charge is the same.

7. Repeat steps 1 through 4 for each half-reaction.

8. If electrons lost do not equal the number gained, multiply each half-reaction by the necessary factor.

9. Sum half-reactions to obtain the balanced net ionic reaction. Inspect H^+ ions and H_2O molecules to cancel.

10. Check the final equation for mass and charge balance.

Important: If expressing H^+ as H_3O^+, change the number of H^+ into the same number of H_3O^+. Add that same number of H_2O to the opposite side of the equation.

Redox reactions can be separated into two half-reactions:

Example 1:

$\quad\quad$ *Oxidation* half-reaction: $\quad Zn\ (s) \;\rightarrow\; Zn^{2+}\ (aq) + 2\ e^-$

$\quad\quad$ *Reduction* half-reaction: $\quad Cu^{2+}\ (aq) + 2\ e^- \;\rightarrow\; Cu\ (s)$

$\quad\quad$ Overall reaction: $Zn\ (s) + Cu^{2+}\ (aq) \rightarrow\; Cu\ (s) + Zn^{2+}\ (aq)$

Example 2:

Oxidation half-reaction:

$$5 \text{ Fe}^{2+}(aq) \rightarrow 5 \text{ Fe}^{3+}(aq) + 5 \text{ e}^-$$

Reduction half-reaction:

$$\text{MnO}_4^-(aq) + 8 \text{ H}^+(aq) + 5 \text{ e}^- \rightarrow \text{Mn}^{2+}(aq) + 4 \text{ H}_2\text{O }(l)$$

Overall reaction:

$$\text{MnO}_4^-(aq) + 5 \text{ Fe}^{2+}(aq) + 8 \text{ H}^+(aq) \rightarrow \text{Mn}^{2+}(aq) + 5 \text{ Fe}^{3+}(aq) + 4 \text{ H}_2\text{O }(l)$$

Galvanic cell half-reactions

Redox reactions are split into two half-reactions in *galvanic cells*, each in two separate compartments as *half-cells*. The chemical energy drives electrons through the external circuit connecting the two half-cells, thus producing an electric current.

For example, consider the reaction:

$$\text{Zn }(s) + \text{CuSO}_4(aq) \rightarrow \text{Cu }(s) + \text{ZnSO}_4(aq)$$

If zinc metal is placed in the $CuSO_4$ solution, an exothermic reaction occurs, producing heat. In a galvanic cell, zinc metal is placed in a $ZnSO_4$ solution in one container, a copper metal in $CuSO_4$ solution in another container. A wire connects two metals to complete the circuit; the two solutions are connected by a salt bridge containing strong electrolytes such as KCl (*aq*) or K_2SO_4 (*aq*) that allow ions to flow between the two half-cells. The reactants – zinc metal and copper ions – are not allowed direct contact.

Due to the potential difference between the two half-cells, electrons are forced to flow from the zinc electrode (the anode) to the copper electrode (the cathode).

At the interface between the zinc electrode and $ZnSO_4$ solution, Zn atoms are oxidized to Zn^{2+} ions; at the interface of the copper electrode and the $CuSO_4$ solution, Cu^{2+} ions combine with incoming electrons reduced to Cu atoms.

These oxidation and reduction processes as half-reaction equations:

Anode-reaction (oxidation):

$$\text{Zn }(s) \rightarrow \text{Zn}^{2+}(aq) + 2 \text{ e}^-$$

Cathode-reaction (reduction):

$$\text{Cu}^{2+}(aq) + 2 \text{ e}^- \rightarrow \text{Cu }(s)$$

Cell notation

The two half-cells are represented by notations $Zn|Zn^{2+}$ and $Cu^{2+}|Cu$.

In a galvanic cell, the more readily oxidized metal serves as an anode and the other as a cathode.

Determine the species' tendency to oxidize from the standard reduction potentials table.

Zinc is more easily oxidized than copper, serving as the anode, while copper forms the cathode.

> Oxidation half-reaction occurs at the anode
>
> Reduction half-reaction at the cathode.

This galvanic cell is represented using the following cell notation:

$$Zn|ZnSO_4\,(aq)||CuSO_4\,(aq)|Cu$$

Cell notation has the anode on the left side, with the cathode on the right.

At the anode half-cell, Zn^{2+} ions are continuously formed, creating an excess of positive ions.

Cu2+ ions are continuously reduced to Cu at the cathode half-cell, causing a decrease in cation concentration.

To maintain electrically neutral solutions in half-cells, anions flow into the anode half-cell, and cations flow into the cathode half-cell. Thus, an electric current is the flow of charged particles. Electrons flow from anode to cathode through the wire; cations and anions flow in the opposite direction through the salt bridge.

Notes for active learning

Reduction Potential

Standard reduction potential

The *standard half-cell potential* (or *reduction potential*) is determined by connecting the half-cell (under standard conditions) to the *standard hydrogen electrode* (SHE) as a reference half-cell.

This reference half-cell consists of a Pt-electrode in a solution containing 1 M H^+, into which H_2 gas is purged at a constant pressure of 1 atm.

Since the reference half-cell is assigned zero potential, the cell potential measured under these conditions is the substance's standard half-cell potential.

For example, when a $Zn|Zn^{2+}(aq, 1\ M)$ half-cell is connected to the reference half-cell (SHE), the voltage measured at 25 °C is found to be 0.763 V.

Electron flows from $Zn|Zn^{2+}$ to SHE; relative to SHE, the reduction potential for $Zn|Zn^{2+}$ half-cell is the negative value of the measured voltage. (E° = reduction potential)

$$Zn^{2+}(aq) + 2\ e^- \rightarrow Zn\ (s) \qquad E^\circ = -076\ V$$

$$2\ H^+(aq) + 2\ e^- \rightarrow H_2\ (s) \qquad E^\circ = 0.000\ V$$

$$Cu^{2+}(aq) + 2\ e^- \rightarrow Cu\ (s) \qquad E^\circ = 0.34\ V$$

Since values are obtained against the standard hydrogen potential, species with positive reduction potentials are easier to reduce than H^+ ions.

In contrast, those with negative reduction potentials are more difficult to reduce.

Therefore, relative to H^+, Cu^{2+} is easier to reduce, whereas Zn^{2+} is challenging. When $Zn|Zn^{2+}$ is connected to $Cu|Cu^{2+}$ half-cells, the spontaneous process is the flow of e^- from $Zn|Zn^{2+}$ to $Cu|Cu^{2+}$ half-cell.

Half-reactions

A half-reaction is the oxidation or reduction reaction component of a redox reaction. A half-reaction is derived from the change in oxidation states of individual substances in the redox reaction.

Balancing redox reactions requires splitting the equation into the two half-reactions of reduction and oxidation.

In general, the half-reactions with the most positive (or the least negative) reduction potential occurs before others.

The *kinetic factor* drives the reaction; if the difference in the standard reduction potentials between two half-cell reactions is small, the concentration may determine the outcome of electrolysis.

For example, the following reactions are possible for the electrolysis of 1 M NaCl (*aq*):

At anode:	$2\ Cl^- (aq) \rightarrow Cl_2 (g) + 2\ e^-$	$E° = -1.36$ V
	$2\ H_2O \rightarrow O_2 (g) + 4\ H^+ (aq) + 4\ e^-$	$E° = -1.23$ V
At cathode:	$2\ H_2O + 2\ e^- \rightarrow H_2 (g) + 2\ OH^- (aq)$	$E° = -0.83$ V
	$Na^+ (aq) + e^- \rightarrow Na (s)$	$E° = -2.71$ V

In the electrolysis of aqueous sodium chloride, Cl_2 forms at the anode instead of O_2, although $E°$ (-1.23 V) for the formation of O_2 from H_2O is less negative than for the formation of Cl_2 from Cl^- (-1.36 V).

The formation of O_2 involves higher activation energy (overvoltage); thus, kinetically less favorable.

Therefore, H_2 is formed at the cathode since it requires less voltage than the reduction of $Na^+(aq)$.

For example, suppose a solution containing metal cations that require lower potential (or one with a positive reduction potential) is electrolyzed.

The metal deposits on the cathode, and no hydrogen are produced. The electrolysis of aqueous copper(II) chloride solution yields Cl_2 gas at the cathode's anode and copper metal deposits.

If the anode is copper, it is oxidized because of the lower voltage requirements relative to water.

Anode reaction:	$Cu (s) \rightarrow Cu^{2+} (aq) + 2\ e^-$	$E° = -0.34$ V
Cathode reaction:	$Cu^{2+} (aq) + 2\ e^- \rightarrow Cu (s)$	$E° = +0.34$ V

Electrical Potential Difference

Cell potential

Electrons flow from one electrode to another in a galvanic cell because there is an electrical potential difference between the two half-cells, called the *cell potential.*

The magnitude of cell potential depends on the two half-cells' nature, the concentration of electrolytes in each half-cell, and temperature.

> Cell potential for galvanic (voltaic) cells is *positive* because the voltaic cell generates potential.

> Cell potential for electrolytic cells is *negative* because it requires potential input.

The *standard cell potential, $E°_{cell}$,* is measured under standard conditions (1 atm pressure for gas, 1 M of electrolytes, and 25 °C). The parameters for a standard cell require the species in solution to have a concentration of 1 M, the partial pressure of gases to be 1 atm, and the temperature to be 25 °C (298 K).

> Reduction potential = potential of the reduction half-reaction

> Oxidation potential = potential of the oxidation half-reaction

> Cell potential = reduction potential + oxidation potential

In the zinc-copper cell, electrons flow from $Zn|Zn^{2+}$ half-cell to $Cu|Cu^{2+}$ half-cell. The $Zn|Z^{2+}$ half-cell has a higher electrical potential than $Cu|Cu^{2+}$.

Using these conditions makes it possible to describe a standard electrode potential ($E°$) in volts and calculate the cell's electrical potential. The superscript $°$ denotes standard conditions.

Cell potential and Gibbs free energy

Cell potential measures the potential difference between half-cells. It indicates if the cell reaction is spontaneous or nonspontaneous. Spontaneity for a process is often associated with its Gibbs free energy change (ΔG).

The mathematical relationship between cell potential and ΔG is:

> $\Delta G° = -nFE°$ under standard conditions

> $\Delta G = -nFE$ under nonstandard conditions

where ΔG = change in Gibbs free energy in Joules, n = the moles of electrons transferred, the Faraday's constant $F = 96,485$ C/mol e$^-$, E = cell potential

---

Maximum work produced

Faraday's constant (C/mol e$^-$) and volts (V) for cell potential gas ΔG units of coulomb-volts.

Coulomb-volts are equivalent to Joules.

A potential difference of 1 V is equivalent to 1 Joule of *work done per coulomb of charge* that flows between two points in the circuit. (1 V = 1 J/C or 1 J = 1 C·V)

$$\text{Maximum work produced} = \text{charge} \times \text{maximum potential}$$

$$w_{max} = -q\text{E}_{max} = \Delta G$$

Electrical charge

$$q = nF$$

$$\Delta G = -nF\text{E}_{cell}$$

or

$$\Delta G^\circ = -nF\text{E}^\circ_{cell}$$

where n = mole of electrons transferred or that flow through circuit, Faraday's constant F = 96,485 C/mole e$^-$

For example:

$$\text{Zn}\,(s) + \text{Cu}^{2+}\,(aq) \rightarrow \text{Zn}^{2+}\,(aq) + \text{Cu}\,(s) \qquad \text{E}^\circ_{cell} = 1.10 \text{ V}$$

$$\Delta G^\circ = -2 \text{ mol e}^- \times (96,485 \text{ C / mol e}^-) \times 1.10 \text{ V}$$

$$\Delta G^\circ = -2.12 \times 10^5 \text{ J}$$

2.12×10^2 kJ is the maximum work per mole of Zn reacted by Cu^{2+}.

Net cell potential

The *net cell potential* is the sum of the two half-cell potentials.

For the Zn-copper cell, the two half-cell potentials:

Anode half-cell reaction:

$$\text{Zn}\,(s) \rightarrow \text{Zn}^{2+}\,(aq) + 2 \text{ e}^- \qquad \text{E}^\circ_{\text{Zn} \rightarrow \text{Zn2+}} = 0.76 \text{ V}$$

Cathode half-cell reaction:

$$\text{Cu}^{2+}\,(aq) + 2 \text{ e}^- \rightarrow \text{Cu}\,(s) \qquad \text{E}^\circ_{\text{Cu2+} \rightarrow \text{Cu}} = 0.34 \text{ V}$$

Overall cell reaction:

$$\text{Zn}\,(s) + \text{Cu}^{2+}\,(aq) \rightarrow \text{Zn}^{2+}\,(aq) + \text{Cu}\,(s)$$

$$E°_{cell} = E°_{Zn \to Zn2+} + E°_{Cu2+ \to Cu}$$

$$E°_{cell} = 0.76 \text{ V} + 0.34 \text{ V}$$

$$E°_{cell} = 1.10 \text{ V}$$

For $Cr|Cr^{3+}$ and $Ag|Ag^+$ half-cells, cell notation and potential:

$$Cr|Cr^{3+}(aq,1 \text{ M})||Ag^+(aq,1 \text{ M})|Ag$$

(anode)   (cathode)

Anode half-cell reaction:

$$Cr(s) \to Cr^{3+}(aq) + 3 \text{ e}^- \qquad E°_{Cr \to Cr3+} = 0.73 \text{ V}$$

Cathode half-cell reaction:

$$Ag^+(aq) + \text{e}^- \to Ag(s) \qquad E°_{Ag+ \to Ag} = 0.80 \text{ V}$$

Overall cell reaction:

$$Cr(s) + 3 \text{ Ag}^+(aq) \to Cr^{3+}(aq) + 3 \text{ Ag}(s)$$

$$E°_{cell} = E°_{Cr \to Cr3+} + E°_{Ag+ \to Ag}$$

$$E°_{cell} = 0.73 \text{ V} + 0.80 \text{ V}$$

$$E°_{cell} = 1.53 \text{ V}$$

The standard electrode potentials are usually listed as reduction potentials relative to a standard hydrogen electrode (SHE) with 0 volts.

For example, use the following metal ion/metal reaction potentials, calculate the standard cell potential:

$$Co(s) + Cu^{2+}(aq) \to Co^{2+}(aq) + Cu(s)$$

$Cu^{2+}(aq)|Cu(s)$	$Ag^+(aq)|Ag(s)$	$Co^{2+}(aq)|Co(s)$	$Zn^{2+}(aq)|Zn(s)$
+0.34 V	+0.80 V	−0.28 V	−0.76 V

The cell reaction is:

$$Co(s) + Cu^{2+}(aq) \to Co^{2+}(aq) + Cu(s)$$

Separate it into half reactions:

$$Cu^{2+}(aq) \to Cu(s)$$

$$Co(s) \to Co^{2+}(aq)$$

The potentials provided are written in this format:

$Cu^{2+} (aq) \mid Cu (s)$

+0.34 V

It means that for the reduction reaction:

$Cu^{2+} (aq) \rightarrow Cu (s)$, the potential is +0.34 V

The reverse reaction or oxidation reaction is:

$Cu (s) \rightarrow Cu^{2+} (aq)$ has opposing potential value: –0.34 V

Obtain the potential values for each half-reaction.

Reverse sign for potential values of oxidation reactions:

$Cu^{2+} (aq) \rightarrow Cu (s)$ = +0.34 V   (reduction)

$Co (s) \rightarrow Co^{2+} (aq)$ = +0.28 V   (oxidation)

Determine the standard cell potential:

Standard cell potential = sum of half-reaction potential

Standard cell potential = 0.34 V + 0.28 V

Standard cell potential = 0.62 V

Nernst equation

The Nernst Equation is used if the temperature is standard (298 K) and uses the natural logarithm (ln) or common logarithm, base 10 (log).

The Nernst equation is used for a cell not under standard conditions:

$E_{cell} = E_{cell}° - (RT / nF) \ln Q$

where E_{cell} = nonstandard cell potential in volts; $E°_{cell}$ = standard cell potential in volts; R = constant, 8.314 J/mol·K, T = temperature in Kelvin; n = moles of electrons transferred; F = Faraday's constant = 96,485 C/mol e⁻ (C represents coulombs) and Q = the reaction quotient expression

$E_{cell} = E°_{cell} - [(0.0257 \text{ V}) / n] \ln Q$

or

$E_{cell} = E°_{cell} - [(0.0592 \text{ V}) / n] \log Q$

Modifying one of the previous cells, as follows:

Zn (s) | Zn^{2+} (*aq*), 3.00M; SO_4^{2-} (3.00M) ‖ Cu^{2+} (*aq*), 0.0015M; SO_4^{2-} (0.0015M) | Cu (s)

$E°_{cell}$ = +1.10V

For standard temperature (298 K), the Nernst equation expression:

E_{cell} = $E°_{cell}$ − [(0.0257 V) / n] ln $[Zn^{2+}]$ / $[Cu^{2+}]$

E_{cell} = +1.10 V − (0.0257 V / 2 mol e⁻) ln (3.00 M / 0.0015 M)

Evaluating the above expression gives:

E_{cell} = +1.10 V − 0.098 V

E_{cell} = 1.00 V

Table of standard reduction potentials

Reduction potential measures the tendency of a chemical species to acquire electrons and become reduced (i.e., the gain of electrons is a reduction).

Reduction potential is measured in volts (V) or millivolts (mV).

The more positive the E°, the higher the tendency for the species to be reduced (a strong oxidizing agent).

The more negative the E°, the higher the tendency for the species to be oxidized (i.e., strong reducing agent).

Standard Reduction Potentials in Aqueous Solution at 25 °C *	
Acidic Solutions	**E° (V)**
F_2 (*g*) + 2 e⁻ → 2 F⁻ (*aq*)	+2.87
Co^{3+} (*aq*) + e⁻ → Co^{2+} (*aq*)	+1.82
Pb^{4+} (*aq*) + 2 e⁻ → Pb^{2+} (*aq*)	+1.8
H_2O_2 (*aq*) + 2 H⁺ (*aq*) + 2 e⁻ → 2 H_2O	+1.77
NiO_2 (*s*) + 4 H⁺ (*aq*) + 2 e⁻ → Ni^{2+} (*aq*) + 2 H_2O	+1.7
PbO_2 (*s*) + SO_4^{2-} (*aq*) + 4 H⁺ (*aq*) + 2 e⁻ → $PbSO_4$ (*s*) + 2 H_2O	+1.69
Au^+ (*aq*) + e⁻ → Au (*s*)	+1.68
2 HClO (*aq*) + 2 H⁺ (*aq*) + 2 e⁻ → Cl_2 (*g*) + 2 H_2O	+1.63

$Ce^{4+}(aq) + e^- \rightarrow Ce^{3+}(aq)$	+1.61
$NaBiO_3(s) + 6\ H^+(aq) + 2\ e^- \rightarrow Bi^{3+}(aq) + Na^+(aq) + 3\ H_2O$	+1.6
$MnO_4^-(aq) + 8\ H^+(aq) + 5\ e^- \rightarrow Mn^{2+}(aq) + 4\ H_2O$	+1.51
$Au^{3+}(aq) + 3\ e^- \rightarrow Au\ (s)$	+1.5
$ClO_3^-(aq) + 6\ H^+(aq) + 5\ e^- \rightarrow 1/2\ Cl_2(g) + 3\ H_2O$	+1.47
$BrO_3^- + 6\ H^+(aq) + 6\ e^- \rightarrow Br^-(aq) + 3\ H_2O$	+1.44
$Cl_2(g) + 2\ e^- \rightarrow 2\ Cl^-(aq)$	+1.36
$Cr_2O_7^{2-} + 14\ H^+(aq) + 6\ e^- \rightarrow 2\ Cr^{3+}(aq) + 7\ H_2O$	+1.33
$N_2H_5^+(aq) + 3\ H^+(aq) + 2\ e^- \rightarrow 2\ NH_4^+(aq)$	+1.24
$MnO_2(s) + 4\ H^+(aq) + 2\ e^- \rightarrow Mn^{2+}(aq) + 2\ H_2O$	+1.23
$O_2(g) + 4\ H^+(aq) + 4\ e^- \rightarrow 2\ H_2O$	+1.23
$Pt^{2+}(aq) + 2\ e^- \rightarrow Pt\ (s)$	+1.2
$IO_3^-(aq) + 6\ H^+(aq) + 5\ e^- \rightarrow \frac{1}{2}\ I_2(aq) + 3\ H_2O$	+1.19
$ClO_4^-(aq) + 2\ H^+(aq) + 2\ e^- \rightarrow ClO_3^-(aq) + H_2O$	+1.19
$Br_2(l) + 2\ e^- \rightarrow 2\ Br^-(aq)$	+1.07
$AuCl_4^- + 3\ e^- \rightarrow Au\ (s) + 4\ Cl^-(aq)$	+1
$Pd^{2+}(aq) + 2\ e^- \rightarrow Pd\ (s)$	+0.99
$NO_3^-(aq) + 4\ H^+(aq) + 3\ e^- \rightarrow NO\ (g) + 2\ H_2O$	+0.96
$NO_3^-(aq) + 3\ H^+(aq) + 2\ e^- \rightarrow HNO_2(aq) + H_2O$	+0.94
$2\ Hg^{2+}(aq) + 2\ e^- \rightarrow Hg_2^{2+}(aq)$	+0.92
$Hg^{2+}(aq) + 2\ e^- \rightarrow Hg\ (l)$	+0.86
$Ag^+(aq) + e^- \rightarrow Ag\ (s)$	+0.79
$Hg_2^{2+}(aq) + 2\ e^- \rightarrow 2\ Hg\ (l)$	+0.79
$Fe^{3+}(aq) + e^- \rightarrow Fe^{2+}(aq)$	+0.77
$SbCl_6^-(aq) + 2\ e^- \rightarrow SbCl_4^-(aq) + 2\ Cl^-(aq)$	+0.75
$[PtCl_4]^{2-}(aq) + 2\ e^- \rightarrow Pt\ (s) + 4\ Cl^-(aq)$	+0.73
$O_2(g) + 2\ H^+(aq) + 2\ e^- \rightarrow H_2O_2(aq)$	+0.68
$[PtCl_6]^{2-}(aq) + 2\ e^- \rightarrow [PtCl_4]^{2-}(aq) + 2\ Cl^-(aq)$	+0.68
$H_3AsO_4(aq) + 2\ H^+(aq) + 2\ e^- \rightarrow H_3AsO_3(aq) + H_2O$	+0.58
$I_2(s) + 2\ e^- \rightarrow 2\ I^-(aq)$	+0.54

$TeO_2\,(s) + 4\,H^+\,(aq) + 4\,e^- \rightarrow Te\,(s) + 2\,H_2O$	+0.53
$Cu^+\,(aq) + e^- \rightarrow Cu\,(s)$	+0.521
$[RhCl_6]^{3-}\,(aq) + 3\,e^- \rightarrow Rh\,(s) + 6\,Cl^-\,(aq)$	+0.44
$Cu^{2+}\,(aq) + 2\,e^- \rightarrow Cu\,(s)$	+0.34
$HgCl_2\,(s) + 2\,e^- \rightarrow 2\,Hg\,(l) + 2\,Cl^-\,(aq)$	+0.27
$AgCl\,(s) + e^- \rightarrow Ag\,(s) + Cl^-\,(aq)$	+0.22
$SO_4^{2-}\,(aq) + 4\,H^+\,(aq) + 2\,e^- \rightarrow SO_2\,(g) + 2\,H_2O$	+0.2
$SO_4^{2-}\,(aq) + 4\,H^+\,(aq) + 2\,e^- \rightarrow H_2SO_3\,(g) + H_2O$	+0.17
$Cu^{2+}\,(aq) + e^- \rightarrow Cu^+\,(aq)$	+0.15
$Sn^{4+}\,(aq) + 2\,e^- \rightarrow Sn^{2+}\,(aq)$	+0.15
$S\,(s) + 2\,H^+\,(aq) + 2\,e^- \rightarrow H_2S\,(aq)$	+0.14
$AgBr\,(s) + e^- \rightarrow Ag\,(s) + Br^-\,(aq)$	+0.07
$2\,H^+\,(aq) + 2\,e^- \rightarrow H_2\,(g)$ (reference electrode)	0.0
$N_2O\,(g) + 6\,H^+\,(aq) + H_2O + 4\,e^- \rightarrow 2\,NH_3OH^+\,(aq)$	−0.05
$Pb^{2+}\,(aq) + 2\,e^- \rightarrow Pb\,(s)$	−0.13
$Sn^{2+}\,(aq) + 2\,e^- \rightarrow Sn\,(s)$	−0.14
$AgI\,(s) + e^- \rightarrow Ag\,(s) + I^-\,(aq)$	−0.15
$Sn^{4+}\,(aq) + 2\,e^- \rightarrow Sn^{2+}\,(aq)$	+0.15
$S\,(s) + 2\,H^+\,(aq) + 2\,e^- \rightarrow H_2S\,(aq)$	+0.14
$AgBr\,(s) + e^- \rightarrow Ag\,(s) + Br^-\,(aq)$	+0.07
$2\,H^+\,(aq) + 2\,e^- \rightarrow H_2\,(g)$ (reference electrode)	0.0
$N_2O\,(g) + 6\,H^+\,(aq) + H_2O + 4\,e^- \rightarrow 2\,NH_3OH^+\,(aq)$	−0.05
$Pb^{2+}\,(aq) + 2\,e^- \rightarrow Pb\,(s)$	−0.13
$Sn^{2+}\,(aq) + 2\,e^- \rightarrow Sn\,(s)$	−0.14
$AgI\,(s) + e^- \rightarrow Ag\,(s) + I^-\,(aq)$	−0.15
$[SnF_6]^{2-}\,(aq) + 4\,e^- \rightarrow Sn\,(s) + 6\,F^-\,(aq)$	−0.25
$Ni^{2+}\,(aq) + 2\,e^- \rightarrow Ni\,(s)$	−0.25
$Co^{2+}\,(aq) + 2\,e^- \rightarrow Co\,(s)$	−0.28
$Tl^+\,(aq) + e^- \rightarrow Tl\,(s)$	−0.34
$PbSO_4\,(s) + 2\,e^- \rightarrow Pb\,(s) + SO_4^{2-}\,(aq)$	−0.36

	E° (V)
$Se\ (s) + 2\ H^+\ (aq) + 2\ e^- \rightarrow H_2Se\ (aq)$	−0.4
$Cd^{2+}\ (aq) + 2\ e^- \rightarrow Cd\ (s)$	−0.40
$Cr^{3+}\ (aq) + e^- \rightarrow Cr^{2+}\ (aq)$	−0.41
$Fe^{2+}\ (aq) + 2\ e^- \rightarrow Fe\ (s)$	−0.44
$2\ CO_2\ (g) + 2\ H^+\ (aq) + 2\ e^- \rightarrow (COOH)_2\ (aq)$	−0.49
$Ga^{3+}\ (aq) + 3\ e^- \rightarrow Ga\ (s)$	−0.53
$HgS\ (s) + 2\ H^+\ (aq) + 2\ e^- \rightarrow Hg\ (l) + H_2S\ (g)$	−0.72
$Cr^{3+}\ (aq) + 3\ e^- \rightarrow Cr\ (s)$	−0.74
$Zn^{2+}\ (aq) + 2\ e^- \rightarrow Zn\ (s)$	−0.76
$2\ H_2O\ (l) + 2\ e^- \rightarrow H_2\ (g) + 2\ OH^-\ (aq)$	−0.83
$Cr^{2+}\ (aq) + 2\ e^- \rightarrow Cr\ (s)$	−0.91
$Mn^{2+}\ (aq) + 2\ e^- \rightarrow Mn\ (s)$	−1.18
$V^{2+}\ (aq) + 2\ e^- \rightarrow V\ (s)$	−1.18
$Zr^{4+}\ (aq) + 4\ e^- \rightarrow Zr\ (s)$	−1.53
$Al^{3+}\ (aq) + 3\ e^- \rightarrow Al\ (s)$	−1.66
$H_2\ (g) + 2\ e^- \rightarrow 2\ H^-\ (aq)$	−2.25
$Mg^{2+}\ (aq) + 2\ e^- \rightarrow Mg\ (s)$	−2.37
$Na^+\ (aq) + e^- \rightarrow Na\ (s)$	−2.71
$Ca^{2+}\ (aq) + 2\ e^- \rightarrow Ca\ (s)$	−2.87
$Sr^{2+}\ (aq) + 2\ e^- \rightarrow Sr\ (s)$	−2.89
$Ba^{2+}\ (aq) + 2\ e^- \rightarrow Ba\ (s)$	−2.9
$Rb^+\ (aq) + e^- \rightarrow Rb\ (s)$	−2.93
$K^+\ (aq) + e^- \rightarrow K\ (s)$	−2.93
$Li^+\ (aq) + e^- \rightarrow Li\ (s)$	−3.05
Basic Solutions	**E° (V)**
$ClO^-\ (aq) + H_2O + 2\ e^- \rightarrow Cl^-\ (aq) + 2\ OH^-\ (aq)$	+0.89
$OOH^-\ (aq) + H_2O + 2\ e^- \rightarrow 3\ OH^-\ (aq)$	+0.88
$2\ NH_2OH\ (aq) + 2\ e^- \rightarrow N_2H_4\ (aq) + 2\ OH^-\ (aq)$	+0.74
$ClO_3^-\ (aq) + 3\ H_2O + 6\ e^- \rightarrow Cl^-\ (aq) + 6\ OH^-\ (aq)$	+0.62
$MnO_4^-\ (aq) + 2\ H_2O + 3\ e^- \rightarrow MnO_2\ (s) + 4\ OH^-\ (aq)$	+0.59

$MnO_4^-(aq) + e^- \rightarrow MnO_4^{2-}(aq)$	+0.56
$NiO_2(s) + 2\,H_2O + 2\,e^- \rightarrow Ni(OH)_2(s) + 2\,OH^-(aq)$	+0.49
$Ag_2CrO_4(s) + 2\,e^- \rightarrow 2\,Ag(s) + CrO_4^{2-}(aq)$	+0.45
$O_2(g) + 2\,H_2O + 4\,e^- \rightarrow 4\,OH^-(aq)$	+0.4
$ClO_4^-(aq) + H_2O + 2\,e^- \rightarrow ClO_3^-(aq) + 2\,OH^-(aq)$	+0.36
$Ag_2O(s) + H_2O + 2\,e^- \rightarrow 2\,Ag(s) + 2\,OH^-(aq)$	+0.34
$2\,NO_2^-(aq) + 3\,H_2O + 4\,e^- \rightarrow N_2O(g) + 6\,OH^-(aq)$	+0.15
$N_2H_4(aq) + 2\,H_2O + 2\,e^- \rightarrow 2\,NH_3(aq) + 2\,OH^-(aq)$	+0.1
$[Co(NH_3)_6]^{3+}(aq) + e^- \rightarrow [Co(NH_3)_6]^{2+}(aq)$	+0.1
$HgO(s) + H_2O + 2\,e^- \rightarrow Hg(l) + 2\,OH^-(aq)$	+0.09
$O_2(g) + H_2O + 2\,e^- \rightarrow OOH^-(aq) + OH^-(aq)$	+0.08
$NO_3^-(aq) + H_2O + 2\,e^- \rightarrow NO_2^-(aq) + 2\,OH^-(aq)$	+0.01
$MnO_2(s) + 2\,H_2O + 2\,e^- \rightarrow Mn(OH)_2(s) + 2\,OH^-(aq)$	−0.05
$CrO_4^{2-}(aq) + 4\,H_2O + 3\,e^- \rightarrow Cr(OH)_3(s) + 5\,OH^-(aq)$	−0.12
$Cu(OH)_2(s) + 2\,e^- \rightarrow Cu(s) + 2\,OH^-(aq)$	−0.36
$Fe(OH)_3(s) + e^- \rightarrow Fe(OH)_2(s) + OH^-(aq)$	−0.56
$2\,H_2O + 2\,e^- \rightarrow H_2(g) + 2\,OH^-(aq)$	−0.83
$2\,NO_3^-(aq) + 2\,H_2O + 2\,e^- \rightarrow N_2O_4(g) + 4\,OH^-(aq)$	−0.85
$Fe(OH)_2(s) + 2\,e^- \rightarrow Fe(s) + 2\,OH^-(aq)$	−0.88
$SO_4^{2-}(aq) + H_2O + 2\,e^- \rightarrow SO_3^{2-}(aq) + 2\,OH^-(aq)$	−0.93
$N_2(g) + 4\,H_2O + 4\,e^- \rightarrow N_2H_4(aq) + 4\,OH^-(aq)$	−1.15
$[Zn(OH)_4]^{2-}(aq) + 2\,e^- \rightarrow Zn(s) + 4\,OH^-(aq)$	−1.22
$Zn(OH)_2(s) + 2\,e^- \rightarrow Zn(s) + 2\,OH^-(aq)$	−1.25
$[Zn(CN)_4]^{2-}(aq) + 2\,e^- \rightarrow Zn(s) + 4\,CN^-(aq)$	−1.26
$Cr(OH)_3(s) + 3\,e^- \rightarrow Cr(s) + 3\,OH^-(aq)$	−1.3
$SiO_3^{2-}(aq) + 3\,H_2O + 4\,e^- \rightarrow Si(s) + 6\,OH^-(aq)$	−1.7

table is for reference only, and values are approximate

Notes for active learning

Electron Flow

Direction of electron flow

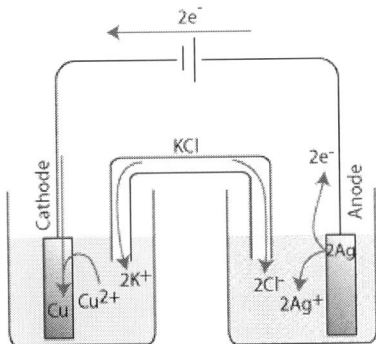

Electrolytic charge flow

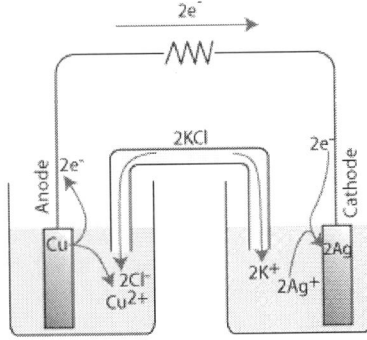

Galvanic (voltaic) charge flow

Electrons flow from anode to cathode.

> Mnemonic: *A to C in alphabetical order*.

Alternatively, AC power as *the A comes first and stands for Anode*.

Oxidation (at the anode) produces electrons (and cations) and releases electrons toward the cathode.

The cathode receives those electrons and uses them for reduction.

The species with the highest oxidation potential (lowest reduction potential) is the anode, and the species with the highest reduction potential is the cathode.

In the diagram above, the galvanic (voltaic) cell shows a natural flow because Cu (higher oxidation potential/lower reduction potential) is the anode, and Ag (higher reduction potential) is the cathode.

The electrolytic cell shows the opposite; a battery drives the reaction to force Cu to be the cathode and Ag to be the anode. Electrons flow in wires and electrodes, while ions flow in an electrolyte solution, thus creating a completed circuit.

The oxidation-reduction reaction continues until the electrolyte concentrations in the half-cells become equal.

At *anode half-cell*:

$$Cu\ (s) \rightarrow Cu^{2+}\ (aq) + 2\ e^-\ (in\ 0.0010\ M\ Cu^{2+})$$

At *cathode half-cell*:

$$Cu^{2+}\ (aq) + 2\ e^- \rightarrow Cu\ (s)\ (in\ 0.50\ M\ Cu^{2+})$$

Electromotive force and voltage

Electromotive force (emf), or cell potential (E_{cell}), is the driving force that enables electrons to flow from one electrode to the other, which has the unit volt (V).

A volt is one Joule/coulomb, where the coulomb is the unit of charge.

Voltage is the driving force for reactions, often called the EMF (electromotive force).

The higher the voltage, the greater the driving force.

A voltmeter must have its negative terminal connected to the anode and positive terminal to the cathode for the correct polarity (± sign).

Voltage is positive for a spontaneous reaction.

Batteries

Batteries as galvanic cells

Batteries are galvanic cells connected in series, where the battery potential equals the cells' potentials.

Three types of batteries:

> *Primary batteries* are not rechargeable.

> *Secondary batteries* are rechargeable.

> *Fuel cells* need an ample supply of fuel to provide the energy.

Primary batteries include common (acidic) dry, alkaline, and mercury batteries. A new dry cell battery has a potential of about 1.5 V regardless of size, but the amount of energy a battery can deliver depends on its size. A D battery delivers more current (greater amperes) than an AAA battery.

Standard dry batteries use aqueous NH_4Cl paste as an electrolyte and are acidic batteries due to the ionization:

$$NH_4^+ (aq) \rightarrow NH_3 (aq) + H^+ (aq)$$

Acidic dry batteries consist of Zinc casing as an anode, the reducing agent; graphite rod as an (inert) cathode; aqueous NH_4Cl paste as an electrolyte, and MnO_2 powder as the oxidizing agent.

$$Zn|Zn^{2+},NH_4^+,NH_3 (aq)||Mn_2O_3,MnO_2|C (s)$$

Anode reaction:

$$Zn (s) \rightarrow Zn^{2+} (aq) + 2 e^-$$

Cathodic reaction:

$$2 MnO_2 (s) + 2 NH_4^+ (aq) + 2 e^- \rightarrow Mn_2O_3 (s) + 2 NH_3 (aq) + H_2O (l)$$

Net reaction:

$$Zn (s) + 2 MnO_2 (s) + 2 NH_4^+ (aq) \rightarrow Zn^{2+} (aq) + Mn_2O_3 (s) + 2 NH_3 (aq) + H_2O (l)$$

The reverse reaction is prevented by forming $[Zn(NH_3)_4]^{2+}$ ions.

Alkaline batteries use zinc (reducing agent) and MnO_2 (oxidizing agent), but an aqueous KOH paste (not NH_4Cl) as the electrolyte.

Reaction at anode: $Zn\ (s) + 2\ OH^-\ (aq) \rightarrow ZnO\ (s) + H_2O\ (l) + 2\ e^-$

Reaction a cathode: $2\ MnO_2\ (s) + H_2O\ (l) + 2\ e^- \rightarrow Mn_2O_3\ (s) + 2\ OH^-\ (aq)$

Net reaction: $Zn\ (s) + 2\ MnO_2\ (s) \rightarrow ZnO\ (s) + Mn_2O_3\ (s)$

Since the reactants involved are solid, alkaline batteries deliver a constant voltage until the limiting reactant is exhausted. They last longer because zinc metal corrodes more slowly under basic conditions.

Mercury batteries are used in calculators and watches.

The following reactions occur at the anode and cathode sections of the cell:

Anode reaction: $Zn\ (s) + 2\ OH^-\ (aq) \rightarrow ZnO\ (s) + H_2O + 2\ e^-$

Cathode reaction: $HgO\ (s) + H_2O + 2\ e^- \rightarrow Hg\ (l) + 2\ OH^-\ (aq)$

Net cell reaction: $Zn\ (s) + HgO\ (s) \rightarrow ZnO\ (s) + Hg\ (l)$

Lead-storage batteries

Lead-storage batteries are used in automobiles and contain sulfuric acid as an electrolyte. Each battery cell contains several lead alloy grids with alternating grids packed with lead metal and lead(IV) oxide, PbO_2. The standard 12-V car battery contains six cells connected in series.

Each set of grids, which are electrodes, is connected in parallel, enabling the cell to deliver more current. The amount of current delivered depends on the electrode's surface area. Each cell in a lead storage battery has a potential of about 2.01 V.

Spontaneous reactions occurring in a lead storage battery:

Anode reaction:

$$Pb\ (s) + HSO_4^-\ (aq) \rightarrow PbSO_4\ (s) + H^+\ (aq) + 2\ e^-$$

Cathode reaction:

$$PbO_2\ (s) + 3\ H^+\ (aq) + HSO_4^-\ (aq) + 2\ e^- \rightarrow PbSO_4\ (s) + 2\ H_2O\ (m)$$

Net reaction:

$Pb\ (s) + PbO_2\ (s) + 2\ H^+\ (aq) + 2\ HSO_4^-\ (aq) \rightarrow 2\ PbSO_4\ (s) + 2\ H_2O\ (l) \rightarrow$ discharging

The greatest advantage of lead-storage batteries (i.e., secondary batteries) is being rechargeable. However, while the car is driven, the battery obtains energy from the motor through the alternator. Starting the engine causes the discharge reaction to occur.

The following re-charging reaction occurs:

$$2 \text{ PbSO}_4 (s) + 2 \text{ H}_2\text{O} (l) \rightarrow \text{Pb} (s) + \text{PbO}_2 (s) + 2 \text{ H}^+ (aq) + 2 \text{ HSO}_4^- (aq)$$

Lead-storage batteries have a longer lifetime than other batteries and deliver a large amount of current and electrical energy within a short time.

The major disadvantages of lead-storage batteries are:

    1) they are heavy, bulky, and relatively non-portable

    2) lead is a toxic metal, and disposal is an environmental problem

    3) the battery must be kept upright because H_2SO_4 is corrosive

Nickel-cadmium batteries

Nickel-cadmium batteries are rechargeable batteries used in small appliances such as cordless phones. They contain cadmium as the anode and hydrated nickel oxide as the cathode.

The electrolyte is an aqueous KOH paste.

Anode reaction:

$$\text{Cd} (s) + 2 \text{ OH}^- (aq) \rightarrow \text{Cd(OH)}_2 (s) + 2 \text{ e}^-$$

Cathode reaction:

$$2 \text{ NiO(OH)} (s) + 2 \text{ H}_2\text{O} (l) + 2 \text{ e}^- \rightarrow 2 \text{ Ni(OH)}_2 (s) + 2 \text{ OH}^- (aq)$$

Net cell reaction:

$$\text{Cd} (s) + 2 \text{ NiO(OH)} (s) + 2 \text{ H}_2\text{O} (l) \rightarrow \text{Cd(OH)}_2 (aq) + 2 \text{ Ni(OH)}_2 (aq)$$

Notes for active learning

Practice Questions

1. Nickel-cadmium batteries are used in rechargeable electronic calculators. Given the following reaction for a discharging NiCad battery, which substance is reduced?

$$Cd\ (s) + NiO_2\ (s) + 2H_2O\ (l) \rightarrow Cd(OH)_2\ (s) + Ni(OH)_2\ (s)$$

A. $Cd(OH)_2$

B. H_2O

C. NiO_2

D. Cd

E. $Ni(OH)_2$

2. How long must a current of 2 amps run for to liberate 3 moles of H_2O when electrolyzed in the following reaction? (Use Faraday's constant = 96,500C/mol and the conversion factor of 1 amp = 1 coulomb/sec)

$$H_2O_2 + 2\ H^+ + 2\ e^- \rightleftarrows 2\ H_2O$$

A. 4.83×10^4 sec

B. 4.83×10^5 sec

C. 1.45×10^5 sec

D. 2.41×10^5 sec

E. 9.65×10^3 sec

3. Which is NOT true for the following redox reaction occurring in an electrolytic cell?

$$Cd\ (s) + Zn(NO_3)_2\ (aq) \xrightarrow{\text{Electricity}} Zn\ (s) + Cd(NO_3)_2\ (aq)$$

A. Cd^{2+} is produced at the anode

B. Zn metal is produced at the anode

C. Oxidation half-reaction: $Cd \rightarrow Cd^{2+} + 2\ e^-$

D. Reduction half-reaction: $Zn^{2+} + 2\ e^- \rightarrow Zn$

E. Zn metal is produced at the cathode

4. Which statements is true of the following redox reaction occurring in a nonspontaneous electrolytic cell?

$$Br_2\ (l) + 2\ NaCl\ (aq) \xrightarrow{\text{Electricity}} Cl_2\ (g) + 2\ NaBr\ (aq)$$

A. Br_2 liquid is produced at the cathode

B. Cl_2 gas is produced at the cathode

C. Reduction half–reaction: $2\ Cl^- \rightarrow Cl_2 + 2\ e^-$

D. Cl_2 gas is produced at the anode

E. Oxidation half-reaction: $Br_2 + 2\ e^- \rightarrow 2\ Br^-$

5. Which is true regarding the redox reaction occurring in a spontaneous electrochemical cell?

$$Sn\ (s) + Cu^{2+}\ (aq) \rightarrow Cu\ (s) + Sn^{2+}\ (aq)$$

A. Electrons flow from the Sn electrode to the Cu electrode

B. Anions in the salt bridge flow from the Sn half-cell to the Cu half-cell

C. Sn is oxidized at the cathode

D. Cu^{2+} is reduced at the anode

E. Cl_2 gas is produced at the anode

6. Which of the following is a unit of electrical charge?

A. joule

B. volt

C. coulomb

D. ampere

E. watt

7. If a galvanic cell has two electrodes, which is correct?

A. Oxidation occurs at the negatively charged cathode

B. Oxidation occurs at the positively charged cathode

C. Oxidation occurs at the positively charged anode

D. Oxidation occurs at the negatively charged anode

E. Oxidation occurs at the uncharged dynode

8. A battery operates by:

A. oxidation

B. reduction

C. both oxidation and reduction

D. neither oxidation nor reduction

E. none of the above

9. What is the term for the electrode in an electrochemical cell where oxidation occurs?

A. oxidation electrode

B. reduction electrode

C. anode

D. cathode

E. none of the above

10. What is the general term for an apparatus containing two solutions with electrodes in separate compartments connected by a wire and salt bridge?

A. electrolytic cell

B. voltaic cell

C. battery

D. electrochemical cell

E. dry cell

11. The electrode with the standard reduction potential of 0 V is assigned as the standard reference electrode and uses the half-reaction:

A. $2 \, NH_4^+ \,(aq) + 2 \, e^- \leftrightarrows H_2 \,(g) + 2 \, NH_3 \,(g)$

B. $Ag^+ \,(aq) + e^- \leftrightarrows Ag \,(s)$

C. $Cu^{2+} \,(aq) + 2 \, e^- \leftrightarrows Cu \,(s)$

D. $Zn^{2+} \,(aq) + 2 \, e^- \leftrightarrows Zn \,(s)$

E. $2 \, H^+ \,(aq) + 2 \, e^- \leftrightarrows H_2 \,(g)$

12. In which type of cell does the following reaction occur when electrons are forced into a system by applying an external voltage?

$$Fe^{2+} + 2 \, e^- \rightarrow Fe \,(s) \qquad E^\circ = -0.44 \, V$$

A. concentration cell

B. battery

C. electrochemical cell

D. galvanic cell

E. electrolytic cell

13. What is the term for a reaction representing separate oxidation or reduction processes?

A. reduction reaction

B. redox reaction

C. oxidation reaction

D. half-reaction

E. none of the above

14. What is the term for providing electricity to a nonspontaneous redox process to cause a reaction?

A. hydrolysis

B. protolysis

C. electrochemistry

D. electrolysis

E. none of the above

15. How is electrolysis different from the chemical process inside a battery?

A. Pure compounds cannot be generated *via* electrolysis

B. Electrolysis only uses electrons from a cathode

C. Electrolysis does not use electrons

D. They are the same process in reverse

E. Chemical changes do not occur in electrolysis

16. How many electrons are needed to balance the following half-reaction $H_2S \rightarrow S_8$ in an acidic solution?

A. 16 electrons to the right side

B. 6 electrons to the right side

C. 12 electrons to the left side

D. 8 electrons to the right side

E. 14 electrons to the left side

17. How are photovoltaic cells different from many other forms of solar energy?

A. Light is reflected, and the coolness of the shade provides a temperature differential

B. Light is passively converted into heat

C. Light is converted directly to electricity

D. Light is converted into heat and then into electricity

E. Light is converted into heat and then into steam

18. Which is true regarding the following redox reaction occurring in a spontaneous electrochemical cell?

$$Sn\ (s) + Cu^{2+}\ (aq) \rightarrow Cu\ (s) + Sn^{2+}\ (aq)$$

A. Anions in the salt bridge flow from the Cu half-cell to the Sn half-cell

B. Electrons flow from the Cu electrode to the Sn electrode

C. Cu^{2+} is oxidized at the cathode

D. Sn is reduced at the anode

E. None of the above

19. What is the term for converting chemical energy to electrical energy from redox reactions?

A. Redox chemistry

B. Electrochemistry

C. Cell chemistry

D. Battery chemistry

E. Electrolysis

Detailed Explanations

1. C is correct.

Determine the *oxidation number* of each species.

The reduced substance has a decrease in the oxidation number.

In NiO_2, the oxidation number of $Ni = +4$.

In $Ni(OH)_2$, the oxidation number of $Ni = +2$.

2. C is correct.

3 moles of H_2O need 3 moles of e^-.

Calculate time needed for 3 moles of H_2O:

$$(3 \text{ moles } e^-) \times (96,500 \text{ coulomb} / 1 \text{ mole } e^-) \times (1 \text{ sec} / 2 \text{ coulomb})$$
$$\text{time (for 3 moles of } H_2O) = 1.45 \times 10^5 \text{ sec}$$

3. B is correct.

An *electrolytic cell* is a nonspontaneous electrochemical cell requiring electrical energy to be supplied (e.g., battery) to initiate the reaction.

The *anode* is positive, and the *cathode* is the negative electrode.

For electrolytic and galvanic cells, *oxidation occurs at the anode and reduction at the cathode*.

Therefore, Zn metal is produced at the cathode because it is a reduction product (from an oxidation number of +3 on the left to 0 on the right). Zn will not be produced at the anode.

4. D is correct.

In electrochemical (i.e., galvanic) cells, oxidation occurs at the anode and reduction at the cathode.

Cl_2 gas is produced at the anode because it is an oxidation product (from–1 on the left to 0 on the right).

5. A is correct.

A *spontaneous* electrochemical cell is known as a *galvanic cell*.

6. C is correct.

A *coulomb* is a unit of electrical charge.

7. D is correct.

Oxidation (i.e., loss of electrons) occurs at the *anode*.

8. C is correct.

A battery is an electrochemical (i.e., galvanic) cell with reduction and oxidation reactions to generate electricity.

9. C is correct.

Oxidation (i.e., loss of electrons) occurs at the anode.

10. D is correct.

The salt bridge should be the most apparent indication of an electrochemical cell – electrolytic cells do not have salt bridges.

11. E is correct.

By convention, the reference standard for potential is hydrogen reduction.

12. E is correct.

Electrolytic cells do not occur spontaneously; the reaction requires *external electrical energy*.

13. D is correct.

A redox reaction, or oxidation-reduction reaction, involves transferring electrons between two reacting substances.

An oxidation reaction refers explicitly to the substance that is losing electrons, and a reduction reaction refers explicitly to the substance that is gaining reactions.

The oxidation and reduction reactions alone are *half-reactions* because they occur together to form a whole reaction.

Therefore, half-reaction is the answer because it can represent a separate oxidation or reduction process.

14. D is correct.

Electrolysis uses a direct electric current to provide a nonspontaneous redox process to drive the reaction.

The direct electric current must be passed through an ionic substance or solution that contains electrolytes.

15. D is correct.

Electrolysis is the same process in reverse for the chemical process inside a battery.

Electrolysis is used to separate elements.

16. A is correct.

Half-reaction:

$$H_2S \rightarrow S_8$$

Balancing half-reaction in acidic conditions:

Step 1: Balance atoms except for H and O

$$8\ H_2S \rightarrow S_8$$

Step 2: To balance oxygen, add H_2O to the side with fewer oxygen atoms

There is no oxygen at all, so skip this step.

Step 3: To balance hydrogen, add H^+:

$$8\ H_2S \rightarrow S_8 + 16\ H^+$$

Balance charges by adding electrons to the side with a greater/more positive total charge

Total charge on left side: 0

Total charge on right side: $16(+1) = +16$

Add 16 electrons to right side:

$$8\ H_2S \rightarrow S_8 + 16\ H^+ + 16\ e^-$$

17. C is correct.

The operation of a photovoltaic (PV) cell has the following requirements:

1) Light is absorbed, which excites electrons.

2) Separation of charge carries opposite types.

3) The separated charges are transferred to an external circuit.

In contrast, solar panels supply heat by absorbing sunlight.

A photoelectrolytic (or photoelectrochemical) cell is a photovoltaic cell that splits water directly into hydrogen and oxygen using only solar illumination.

18. A is correct.

Cu^{2+} is reduced (i.e., gains electrons), while Sn^{2+} is oxidized (i.e., loses electrons).

Salt bridge contains cations (positive ions) and anions (negative ions).

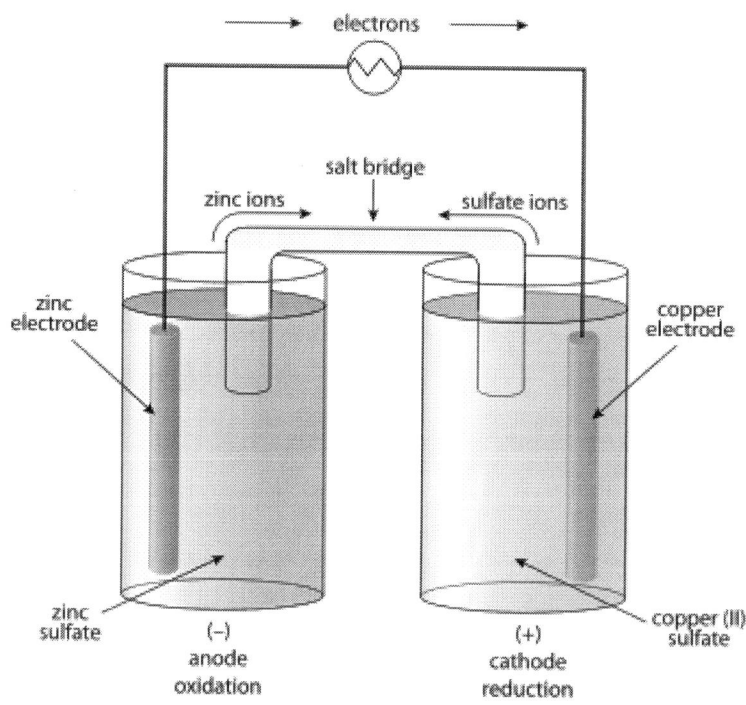

Schematic of a sample Zn–Cu galvanic cell

Anions flow towards the oxidation half-cell because the oxidation product is positively charged, and the anions are required to balance the charges within the cell. The opposite is true for cations; they flow towards the reduction half-cell.

19. B is correct.

Electrochemistry is the branch of physical chemistry that studies chemical reactions at the interface of an ionic conductor (i.e., the electrolyte) and an electrode.

Electric charges move between the electrolyte and the electrode through a series of redox reactions, and chemical energy is converted to electrical energy.

Notes for active learning

Notes for active learning

APPENDIX

Periodic Table of the Elements

Legend:
- Atomic Number
- Valence
- **Symbol**
- Name
- Atomic Mass

1 IA 1A	2 IIA 2A	3 IIIB 3B	4 IVB 4B	5 VB 5B	6 VIB 6B	7 VIIB 7B	8 VIII 8	9 VIII 8	10 VIII 8	11 IB 1B	12 IIB 2B	13 IIIA 3A	14 IVA 4A	15 VA 5A	16 VIA 6A	17 VIIA 7A	18 VIIIA 8A
1 -1,+1 **H** Hydrogen 1.008																	2 0 **He** Helium 4.003
3 +1 **Li** Lithium 6.941	4 +2 **Be** Beryllium 9.012											5 +3 **B** Boron 10.811	6 +4,+3,+2,+1 -4,-3 **C** Carbon 12.011	7 +5,+4 -3 **N** Nitrogen 14.007	8 -2 **O** Oxygen 15.999	9 -1 **F** Fluorine 18.998	10 0 **Ne** Neon 20.180
11 +1 **Na** Sodium 22.990	12 +2 **Mg** Magnesium 24.305											13 +3 **Al** Aluminum 26.982	14 +4,+2,-4 **Si** Silicon 28.086	15 +5,+3,-3 **P** Phosphorus 30.974	16 +6,+4,-2,-2 **S** Sulfur 32.066	17 +7,+5,+3,+1 -1 **Cl** Chlorine 35.453	18 0 **Ar** Argon 39.948
19 +1 **K** Potassium 39.098	20 +2 **Ca** Calcium 40.078	21 +3 **Sc** Scandium 44.956	22 +4 **Ti** Titanium 47.88	23 +5 **V** Vanadium 50.942	24 +6,+3 **Cr** Chromium 51.996	25 +7,+4,+2,+3 **Mn** Manganese 54.938	26 +6,+3,+2 **Fe** Iron 55.845	27 +3,+2 **Co** Cobalt 58.933	28 +2 **Ni** Nickel 58.693	29 +2,+1 **Cu** Copper 63.546	30 +2 **Zn** Zinc 65.38	31 +3 **Ga** Gallium 69.723	32 +4,+2 **Ge** Germanium 72.631	33 +5,+3,-3 **As** Arsenic 74.922	34 +6,+4,+2,-2 **Se** Selenium 78.971	35 +5,+3,+1 -1 **Br** Bromine 79.904	36 +2,0 **Kr** Krypton 84.798
37 +1 **Rb** Rubidium 85.468	38 +2 **Sr** Strontium 87.62	39 +3 **Y** Yttrium 88.906	40 +4 **Zr** Zirconium 91.224	41 +5 **Nb** Niobium 92.906	42 +6,+4 **Mo** Molybdenum 95.95	43 +7,+4 **Tc** Technetium 98.507	44 +4,+3 **Ru** Ruthenium 101.07	45 +3,+2 **Rh** Rhodium 102.906	46 +4,+2 **Pd** Palladium 106.42	47 +1 **Ag** Silver 107.868	48 +2 **Cd** Cadmium 112.414	49 +3 **In** Indium 114.818	50 +4,+2 **Sn** Tin 118.711	51 +5,+3,-3 **Sb** Antimony 121.760	52 +6,+4,+2,-2 **Te** Tellurium 127.6	53 +7,+5,+3,+1 -1 **I** Iodine 126.904	54 +4,+2,0 **Xe** Xenon 131.294
55 +1 **Cs** Cesium 132.905	56 +2 **Ba** Barium 137.328	57-71	72 +4 **Hf** Hafnium 178.49	73 +5 **Ta** Tantalum 180.948	74 +6,+4 **W** Tungsten 183.85	75 +7 **Re** Rhenium 186.207	76 +8 **Os** Osmium 190.23	77 +4,+3 **Ir** Iridium 192.22	78 +4,+2 **Pt** Platinum 195.08	79 +3,+1 **Au** Gold 196.967	80 +2,+1 **Hg** Mercury 200.59	81 +3,+1 **Tl** Thallium 204.383	82 +4,+2 **Pb** Lead 207.2	83 +5,+3 **Bi** Bismuth 208.980	84 +4,+2,+2 **Po** Polonium [208.982]	85 +1,-1 **At** Astatine 209.987	86 +2,0 **Rn** Radon 222.018
87 +1 **Fr** Francium 223.020	88 +2 **Ra** Radium 226.025	89-103	104 unknown **Rf** Rutherfordium [261]	105 unknown **Db** Dubnium [262]	106 unknown **Sg** Seaborgium [266]	107 unknown **Bh** Bohrium [264]	108 unknown **Hs** Hassium [269]	109 unknown **Mt** Meitnerium [278]	110 unknown **Ds** Darmstadtium [281]	111 unknown **Rg** Roentgenium [280]	112 unknown **Cn** Copernicium [285]	113 unknown **Nh** Nihonium [286]	114 unknown **Fl** Flerovium [289]	115 unknown **Mc** Moscovium [289]	116 unknown **Lv** Livermorium [293]	117 unknown **Ts** Tennessine [294]	118 unknown **Og** Oganesson [294]

Lanthanide Series

57 +3 **La** Lanthanum 138.905	58 +4,+3 **Ce** Cerium 140.116	59 +3 **Pr** Praseodymium 140.908	60 +3 **Nd** Neodymium 144.243	61 +3 **Pm** Promethium 144.913	62 +3 **Sm** Samarium 150.36	63 +3,+2 **Eu** Europium 151.964	64 +3 **Gd** Gadolinium 157.25	65 +3 **Tb** Terbium 158.925	66 +3 **Dy** Dysprosium 162.500	67 +3 **Ho** Holmium 164.930	68 +3 **Er** Erbium 167.259	69 +3 **Tm** Thulium 168.934	70 +3 **Yb** Ytterbium 173.055	71 +3 **Lu** Lutetium 174.967

Actinide Series

89 +3 **Ac** Actinium 227.028	90 +4 **Th** Thorium 232.038	91 +5 **Pa** Protactinium 231.036	92 +6 **U** Uranium 238.029	93 +5 **Np** Neptunium 237.048	94 +4 **Pu** Plutonium 244.064	95 +3 **Am** Americium 243.061	96 +3 **Cm** Curium 247.070	97 +3 **Bk** Berkelium 247.070	98 +3 **Cf** Californium 251.080	99 +3 **Es** Einsteinium [254]	100 +3 **Fm** Fermium 257.095	101 +3 **Md** Mendelevium 258.1	102 +3 **No** Nobelium 259.101	103 +3 **Lr** Lawrencium [262]

Notes for active learning

Common Chemistry Equations

Throughout the test the following symbols have the definitions specified unless otherwise noted.

L, mL	= liter(s), milliliter(s)	mm Hg	= millimeters of mercury
g	= gram(s)	J, kJ	= joule(s), kilojoule(s)
nm	= nanometer(s)	V	= volt(s)
atm	= atmosphere(s)	mol	= mole(s)

ATOMIC STRUCTURE

$E = h\nu$

$c = \lambda\nu$

E = energy
ν = frequency
λ = wavelength

Planck's constant, $h = 6.626 \times 10^{-34}$ J s

Speed of light, $c = 2.998 \times 10^8$ m s^{-1}

Avogadro's number $= 6.022 \times 10^{23}$ mol^{-1}

Electron charge, $e = -1.602 \times 10^{-19}$ coulomb

EQUILIBRIUM

$K_c = \dfrac{[C]^c[D]^d}{[A]^a[B]^b}$, where $a\,A + b\,B \rightleftarrows c\,C + d\,D$

$K_p = \dfrac{(P_C)^c(P_D)^d}{(P_A)^a(P_B)^b}$

$K_a = \dfrac{[H^+][A^-]}{[HA]}$

$K_b = \dfrac{[OH^-][HB^+]}{[B]}$

$K_w = [H^+][OH^-] = 1.0 \times 10^{-14}$ at 25°C

$\quad = K_a \times K_b$

$pH = -\log[H^+]$, $pOH = -\log[OH^-]$

$14 = pH + pOH$

$pH = pK_a + \log\dfrac{[A^-]}{[HA]}$

$pK_a = -\log K_a$, $pK_b = -\log K_b$

Equilibrium Constants

K_c (molar concentrations)
K_p (gas pressures)
K_a (weak acid)
K_b (weak base)
K_w (water)

KINETICS

$\ln[A]_t - \ln[A]_0 = -kt$

$\dfrac{1}{[A]_t} - \dfrac{1}{[A]_0} = kt$

$t_{1/2} = \dfrac{0.693}{k}$

k = rate constant
t = time
$t_{1/2}$ = half-life

GASES, LIQUIDS, AND SOLUTIONS

$$PV = nRT$$

$$P_A = P_{total} \times X_A, \text{ where } X_A = \frac{\text{moles A}}{\text{total moles}}$$

$$P_{total} = P_A + P_B + P_C + \ldots$$

$$n = \frac{m}{M}$$

$$K = {}^\circ C + 273$$

$$D = \frac{m}{V}$$

$$KE \text{ per molecule} = \frac{1}{2}mv^2$$

Molarity, M = moles of solute per liter of solution

$$A = abc$$

P = pressure
V = volume
T = temperature
n = number of moles
m = mass
M = molar mass
D = density
KE = kinetic energy
v = velocity
A = absorbance
a = molar absorptivity
b = path length
c = concentration

Gas constant, R = 8.314 J mol^{-1} K^{-1}

= 0.08206 L atm mol^{-1} K^{-1}

= 62.36 L torr mol^{-1} K^{-1}

1 atm = 760 mm Hg

= 760 torr

STP = 0.00 $^\circ$C and 10^5 Pa

THERMOCHEMISTRY/ ELECTROCHEMISTRY

$$q = mc\Delta T$$

$$\Delta S^\circ = \sum S^\circ \text{ products} - \sum S^\circ \text{ reactants}$$

$$\Delta H^\circ = \sum \Delta H_f^\circ \text{ products} - \sum \Delta H_f^\circ \text{ reactants}$$

$$\Delta G^\circ = \sum \Delta G_f^\circ \text{ products} - \sum \Delta G_f^\circ \text{ reactants}$$

$$\Delta G^\circ = \Delta H^\circ - T\Delta S^\circ$$

$$= -RT \ln K$$

$$= -nFE^\circ$$

$$I = \frac{q}{t}$$

q = heat
m = mass
c = specific heat capacity
T = temperature
S° = standard entropy
H° = standard enthalpy
G° = standard free energy
n = number of moles
E° = standard reduction potential
I = current (amperes)
q = charge (coulombs)
t = time (seconds)

Faraday's constant, F = 96,485 coulombs per mole of electrons

1 volt = $\dfrac{1 \text{ joule}}{1 \text{ coulomb}}$

Glossary of Chemistry Terms

A

Absolute entropy (of a substance) – the increase in the entropy of a substance as it goes from a perfectly ordered crystalline form at 0 K (where its entropy is zero) to the temperature in question.

Absolute zero – the zero point on the absolute temperature scale; –273.15 °C or 0 K; theoretically, the temperature at which molecular motion ceases (i.e., the system does not emit or absorb energy, atoms at rest).

Absorption spectrum – spectrum associated with absorption of electromagnetic radiation by atoms (or other species), resulting from transitions from lower to higher energy states.

Accuracy – how close a value is to the actual value; see *precision*.

Acid – a substance that produces H^+ (*aq*) ions in an aqueous solution and gives a pH of less than 7.0; strong acids ionize entirely or almost entirely in dilute aqueous solution; weak acids ionize only slightly. It turns litmus red.

Acid dissociation constant – an equilibrium constant for dissociating a weak acid.

Acid rain – rainwater with a pH of less than 5.7; caused by the gases NO_2 (vehicle exhaust fumes) and SO_2 (from burning fossil fuels) dissolving in the rain. It kills fish, wildlife, and trees and destroys buildings and lakes.

Acidic salt – contains an ionizable hydrogen atom; does not necessarily produce acidic solutions.

Actinides – the fifteen chemical elements that are between actinium (89) and lawrencium (103).

Activated complex – a structure forming because of a collision between molecules while new bonds form.

Activation energy – the amount of energy that reactants must absorb in their ground states to reach the transition state needed for a reaction can occur.

Active metal – a metal with low ionization energy that loses electrons readily to form cations.

Activity (of a component of ideal mixture) – a dimensionless quantity whose magnitude is equal to the molar concentration in an ideal solution; equal to partial pressure in an ideal gas mixture; 1 for pure solids or liquids.

Activity series – a listing of metals (and hydrogen) in order of decreasing activity.

Actual yield – the amount of a specified pure product obtained from a given reaction; see *theoretical yield.*

Addition reaction – a reaction in which two atoms or groups of atoms are added to a molecule, one on each side of a double or triple bond.

Adhesive forces – forces of attraction between a liquid and another surface.

Adsorption – the adhesion of a species onto the surfaces of particles.

Aeration – the mixing of air into a liquid or a solid.

Alcohol – hydrocarbon derivative containing a ~OH group attached to a carbon atom, not in an aromatic ring.

Alkali metals – the elements of Group IA on the periodic table (e.g., Na, K, Rb).

Alkaline battery – a dry cell in which the electrolyte contains KOH.

Alkaline earth metals – group IIA metals on the periodic table; see *earth metals*.

Allomer – a substance that has a different composition than another but the same crystalline structure.

Allotropes – elements with different structures (therefore different forms), such as carbon (e.g., diamonds, graphite, and fullerene).

Allotropic modifications (allotropes) – different forms of the same element in the same physical state.

Alloy – a mixture of metals. For example, bronze is an alloy formed from copper and tin.

Alloying – mixing metal with other substances (usually other metals) to modify its properties.

Alpha (α) particle – a helium nucleus; helium ion with 2+ charge; an assembly of two protons and two neutrons.

Amorphous solid – a non-crystalline solid with no well-defined ordered structure.

Ampere – unit of electrical current; one ampere equals one coulomb per second.

Amphiprotism – the ability of a substance to exhibit amphiprotic by accepting donated protons.

Amphoterism – the ability to react with both acids and bases; to act as either an acid or a base.

Amplitude – the maximum distance that medium particles carrying the wave move from their rest position.

Anion – a negative ion; an atom or group of atoms that has gained one or more electrons.

Anode – in a cathode ray tube, the positive electrode (electrode at which oxidation occurs); the positive side of a dry cell battery or a cell.

Antibonding orbital – a molecular orbital higher in energy than any of the atomic orbitals from which it is derived; lends instability to a molecule or ion when populated with electrons; denoted with star (*) superscript.

Artificial transmutation – an artificially induced nuclear reaction caused by the bombardment of a nucleus with subatomic particles or small nuclei.

Associated ions – short-lived species formed by the collision of dissolved ions of opposite charges.

Atmosphere – a unit of pressure; the pressure supports a column of mercury 760 mm high at 0 °C.

Atom – a chemical element in its smallest form; it comprises neutrons and protons within the nucleus and electrons circling the nucleus; it is the smallest part of an element that can exist.

Atomic mass unit (amu) – one-twelfth of the mass of an atom of the carbon-12 isotope; used for stating atomic and formula weights; known as a dalton.

Atomic number – represents an element corresponding with the number of protons within the nucleus; the number of protons in the nucleus of the atom.

Atomic orbital (*AO*) – a region or volume in space where the probability of finding electrons is highest.

Atomic radius – radius of an atom.

Atomic weight – weighted average of the masses of the constituent isotopes of an element; the relative masses of atoms of different elements.

Aufbau (or *building up*) principle – describes the order in which electrons fill orbitals in atoms.

Autoionization – an ionization reaction between identical molecules.

Avogadro's Law – equal volumes of gases contain the same number of molecules at the same temperature and pressure.

Avogadro's number (N_A) – the number (6.022×10^{23}) of atoms, molecules, or particles found in precisely 1 mole of a substance.

B

Background radiation – extraneous to an experiment; usually the low-level natural radiation from cosmic rays and trace radioactive substances present in the environment.

Band – a series of very closely spaced nearly continuous molecular orbitals that belong to the crystal as a whole.

Band of stability – band containing nonradioactive nuclides in a plot of neutrons *vs.* their atomic number.

Band theory of metals – the theory that accounts for the bonding and properties of metallic solids.

Barometer – a device used to measure the pressure in the atmosphere.

Base – a substance that produces ⁻OH (*aq*) ions in an aqueous solution; accepts a proton and has a high pH; strongly soluble bases are soluble in water and are entirely dissociated; weak bases ionize only slightly; a typical example of a base is sodium hydroxide (NaOH). It turns litmus blue.

Basic anhydride – the oxide of a metal that reacts with water to form a base.

Basic salt – a salt containing an ionizable OH group.

Beta (β) particle – an electron emitted from the nucleus when a neutron decays to a proton and an electron.

Binary acid – a binary compound in which H is bonded to one or more electronegative nonmetals.

Binary compound – consists of two elements; it may be ionic or covalent.

Binding energy (nuclear binding energy) – the energy equivalent ($E = mc^2$) of the mass deficiency of an atom (where E is the energy in joules, m is the mass in kilograms, and c is the speed of light in m/s^2).

Boiling – the phase transition of liquid vaporizing.

Boiling point – the temperature at which the vapor pressure of a liquid is equal to the applied pressure; the *condensation point*.

Boiling point elevation – the increase in the boiling point of a solvent caused by the dissolution of a nonvolatile solute.

Bomb calorimeter – a device used to measure the heat transfer between a system and its surroundings at constant volume.

Bond – the attraction and repulsion between atoms and molecules is a cornerstone of chemistry.

Bond energy – the amount of energy necessary to break one mole of bonds in a substance, dissociating the substance in its gaseous state into atoms of its elements in the gaseous state.

Bond order – half the number of electrons in bonding orbitals minus half the electrons in antibonding orbitals.

Bonding orbital – a molecular orbit lower in energy than any of the atomic orbitals from which it is derived; lends stability to a molecule or ion when populated with electrons.

Bonding pair – pair of electrons involved in a covalent bond.

Boron hydrides – binary compounds of boron and hydrogen.

Born-Haber cycle – a series of reactions (and the accompanying enthalpy changes) which, when summed, represents the hypothetical one-step reaction by which elements in their standard states are converted into crystals of ionic compounds (and the accompanying enthalpy changes).

Boyle's Law – at a constant temperature, the volume occupied by a definite mass of a gas is inversely proportional to the applied pressure.

Breeder reactor – a nuclear reactor that produces more fissionable nuclear fuel than it consumes.

Brønsted-Lowrey acid – a chemical species that donates a proton.

Brønsted-Lowrey base – a chemical species that accepts a proton.

Buffer solution – resists change in pH; contains either a weak acid and a soluble ionic salt of the acid or a weak base and a soluble ionic salt of the base.

Buret – a piece of volumetric glassware, usually graduated in 0.1 mL intervals, used to deliver solutions for titrations in a quantitative (drop-like) manner; also spelled *burette*.

C

Calorie – the amount of heat required to raise the temperature of one gram of water from 14.5 °C to 15.5 °C; 1 calorie = 4.184 joules.

Calorimeter – a device used to measure the heat transfer between a system and its surroundings.

Canal ray – a stream of positively charged particles (cations) that moves toward the negative electrode in cathode ray tubes; observed to pass through canals in the negative electrode.

Capillary – a tube having a very small inside diameter.

Capillary action – the drawing of a liquid up the inside of a small-bore tube when adhesive forces exceed cohesive forces; the depression of the surface of the liquid when cohesive forces exceed the adhesive forces.

Catalyst – a chemical compound used to change the rate (to speed or slow it) of a regenerated reaction (i.e., not consumed) at the end of the reaction.

Catenation – the bonding of atoms of the same element into chains or rings (i.e., the ability of an element to bond with itself).

Cathode – the electrode at which reduction occurs; in a cathode ray tube, the negative electrode.

Cathodic protection – protection of a metal (making a cathode) against corrosion by attaching it to a sacrificial anode of more easily oxidized metal.

Cathode ray tube – a closed glass tube containing gas under low pressure, with electrodes near the ends and a luminescent screen near the positive electrode; produces cathode rays when a high voltage is applied.

Cation – a positive ion; an atom or group of atoms that lost one or more electrons.

Cell potential – the potential difference, E_{cell}, between oxidation and reduction half-cells under nonstandard conditions; the force in a galvanic cell pulls electrons through a reducing agent to an oxidizing agent.

Central atom – an atom in a molecule or polyatomic ion bonded to more than one other atom.

Chain reaction – a reaction that, once initiated, sustains itself and expands; a reaction in which reactive species, such as radicals, are produced in more than one step; these reactive species propagate the chain reaction.

Charles' Law – at constant pressure, the volume occupied by a definite mass of gas is directly proportional to its absolute temperature.

Chemical bonds – the attractive forces holding atoms together in elements or compounds.

Chemical change – when one or more new substances are formed.

Chemical equation – description of a chemical reaction by placing the formulas of the reactants on the left of an arrow and the formulas of the products on the right.

Chemical equilibrium – a state of dynamic balance in which the rates of forward and reverse reactions are equal; there is no net change in concentrations of reactants or products while a system is at equilibrium.

Chemical kinetics – the study of rates and mechanisms of chemical reactions and factors they depend on.

Chemical periodicity – the variations in properties of elements with their position in the periodic table.

Chemical reaction – the change of one or more substances into another or multiple substances.

Cloud chamber – a device for observing the paths of speeding particles as vapor molecules condense on them to form fog-like tracks.

Cobalt chloride paper – water test; water changes the color from blue to pink.

Coefficient of expansion – the ratio of the change in the length or the volume of a body to the original length or volume for a unit change in temperature.

Cohesive forces – the forces of attraction among particles of a liquid.

Colligative properties – physical properties of solutions that depend upon the number but not the kind of solute particles present.

Collision theory – theory of reaction rates that states that effective collisions between reactant molecules must occur for the reaction to occur.

Colloid – a heterogeneous mixture in which solute-like particles do not settle out (e.g., many kinds of milk).

Combination reaction – two substances (elements or compounds) combine to form one compound.

Combustible – classification of liquid substances that burn based on flashpoints; any liquid having a flashpoint at or above 37.8 °C (100 °F) but below 93.3 °C (200 °F), except any mixture having components with flashpoints of 93.3 °C (200 °F) or higher, the total of which makes up 99% or more of the volume of the mixture.

Combustion (or *burning*) – an exothermic reaction between an oxidant and fuel with heat and often light.

Common ion effect – suppression of ionization of a weak electrolyte by the presence in the same solution of a strong electrolyte containing one of the same ions as the weak electrolyte.

Complex ions – ions resulting from coordinating covalent bonds between simple ions and other ions or molecules.

Composition stoichiometry – describes the quantitative (mass) relationships among elements in compounds.

Compound – a substance of two or more chemically bonded elements in fixed proportions; can be decomposed into constituent elements.

Compressed gas – a single or mixture of gases having (in a container) an absolute pressure exceeding 40 psi at 21.1 °C (70 °F).

Compression – an area in a longitudinal wave where the particles are closer and pushed in.

Concentration – the amount of solute per unit volume, the mass of solvent or solution.

Condensation – the phase change from gas to liquid.

Condensed phases – the liquid and solid phases; phases in which particles interact strongly.

Condensed states – the solid and liquid states.

Conduction – heat transfer between substances in direct contact with each other (i.e., must be touching); when particles of a hotter substance vibrate, these molecules bump into nearby particles and transfer some energy.

Conduction band – a vacant or partially filled band of energy levels just higher in energy than a filled band; a band within which, or into which, electrons must be promoted to allow electrical conduction to occur in a solid.

Conductor – material that allows electric flow more freely.

Conjugate acid-base pair – in Brønsted-Lowry terms, a reactant and a product that differ by a proton, H^+.

Conformations – structures of a compound that differ by the extent of their rotation about a single bond.

Continuous spectrum – contains wavelengths in a specified region of the electromagnetic spectrum.

Control rods – rods of materials such as cadmium or boron steel that act as neutron absorbers (not merely moderators), used in nuclear reactors to control neutron fluxes and therefore fission rates.

Conjugated double bonds – double bonds separated from each other by one single bond – C=C–C=C–

Contact process – the industrial process for sulfur trioxide and sulfuric acid production from sulfur dioxide.

Convection – the physical flow of matter when heat flows by energized molecules from one place to another through the movement of fluids. The transfer of heat through a liquid or a gas when molecules of the liquid or gas move and carry the heat.

Coordinate covalent bond – a covalent bond with shared electrons furnished by the same species; a bond between a Lewis acid and a Lewis base.

Coordination compound or complex – a compound containing coordinate covalent bonds.

Coordination number – the number of donor atoms coordinated to metal; the number of nearest neighbors of an atom or ion in describing crystals.

Coordination sphere – the metal ion and its coordinating ligands but no uncoordinated counter-ions.

Corrosion – oxidation of metals (e.g., rusting) in the presence of air and moisture.

Coulomb – the SI unit of electrical charge; unit symbol – C.

Covalent bond – a force of attraction (chemical bond) formed by the sharing of electron pairs between two atoms.

Covalent compounds – compounds made of two or more nonmetal atoms bonded by sharing valence electrons.

Critical mass – the minimum mass of a particular fissionable nuclide in a given volume required to sustain a nuclear chain reaction.

Critical point – the combination of critical temperature and critical pressure of a substance.

Critical pressure – the pressure required to liquefy a gas (vapor) at its *Critical temperature.*

Critical temperature – the temperature above which a gas cannot be liquefied; the temperature above which a substance cannot exhibit distinct gas and liquid phases.

Crystal – a solid packed with ions, molecules, or atoms in an orderly fashion.

Crystal field stabilization energy – a measure of the net energy of stabilization gained by a metal ion's nonbonding d electrons due to complex formation.

Crystal field theory – bonding in transition metal complexes in which ligands and metal ions are treated as point charges; a purely ionic model; ligand point charges represent the crystal (electrical) field perturbing the metal's d orbitals containing nonbonding electrons.

Crystal lattice – a pattern of arrangement of particles in a crystal.

Crystal lattice energy – the amount of energy that holds a crystal together; the energy change when a mole of solid forms fom its constituent molecules or ions (for ionic compounds) in their gaseous state (always negative).

Crystalline solid – a solid characterized by a regular, ordered arrangement of particles.

Curie (Ci) – the basic unit to describe the intensity of radioactivity in a sample of material; one curie equals 37 billion disintegrations per second or approximately the amount of radioactivity given off by 1 gram of radium.

Current – a flow of charged particles, such as electrons or ions, moving through an electrical space or conductor. It is measured as the net rate of flow of electric charge; the unit is Ampere (A).

Cuvette – glassware used in spectroscopic experiments; usually made of plastic, glass, or quartz and should be as clean and transparent as possible.

Cyclotron – a device for accelerating charged particles along a spiral path.

D

Dalton's Law (or the *law of partial pressures*) – the pressure exerted by a mixture of gases is the sum of the partial pressures of the individual gases.

Daughter nuclide – nuclide produced in nuclear decay.

Debye – the unit used to express dipole moments.

Degenerate – in orbitals, describes orbitals of the same energy.

Deionization – the removal of ions; in the case of water, mineral ions such as sodium, iron and calcium.

Deliquescence – substances that absorb water from the atmosphere to form liquid solutions.

Delocalization – in reference to electrons, bonding electrons distributed among more than two atoms bonded; occurs in species that exhibit resonance.

Density – mass per unit volume; $D = M \times V$.

Deposition – settling particles within a solution; the direct solidification of vapor by cooling; see *sublimation*.

Derivative – a compound that can be imagined arising from a parent compound by replacing one atom with another atom or group of atoms; used extensively in organic chemistry to identify compounds.

Detergent – a soap-like emulsifier with a sulfate, SO_3, or a phosphate group instead of a carboxylate group.

Deuterium – an isotope of hydrogen whose atoms are twice as massive as ordinary hydrogen; deuterium atoms contain a proton and a neutron in the nucleus.

Dextrorotatory – refers to an optically active substance that rotates plane-polarized light clockwise, also known as *dextro*.

Diagonal similarities – chemical similarities in the Periodic Table of Elements of Period 2 to elements of Period 3 one group to the right, especially evident toward the left of the periodic table.

Diamagnetism – weak repulsion by a magnetic field.

Differential Scanning Calorimetry (DSC) – a technique for measuring temperature, direction, and magnitude of thermal transitions in a sample material by heating/cooling and comparing the amount of energy required to maintain its rate of temperature increase or decrease with an inert reference material under similar conditions.

Differential Thermal Analysis (DTA) – a technique for observing the temperature, direction and magnitude of thermally induced transitions in a material by heating/cooling a sample and comparing its temperature with an inert reference material under similar conditions.

Differential thermometer – a thermometer used to measure very small temperature changes accurately.

Dilution – the process of reducing the concentration of a solute in a solution, usually by mixing with more solvent.

Dimer – molecule formed by combining two smaller (identical) molecules.

Dipole – electric or magnetic separation of charge; charge separation between two covalently bonded atoms.

Dipole-dipole interactions – attractive electrostatic forces between polar molecules (i.e., between molecules with permanent dipoles).

Dipole moment – the product of the distance separating opposite charges of an equal magnitude of charge; a measure of the polarity of a bond or molecule; a measured dipole refers to the dipole moment of an entire molecule.

Dispersing medium – the solvent-like phase in a colloid.

Dispersed phase – the solute-like species in a colloid.

Displacement reactions – reactions in which one element displaces another from a compound.

Disproportionation reactions – redox reactions in which the oxidizing agent and the reducing agent are the same species.

Dissociation – in an aqueous solution, the process by which a solid ionic compound separates into its ions.

Dissociation constant – equilibrium constant for dissociating a complex ion into a simple ion and coordinating species (ligands).

Dissolution or solvation – the spread of ions in a monosaccharide.

Distilland – the material in a distillation apparatus that is to be distilled.

Distillate – the material in a distillation apparatus collected in the receiver.

Distillation – separating a liquid mixture into its components based on differences in boiling points; the process in which components of a mixture are separated by boiling away the more volatile liquid; the vaporization of a liquid by heating and then the condensation of the vapor by cooling.

Domain – a cluster of atoms in a ferromagnetic substance, which align in the same direction in the presence of an external magnetic field.

Donor atom – a ligand atom whose electrons are shared with a Lewis acid.

***d*-orbitals** – beginning in the third energy level, a set of five degenerate orbitals per energy level, higher in energy than s and p orbitals of the same energy level.

Dosimeter – a small, calibrated electroscope worn by laboratory personnel to measure incident ionizing radiation or chemical exposure.

Double bond – covalent bond resulting from the sharing of four electrons (two pairs) between two atoms.

Double salt – solid consisting of two co-crystallized salts.

Doublet – two peaks or bands of about equal intensity appearing close on a spectrogram.

Downs cell – electrolytic cell for the commercial electrolysis of molten sodium chloride.

DP number – the degree of polymerization; the average number of monomer units per polymer unit.

Dry cells (voltaic cells) – ordinary batteries for appliances (e.g., flashlights, radios).

Dumas method – a method used to determine the molecular weights of volatile liquids.

Dynamic equilibrium – an equilibrium in which the processes occur continuously with no net change.

E

Earth metal – highly reactive elements in group IIA of the periodic table (includes beryllium, magnesium, calcium, strontium, barium, and radium); see *alkaline earth metal*.

Effective collisions – a collision between molecules resulting in a reaction; one in which the molecules collide with proper relative orientations and sufficient energy to react.

Effective molality – the sum of the molalities of solute particles in a solution.

Effective nuclear charge – the nuclear charge experienced by the outermost electrons of an atom; the actual nuclear charge minus the effects of shielding due to inner-shell electrons (e.g., a set of dx_2-y_2 and dz_2 orbitals); those d orbitals within a set with lobes directed along the x, y and z-axes.

Electrical conductivity – the measure of how easily an electric current can flow through a substance.

Electric charge – a measured property (coulombs) that determines electromagnetic interaction.

Electrochemical cell – using a chemical reaction's current; electromotive force is made.

Electrochemistry – the study of chemical changes produced by electrical current and electricity production by chemical reactions.

Electrodes – surfaces upon which oxidation and reduction half-reactions occur in electrochemical cells; a conductor dips into an electrolyte and allows the electrons to flow to and from the electrolyte.

Electrode potentials – potentials, E, of half-reactions as reductions versus the standard hydrogen electrode.

Electrolysis – 1) occurs in electrolytic cells; chemical decomposition occurs by passing an electric current through a solution containing ions. 2) producing a chemical change using electricity; used to split up water into H and O_2.

Electrolyte – a substance (i.e., anions, cations) which, when dissolved in water, conducts electricity. An ionic solution that conducts a certain amount of current and split categorically as weak and strong.

Electrolytic cells – electrochemical cells in which electrical energy causes nonspontaneous redox reactions to occur (i.e., forced to occur by applying an outside source of electrical energy).

Electrolytic conduction – electrical current passes by ions through a solution or pure liquid.

Electromagnetic radiation – energy propagated using electric and magnetic fields that oscillate in directions perpendicular to the direction of travel of the energy; a type of wave that can go through vacuums as well as material; classified as a "self-propagating wave."

Electromagnetism – fields of an electric charge and electric properties that change how particles move and interact.

Electromotive force – a device that gains energy as electric charges pass through it.

Electromotive series – the relative order of tendencies for elements and their simple ions to act as oxidizing or reducing agents; also known as the "activity series."

Electron – a subatomic particle having a mass of 0.00054858 amu and a charge of 1−.

Electron affinity – the amount of energy absorbed in the process in which an electron is added to a neutral isolated gaseous atom to form a gaseous ion with a 1− charge; it has a negative value if energy is released.

Electron configuration – the specific distribution of electrons in atomic orbitals of atoms or ions.

Electron-deficient compounds –contain at least one atom (other than H) that shares fewer than eight electrons.

Electron shells – an orbital around the atom's nucleus with a fixed number of electrons (usually two or eight).

Electronic transition – the transfer of an electron from one energy level to another.

Electronegativity – a measure of the relative tendency of an atom to attract electrons to itself when chemically combined with another atom.

Electronic geometry – the geometric arrangement of orbitals containing the shared and unshared electron pairs surrounding the central atom or polyatomic ion.

Electrophile – positively charged or electron-deficient.

Electrophoresis – a technique for separating ions by migration rate and direction of migration in an electric field.

Electroplating – a metal is covered with another metal layer using electricity; plating a metal onto a (cathodic) surface by electrolysis.

Element – a substance that cannot be decomposed into simpler substances by chemical means; defined by its *atomic number*. A substance that cannot be split into simpler substances by chemical means.

Eluant (or eluent) – the solvent used in the process of elution, as in liquid chromatography.

Eluate – a solvent (or mobile phase) which passes through a chromatographic column and removes the sample components from the stationary phase.

Emission spectrum – the emission of electromagnetic radiation by atoms (or other species) resulting from electronic transitions from higher to lower energy states.

Empirical formula – gives the simplest whole-number ratio of atoms of each element present in a compound; also known as the simplest formula.

Emulsifying agent – a substance that coats the particles of the dispersed phase and prevents coagulation of colloidal particles; an emulsifier.

Emulsion – colloidal suspension of a liquid in a liquid.

Endothermic – describes processes that absorb heat energy (*H*).

Endothermicity – the absorption of heat by a system as the process occurs.

Endpoint – the point at which an indicator changes color and a titration is stopped.

Energy – a system's ability to do work.

Enthalpy (*H*) – the heat content of a specific amount of substance; $E = PV$.

Entropy *(S)* – a thermodynamic state or property that measures the degree of disorder (i.e., randomness) of a system; the amount of energy not available for work in a closed thermodynamic system (usually denoted by S).

Enzyme – a protein that acts as a catalyst in biological systems.

Equation of state – an expression describes the behavior of matter in a given state; the van der Waals equation describes the behavior of the gaseous state.

Equilibrium or chemical equilibrium – a state of dynamic balance with the rates of forward and reverse reactions equal; the state of a system when neither forward nor reverse reaction is thermodynamically favored.

Equilibrium constant – a quantity that characterizes the equilibrium position for a reversible reaction; its magnitude is equal to the mass action expression at equilibrium; equilibrium, "K," varies with temperature.

Equivalence point – the point when chemically equivalent amounts of reactants have reacted.

Equivalent weight – an oxidizing or reducing agent whose mass gains (oxidizing agents) or loses (reducing agents) 6.022×10^{23} electrons in a redox reaction.

Evaporation – vaporization of a liquid below its boiling point.

Evaporation rate – the rate at which a particular substance will vaporize (evaporate) compared to the rate of a known substance such as ethyl ether, especially useful for health and fire-hazard considerations.

Excited state – any state other than the ground state of an atom or molecule; see *ground state*.

Exothermic – describes processes that release heat energy (*H*).

Exothermicity – the release of heat by a system as a process occurs.

Explosive – a chemical or compound that causes a sudden, almost instantaneous release of pressure, gas, heat, and light when subjected to sudden shock, pressure, high temperature, or applied potential.

Explosive limits – the range of concentrations over which a flammable vapor mixed with the proper ratios of air will ignite or explode if a source of ignition is provided.

Extensive property – a property that depends upon the amount of material in a sample.

Extrapolate – to estimate the value of a result outside the range of a series of known values; a technique used in standard additions calibration procedure.

F

Faraday constant – a unit of electrical charge widely used in electrochemistry and equal to ~ 96,500 coulombs; represents 1 mole of electrons, or the Avogadro number of electrons: 6.022×10^{23} electrons.

Faraday's law of electrolysis – a two-part law that Michael Faraday published about electrolysis. 1. the mass of a substance altered at an electrode during electrolysis is directly proportional to the quantity of electricity transferred at that electrode. 2. the mass of an elemental material altered at an electrode is directly proportional to the element's equivalent weight; one equivalent weight of a substance is produced at each electrode during the passage of 96,487 coulombs of charge through an electrolytic cell.

Fast neutron – a neutron ejected at high kinetic energy in a nuclear reaction.

Ferromagnetism – the ability of a substance to become permanently magnetized by exposure to an external magnetic field.

Flashpoint – the temperature at which a liquid will yield enough flammable vapor to ignite; there are various recognized industrial testing methods; therefore, the method used must be stated.

Fluorescence – absorption of high energy radiation by a substance and subsequent emission of visible light.

First Law of Thermodynamics – the amount of energy in the universe is constant (i.e., energy is neither created nor destroyed in ordinary chemical reactions and physical changes); known as the Law of Conservation of Energy.

Fluids – substances that flow freely; gases and liquids.

Flux – a substance added to react with the charge or a product of its reduction; in metallurgy, it is usually added to lower a melting point.

Foam – colloidal suspension of a gas in a liquid.

Formal charge – a method of counting electrons in a covalently bonded molecule or ion; it counts bonding electrons as though they were equally shared between the two atoms.

Formula – a combination of symbols that indicates the chemical composition of a substance.

Formula unit – the smallest repeating unit of a substance; the molecule for nonionic substances.

Formula weight – the mass of one formula unit of a substance in atomic mass units.

Fossil fuels – formed from the remains of plants and animals that lived millions of years ago.

Fractional distillation – when a fractioning column is used in a distillation apparatus to separate the components of a liquid mixture with different boiling points.

Fractional precipitation – removal of some ions from a solution by precipitation while leaving other ions with similar properties in the solution.

Free energy change – the indicator of the spontaneity of a process at constant T and P (e.g., if ΔG is negative, the process is spontaneous).

Free radical – a highly reactive chemical species carrying no charge and having a single unpaired electron in an orbital.

Freezing – phase transition from liquid to solid.

Freezing point depression – the decrease in the freezing point of a solvent caused by the presence of a solute.

Frequency – the number of repeating points on a wave that passes a given observation point per unit time; the unit is 1 hertz = 1 cycle per 1 second.

Fuel – any substance that burns in oxygen to produce heat.

Fuel cells – a voltaic cell that converts the chemical energy of a fuel and an oxidizing agent directly into electrical energy continuously.

G

Gamma (γ) ray – a highly penetrating type of nuclear radiation similar to x-ray radiation, except that it comes from within the nucleus of an atom and has higher energy; energy-wise, very similar to cosmic rays except that cosmic rays originate from outer space.

Galvanic cell – battery made up of electrochemical with two different metals connected by a salt bridge.

Galvanizing – placing a thin layer of zinc on a ferrous material to protect the underlying surface from corrosion.

Gangue – sand, rock and other impurities surrounding the mineral of interest in an ore.

Gas – a state of matter in which the particles have no definite shape or volume, though they fill their container.

Gay-Lussac's Law – the expression used for each of the two relationships named after the French chemist Joseph Louis Gay-Lussac concerning the properties of gases; more usually applied to his law of combining volumes.

Geiger counter – a gas-filled tube that discharges electrically when ionizing radiation passes through it.

Gel – colloidal suspension of a solid dispersed in a liquid; a semi-rigid solid.

Gibbs (free) energy – the thermodynamic state function of a system that indicates the amount of energy available for the system to do useful work at constant T and P; a value that indicates the spontaneity of a reaction (usually denoted by *G*).

Graham's Law – the rates of effusion of gases are inversely proportional to the square roots of their molecular weights or densities.

Ground state – the lowest energy state or most stable state of an atom, molecule, or ion; see *excited state*.

Group – a vertical column in the periodic table; known as a family.

H

Haber process – a process for the catalyzed industrial production of ammonia from N_2 and H_2 at high temperature and pressure.

Half-cell – the compartment in which the oxidation or reduction half-reaction occurs in a voltaic cell.

Half-life – the time required for half of a reactant to be converted into product(s); the time required for half of a given sample to undergo radioactive decay.

Half-reaction – the oxidation or the reduction part of a redox reaction.

Halogens – group VIIA elements: F, Cl, Br, I; halogens are nonmetals.

Hard water – water high in dissolved minerals that is it difficult to form lather with soap.

Heat – a form of energy that flows between two samples of matter because of their temperature differences.

Heat capacity – the amount of heat required to raise the temperature of a mass one degree Celsius.

Heat of condensation – the amount of heat that must be removed from one gram of vapor at its condensation point to condense the vapor with no change in temperature.

Heat of crystallization – the amount of heat that must be removed from one gram of a liquid at its freezing point to freeze it with no change in temperature.

Heat of fusion – the amount of heat required to melt one gram of a solid at its melting point with no change in temperature; usually expressed in J/g; the molar heat of fusion is the amount of heat required to melt one mole of a solid at its melting point with no change in temperature and is usually expressed in kJ/mol.

Heat of solution – the amount of heat absorbed in forming a solution that contains one mole of the solute; the value is positive if heat is absorbed (endothermic) and negative if heat is released (exothermic).

Heat of vaporization – the amount of heat required to vaporize one gram of a liquid at its boiling point with no change in temperature; usually expressed in J/g; the molar heat of vaporization is the amount of heat required to vaporize one mole of liquid at its boiling point with no change in temperature and is usually expressed as ion kJ/mol.

Heisenberg uncertainty principle – states that it is impossible to accurately determine the *momentum* and *position* of an electron simultaneously.

Henry's Law – the gas pressure above a solution is proportional to the concentration of the gas in the solution.

Hess' Law of heat summation – the enthalpy change for a reaction is the same whether it occurs in one step or a series of steps.

Heterogeneous catalyst – exist in a different phase (solid, liquid, or gas) from the reactants; a contact catalyst.

Heterogeneous equilibria – equilibria involving species in more than one phase.

Heterogeneous mixture – a mixture that does not have uniform composition and properties throughout.

Heteronuclear – consisting of different elements.

High spin complex – crystal field designation for an outer orbital complex; t_{2g} and e_g orbitals are singly occupied before pairing occurs.

Homogeneous catalyst – in the same phase (solid, liquid, or gas) as the reactants.

Homogeneous equilibria – when *reagents* and *products* are of the same phase (i.e., gases, liquids, or solids).

Homogeneous mixture – a mixture which has uniform composition and properties throughout.

Homologous series – compounds with each member differing from the next by a specific number and kind of atoms.

Homonuclear – consisting of only one element.

Hund's rule – single electrons must occupy orbitals of a given sublevel before pairing begins; see *Aufbau* (or *building up*) *principle*.

Hybridization – mixing atomic orbitals to form a new set of atomic orbitals with the same electron capacity and properties and energies intermediate between the original unhybridized orbitals.

Hydrate – a solid compound that contains a definite percentage of bound water.

Hydrate isomers – crystalline complexes that differ in whether water exists inside or outside the coordination sphere.

Hydration – the reaction of a substance with water.

Hydration energy – the energy change accompanying the hydration of a mole of gas and ions.

Hydride – a binary compound of hydrogen.

Hydrocarbons – compounds that contain only carbon and hydrogen.

Hydrogen bond – a relatively strong dipole-dipole interaction (but still considerably weaker than the covalent or ionic bonds) between molecules containing hydrogen directly bonded to a small, highly electronegative atom, such as N, O or F.

Hydrogenation – the reaction in which hydrogen adds across a double or triple bond.

Hydrogen-oxygen fuel cell – hydrogen is the fuel (reducing agent) and oxygen is the oxidizing agent.

Hydrolysis – the reaction of a substance with water or its ions.

Hydrolysis constant – an equilibrium constant for a hydrolysis reaction.

Hydrometer – a device used to measure the densities of liquids and solutions.

Hydrophilic colloids – colloidal particles that repel water molecules.

I

Ideal gas – a hypothetical gas that obeys the postulates of the kinetic-molecular theory.

Ideal gas law – the product of pressure and the volume of an ideal gas is directly proportional to the number of moles of the gas and the absolute temperature. $PV = nRT$

Ideal solution – obeys Raoult's Law strictly.

Immiscible liquids – do not mix to form a solution (e.g., oil and water).

Indicators – for acid-base titrations, organic compounds that exhibit different colors in solutions of different acidities, determine the point at which the reaction between two solutes is complete.

Inert pair effect – characteristic of the post-transition minerals; the tendency of the electrons in the outermost atomic *s* orbital to remain un-ionized or unshared in compounds of post-transition metals.

Inhibitory catalyst – an inhibitor; a catalyst that decreases the rate of reaction.

Inner orbital complex – valence bond designation for a complex in which the metal ion utilizes d orbitals for one shell inside the outermost occupied shell in its hybridization.

Inorganic chemistry – a part of chemistry concerned with inorganic (non carbon-based) compounds.

Insulator – a material that resists the flow of electric current or heat transfer; does not allow heat to flow easily.

Insoluble compound – a substance that will not dissolve in a solvent, even after mixing.

Integrated rate equation – an expression giving the concentration of a reactant remaining after a specified time; has a different mathematical form for different orders of reactants.

Intermolecular forces – forces between individual particles (atoms, molecules, ions) of a substance.

Ion – a molecule that has gained or lost electrons; an atom or a group of atoms carries an electric charge (Na^+).

Ion product for water – equilibrium constant for water ionization; $Kw = [H_3O^+] \cdot [^-OH] = 1.00 \times 10^{-14}$ at 25 °C.

Ionic bond – electrostatic attraction between oppositely charged ions, resulting from a transfer of electrons.

Ionic bonding – chemical bonding resulting from transferring electrons from one atom or group.

Ionic compounds – compounds containing predominantly ionic bonding.

Ionic geometry – arrangement of atoms (not lone pairs of electrons) about the central atom of a polyatomic ion.

Ionization – the breaking up of a compound into separate ions; in an aqueous solution, the process by which a molecular compound reacts with water and forms ions.

Ionization constant – equilibrium constant for the ionization of a weak electrolyte.

Ionization energy – the minimum amount of energy required to remove the most loosely held electron of an isolated gaseous atom or ion.

Ion exchange – a method of removing hardness from water, and it replaces the positive ions that cause the hardness with H^+ ions.

Ionization isomers – result from the interchange of ions inside and outside the coordination sphere.

Isoelectric – having the same electronic configurations.

Isomers – different substances with the same formula.

Isomorphous – refers to crystals having the same atomic arrangement.

Isotopes – two or more forms of atoms of the same element with different masses; atoms containing the same number of protons but different numbers of neutrons.

IUPAC – acronym for "International Union of Pure and Applied Chemistry."

J

Joule (J) – a unit of energy in the SI system; one joule is $1 \text{ kg} \cdot \text{m}^2/\text{s}^2$, which is 0.2390 calories.

K

K capture – absorption of a K shell (n = 1) electron by a proton as it is converted to a neutron.

Kelvin – a unit of measure for temperature based upon an absolute scale.

Kinetics – a sub-field of chemistry specializing in reaction rates.

Kinetic energy (*KE*) – energy that matter processes by its motion.

Kinetic-molecular theory – a theory that attempts to explain macroscopic observations on gases in microscopic or molecular terms.

L

Lanthanides – elements 57 (lanthanum) through 71 (lutetium); grouped because of their similar behavior in chemical reactions.

Lanthanide contraction – a decrease in the radii of the elements following the lanthanides compared to what would be expected if there were no *f*-transition metals.

Latent heat – the energy absorbed or released when a substance changes state without changing temperature.

Lattice – unique arrangement of atoms or molecules in a crystalline liquid or solid.

Law of combining volumes (Gay-Lussac's Law) – at constant temperature and pressure, the volumes of reacting gases (and any gaseous products) can be expressed as ratios of small whole numbers.

Law of conservation of energy – energy cannot be created or destroyed; it can only change form.

Law of conservation of matter – there is no detectable change in the quantity of matter during an ordinary chemical reaction.

Law of conservation of matter and energy – the amount of matter and energy in the universe is fixed.

Law of definite proportions (law of constant composition) – different samples of a pure compound contain the same elements in the same proportions by mass.

Law of partial pressures (or *Dalton's Law*) – the pressure exerted by a mixture of gases is the sum of the partial pressures of the individual gases.

Laws of thermodynamics – physical laws which define quantities of thermodynamic systems describe how they behave and (by extension) set certain limitations such as perpetual motion.

Lead storage battery – secondary voltaic cell used in most automobiles.

Leclanche cell – a common type of *dry cell*.

Le Chatelier's principle – states that a system at equilibrium, or striving to attain equilibrium, responds in such a way as to counteract any stress placed upon it; if stress (change of conditions) is applied to a system at equilibrium, the system will shift in the direction that reduces stress.

Leveling effect – acids stronger than the acid characteristic of the solvent reacts with the solvent produce that acid; a similar statement applies to bases. The strongest acid (base) that can exist in a given solvent is the acid (base) characteristic of the solvent.

Levorotatory – an optically active substance rotates plane-polarized light counterclockwise, known as a *levo*.

Lewis acid – any species that can accept a share in an electron pair.

Lewis base – any species that can make available a share in an electron pair.

Lewis dot formula (electron dot formula) – representation of a molecule, ion or formula unit by showing atomic symbols and only outer shell electrons.

Ligand – a Lewis base in a coordination compound.

Light – that portion of the electromagnetic spectrum visible to the naked eye; known as "visible light."

Limiting reactant – a substance that stoichiometrically limits the number of product(s) that can be formed.

Linear accelerator – a device used for accelerating charged particles along a straight line path.

Line spectrum – an atomic emission or absorption spectrum.

Linkage isomers – a particular ligand bonds to a metal ion through different donor atoms.

Liquid – a state of matter which takes the shape of its container.

Liquid aerosol – colloidal suspension of a liquid in gas.

London dispersion forces – very weak and very short-range attractive forces between short-lived temporary (induced) dipoles; known as "dispersion forces."

Lone pair – pair of electrons residing on one atom and not shared by other atoms; unshared pair.

Low spin complex – crystal field designation for an inner orbital complex; contains electrons paired t_{2g} orbitals before e_g orbitals are occupied in octahedral complexes.

Lubricant – a substance capable of reducing friction (i.e., force that opposes the direction of motion).

M

Magnetic field – a space around a magnet where magnetism can be detected.

Magnetic quantum number – quantum mechanical solution to a wave equation designating the orbital within a given set (s, p, d, f) in which an electron resides.

Manometer – a two-armed barometer.

Mass (m) – a measure of the amount of matter in an object; mass is usually measured in grams or kilograms.

Mass action expression – for a reversible reaction, aA + bB cC + dD; the product of the concentrations of the products (species on right), each raised to the power corresponding to its coefficient in the balanced chemical equation, divided by the product of the concentrations of reactants (species on left), each raised to the power corresponding to its coefficient in the balanced equation; at equilibrium the mass action expression is equal to K.

Mass deficiency – the amount of matter converted into energy when an atom forms from constituent particles.

Mass number – the sum of the numbers of protons and neutrons in an atom; an integer.

Mass spectrometer – an instrument that measures the charge-to-mass ratio of charged particles.

Matter – anything that has mass and occupies space.

Mechanism – the sequence of steps by which reactants are converted into products.

Melting point – the temperature at which liquid and solid coexist in equilibrium.

Meniscus – the shape assumed by the surface of a liquid in a cylindrical container.

Melting – the phase change from a solid to a liquid.

Metal – a chemical element that is a good conductor of electricity and heat and forms cations and ionic bonds with nonmetals; elements below and to the left of the stepwise division (metalloids) in the upper right corner of the periodic table; about 80% of known elements are metals.

Metallic bonding – bonding within metals due to the electrical attraction of positively charged metal ions for mobile electrons that belong to the crystal.

Metallic conduction – conduction of electrical current through a metal or along a metallic surface.

Metalloid – a substance with the properties of metals and nonmetals (B, Al, Si, Ge, As, Sb, Te, Po and At).

Metathesis reactions – reactions in which two compounds react to form two new compounds, with no changes in oxidation number; reactions in which the ions of two compounds exchange partners.

Method of initial rates – method of determining the rate-law expression by carrying out a reaction with different initial concentrations and analyzing the resultant changes in initial rates.

Methylene blue – a heterocyclic aromatic chemical compound with the molecular formula $C_{16}H_{18}N_3SCl$.

Miscible liquids – mix to form a solution (e.g., alcohol and water).

Miscibility – the ability of one liquid to mix with (dissolve in) another liquid.

Mixture – two or more different substances mingled together but not chemically combined. A sample of matter composed of two or more substances, each of which retains its identity and properties.

Moderator – a substance (e.g., deuterium, oxygen, paraffin) capable of slowing fast neutrons upon collision.

Molality – a concentration expressed as a number of moles of solute per kilogram of solvent.

Molarity – the number of moles of solute per liter of solution.

Molar solubility – the number of moles of a solute that dissolves to produce a liter of a saturated solution.

Mole – a measurement of an amount of substance; a single mole contains approximately 6.022 $\times 10^{23}$ units or entities; abbreviated mol.

Molecule – a chemically bonded number of electrically neutral atoms.

Molecular equation – a chemical reaction in which formulas are written as if substances existed as molecules; only complete formulas are used.

Molecular formula – indicates the actual number of atoms present in a molecule of a molecular substance.

Molecular geometry – the arrangement of atoms (not lone pairs of electrons) around a central atom of a molecule or polyatomic ion.

Molecular orbital (*MO*) – resulting from the overlap and mixing of atomic orbitals on different atoms (i.e., a region where an electron can be found in a molecule, as opposed to an atom); an MO belongs to the molecule.

Molecular orbital theory – a theory of chemical bonding based upon postulated molecular orbitals.

Molecular weight – the mass of one molecule of a nonionic substance in atomic mass units.

Molecule – the smallest particle of a compound capable of stable, independent existence.

Mole fraction – the number of moles of a component in a mixture divided by the number of moles in the mixture.

Monoprotic acid – can form only one hydronium ion per molecule; may be strong or weak.

Mother nuclide – nuclide that undergoes nuclear decay.

N

Native state – refers to the occurrence of an element in an uncombined or free state in nature.

Natural radioactivity – spontaneous decomposition of an atom.

Neat – conditions with a liquid reagent or gas performed with no added solvent or co-solvent.

Nernst equation – corrects standard electrode potentials for nonstandard conditions.

Net ionic equation – results from canceling spectator ions and eliminating brackets from a total ionic equation.

Neutralization – the reaction of an acid with a base to form a salt and water; usually, the reaction of hydrogen ions with hydrogen ions to form water molecules.

Neutrino – particle that can travel at speeds close to the speed of light; created due to radioactive decay.

Neutron – a neutral unit or subatomic particle with no net charge and a mass of 1.0087 amu.

Nickel-cadmium cell (NiCd battery) – a dry cell where the anode is Cd, the cathode is NiO_2, and the electrolyte is basic.

Nitrogen cycle – the complex series of reactions by which nitrogen is slowly but continually recycled in the atmosphere, lithosphere, and hydrosphere.

Noble gases – elements of the periodic Group 0; He, Ne, Ar, Kr, Xe, Rn; known as "rare gases;" formerly called "inert gases."

Nodal plane – a region in which the probability of finding an electron is zero.

Nonbonding orbital – a molecular orbital derived only from an atomic orbital of one atom; lends neither stability nor instability to a molecule or ion when populated with electrons.

Nonelectrolyte – a substance whose aqueous solutions do not conduct electricity.

Nonmetal – an element that is not metallic.

Nonpolar bond – a covalent bond in which electron density is symmetrically distributed.

Nuclear – of or about the atomic nucleus.

Nuclear binding energy – the energy equivalent of the mass deficiency; the energy released in forming an atom from the subatomic particles.

Nuclear fission – when a heavy nucleus splits into nuclei of intermediate masses and protons are emitted.

Nuclear magnetic resonance spectroscopy – a technique that exploits the magnetic properties of specific nuclei; helpful in identifying unknown compounds.

Nuclear reaction – involves a change in the composition of a nucleus and can emit or absorb a tremendous amount of energy.

Nuclear reactor – a system in which controlled nuclear fission reactions generate heat energy on a large scale, subsequently converted into electrical energy.

Nucleons – particles comprising the nucleus; protons, and neutrons.

Nucleus – the very small and dense, positively charged center of an atom containing protons and neutrons, as well as other subatomic particles; the net charge is positive.

Nuclides – refers to different atomic forms of elements; in contrast to isotopes, which refer only to different atomic forms of a single element.

Nuclide symbol – designation for an atom A/Z E, in which E is the symbol of an element, Z is its atomic number, and A is its mass number.

Number density – a measure of the concentration of countable objects (e.g., atoms, molecules) in space; the number per volume.

O

Octahedral – molecules and polyatomic ions with one atom in the center and six atoms at the corners of an octahedron.

Octane number – a number that indicates how smoothly a gasoline burns.

Octet rule – during bonding, atoms tend to reach an electron arrangement with eight electrons in the outermost shell. Many representative elements attain at least a share of eight electrons in their valence shells when they form molecular or ionic compounds; there are some limitations.

Open sextet – species with only six electrons in the highest energy level of the central element (many Lewis acids).

Orbital – may refer to an atomic orbital or a molecular orbital.

Organic chemistry – the chemistry of substances that contain carbon-hydrogen bonds.

Organic compound – substances that contain carbon.

Osmosis – when solvent molecules pass through a semi-permeable membrane from a dilute solution into a more concentrated solution.

Osmotic pressure – the hydrostatic pressure produced on the surface of a semi-permeable membrane.

Outer orbital complex – valence bond designation for a complex in which the metal ion utilizes d orbitals in the outermost (occupied) shell in hybridization.

Overlap – the interaction of orbitals on different atoms in the same region of space.

Oxidation – the addition of oxygen or the loss of electrons. An algebraic increase in the oxidation number; may correspond to a loss of electrons.

Oxidation numbers – quantitative values used as mechanical aids in writing formulas and balancing equations; for single-atom ions, they correspond to the charge on the ion; more electronegative atoms are assigned negative oxidation numbers, known as *oxidation states*.

Oxidation-reduction reactions – reactions in which oxidation and reduction occur; known as *redox reactions*.

Oxide – a binary compound of oxygen.

Oxidizing agent – the substance that oxidizes another substance and is reduced.

P

Pairing – a favorable interaction of two electrons with opposite m values in the same orbital.

Pairing energy – the energy required to pair two electrons in the same orbital.

Paramagnetism – attraction toward a magnetic field, stronger than diamagnetism but still weak compared to ferromagnetism.

Partial pressure – the force exerted by one gas in a mixture of gases.

Particulate matter – fine, divided solid particles suspended in polluted air.

Pauli exclusion principle – no electrons in the same atom may have identical sets of four quantum numbers.

Percentage ionization – the percentage of the weak electrolyte that will ionize in a solution of given concentration.

Percent by mass – 100% times the actual yield divided by the theoretical yield.

Percent composition – the mass percent of each element in a compound.

Percent purity – the percent of a specified compound or element in an impure sample.

Period – the elements in a horizontal row of the periodic table.

Periodicity – regular periodic variations of properties of elements with their atomic number (and position in the periodic table).

Periodic Law – the properties of the elements are periodic functions of their atomic numbers.

Periodic table – an arrangement of elements by increasing atomic numbers, emphasizes periodicity.

Peroxide – a compound with oxygen in –1 oxidation state; metal peroxides contain the peroxide ion, O_2^{2-}.

pH – the measure of acidity (or basicity) of a solution; negative logarithm of the concentration (mol/L) of the H_3O^+ [H^+] ion; scale is commonly used over a range 0 to 14.

Phase diagram – shows the equilibrium temperature-pressure relationships for different phases of a substance.

pH scale – a range from 0 to 14. If the pH of a solution is 7 it is neutral; if the pH of a solution is less than 7 it is acidic; if the pH of a solution is greater than 7 it is basic.

Permanent hardness – hardness (relative to lathering soap) in water that cannot be removed by boiling; caused by calcium sulfate.

Photoelectric effect – emission of an electron from the surface of a metal caused by impinging electromagnetic radiation of specific minimum energy; the current increases with increasing radiation intensity.

Photon – a carrier of electromagnetic radiation of all wavelengths, such as gamma rays and radio waves; known as *quantum of light*.

Physical change – when a substance changes from one physical state to another, but no substances with different composition are formed; physical change may involve a phase change (e.g., melting, freezing) or another physical change such as crushing a crystal or separating one volume of liquid into different containers; does not produce a new substance.

Plasma – a physical state of matter that exists at extremely high temperatures in which molecules are dissociated, and most atoms are ionized.

Polar bond – a covalent bond with an unsymmetrical distribution of electron density.

Polarimeter – a device used to measure optical activity.

Polarization – the buildup of a product of oxidation or reduction of an electrode, preventing further reaction.

Polydentate – refers to ligands with more than one donor atom.

Polyene – a compound that contains more than one double bond per molecule.

Polymerization – the combination of many small molecules to form large molecules.

Polymer – a large molecule consisting of chains or rings of linked monomer units, usually characterized by high melting and boiling points.

Polymorphous – refers to substances that can crystallize in more than one crystalline arrangement.

Polyprotic acid – forms two or more hydronium ions per molecule; often, at least one ionization step is weak.

Positron – a nuclear particle with the mass of an electron but opposite charge (positive).

Potential difference (or *voltage*) – the force which moves the electrons around the circuit; the unit is Volt (V).

Potential energy (*PE*) – energy stored in a body or a system due to its position in a force field or configuration.

Power – the rate at which energy is converted from one form to another; the unit is Watts (W). Power = voltage × current (P = VI).

Precipitate – an insoluble solid formed by mixing in solution the constituent ions of a slightly soluble solution.

Precision – how close the results of multiple experimental trials are; see *accuracy*.

Pressure – force per unit area; unit is Pascal (Pa).

Primary standard – a known high degree of purity substance that undergoes one invariable reaction with the other reactant of interest.

Primary voltaic cells – voltaic cells that cannot be recharged; no further chemical reaction is possible once the reactants are consumed.

Products – chemicals produced (from reactants) in a chemical reaction.

Proton – a subatomic particle having a mass of 1.0073 amu and a charge of +1, found in the nuclei of atoms.

Protonation – the addition of a proton (H^+) to an atom, molecule or ion.

Pseudobinaryionic compounds – contain more than two elements but are named like binary compounds.

Q

Quanta – the minimum amount of energy emitted by radiation.

Quantum mechanics – the study of how atoms, molecules, subatomic particles, etc. behave and are structured; a mathematical method of treating particles based on quantum theory, which assumes that energy (of small particles) is not infinitely divisible.

Quantum numbers – numbers that describe the energies of electrons in atoms; derived from quantum mechanical treatment.

Quarks – elementary particles and a fundamental constituent of matter, combining to form hadrons (i.e., protons and neutrons).

R

Radiation – 1) heat transfer through invisible rays, which travel outwards from the hot object without a medium. 2) high-energy particles or rays emitted during the nuclear decay processes.

Radical – an atom or group of atoms that contains one or more unpaired electrons; usually a very reactive species.

Radioactive dating – method of dating ancient objects by determining the ratio of amounts of mother and daughter nuclides present in an object and relating the ratio to the object's age via half-life calculations.

Radioactive tracer – a small amount of radioisotope replacing a nonradioactive isotope of the element in a compound whose path (e.g., in the body) or whose decomposition products are monitored by detection of radioactivity; known as a "radioactive label."

Radioactivity – the spontaneous disintegration of atomic nuclei.

Raoult's Law – the vapor pressure of a solvent in an ideal solution decreases as its mole fraction decreases.

Rate-determining step – the slowest step in a mechanism; the step that determines the overall reaction rate.

Rate-law expression – equation relating the reaction rate to the concentrations of the reactants and the specific rate of the constant.

Rate of reaction – the change in the concentration of a reactant or product per unit time.

Reactants – substances consumed in a chemical reaction; react together in a chemical reaction.

Reaction quotient – the mass action expression under any set of conditions (not necessarily equilibrium); its magnitude relative to K determines the direction in which the reaction must occur to establish equilibrium.

Reaction ratio – the relative amounts of reactants and products involved in a reaction; may be the ratio of moles, millimoles, or masses.

Reaction stoichiometry – describes the quantitative relationships among substances participating in chemical reactions.

Reactivity series (or activity series) – an empirical, calculated, and structurally analytical progression of a series of metals, arranged by "reactivity" from highest to lowest; used to summarize information about the reactions of metals with acids and water, double displacement reactions and the extraction of metals from ores.

Reagent – a substance (or compound) added to a system to cause a chemical reaction or to visualize if a reaction occurs; the terms reactant and reagent are often used interchangeably; however, a *reactant* is more specifically a substance consumed during a chemical reaction.

Reducing agent – a substance that reduces another substance and is itself oxidized.

Reduction – the removal of oxygen or the gaining of electrons.

Resonance – the concept in which two or more equivalent dot formulas for the same arrangement of atoms (resonance structures) are necessary to describe the bonding in a molecule or ion.

Reverse osmosis – forcing solvent molecules to flow through a semi-permeable membrane from a concentrated solution into a dilute solution by applying greater hydrostatic pressure on the concentrated side than the osmotic pressure opposing it.

Reversible reaction – proceses that do not go to completion and occur in the forward and reverse direction.

S

Saline solution – a general term for NaCl (i.e., sodium chloride) in water.

Salt – when a metal replaces the hydrogen of an acid.

Salts – ionic compounds composed of anions and cations.

Salt bridge – a U-shaped tube containing an electrolyte, connects the two half-cells of a voltaic cell.

Saturated solution –no more solute will dissolve at that temperature.

s-block elements – group 1 and 2 elements (alkali and alkaline metals), including hydrogen and helium.

Schrödinger equation – quantum state equation representing the behavior of an electron around an atom; describes the wave function of a physical system evolving.

Second Law of Thermodynamics – the universe tends toward a state of greater disorder in spontaneous processes.

Secondary standard – a solution that has been titrated against a primary standard; a standard solution.

Secondary voltaic cells – voltaic cells that can be recharged; original reactants can be regenerated by reversing the direction of the current flow.

Semiconductor – a substance that does not conduct electricity at low temperatures but will do so at higher temperatures.

Semi-permeable membrane – a thin partition between two solutions through which specific molecules can pass but others cannot.

Shielding effect – electrons in filled sets of s, p orbitals between the nucleus and outer shell electrons shield the outer shell electrons somewhat from the effect of protons in the nucleus; known as the "screening effect."

Sigma (σ) bonds – bonds resulting from the head-on overlap of atomic orbitals. The region of electron sharing is along and (cylindrically) symmetrical to the imaginary line connecting the bonded atoms.

Sigma orbital – molecular orbital resulting from the head-on overlap of two atomic orbitals.

Single bond – covalent bond resulting from the sharing of two electrons (one pair) between two atoms.

Sol – a suspension of solid particles in a liquid; artificial examples include sol-gels.

Solid – one of the states of matter, where the molecules are packed close, resistance to movement/deformation and volume change.

Solubility product constant – equilibrium constant that applies to the dissolution of a slightly soluble compound.

Solubility product principle – the solubility product constant expression for a slightly soluble compound is the product of the concentrations of the constituent ions, each raised to the power that corresponds to the number of ions in one formula unit.

Solute – the dispersed (i.e., dissolved) phase of a solution; the solution is mixed into the solvent (e.g., NaCl in saline water).

Solution – a homogeneous mixture of multiple substances; comprised of solutes and solvents; a mixture of a solute (usually a solid) and a solvent (usually a liquid).

Solvation – the process by which solvent molecules surround and interact with solute ions or molecules.

Solvent – the dispersing medium of a solution (e.g., H_2O in saline water).

Solvolysis – the reaction of a substance with the solvent in which it is dissolved.

s-orbital – a spherically symmetrical atomic orbital; one per energy level.

Specific gravity – the ratio of the density of a substance to the density of water.

Specific heat   the amount of heat required to raise the temperature of one gram of substance one degree Celsius.

Specific rate constant – an experimentally determined (proportionality) constant, which is different for different reactions, and which changes only with temperature; k in the rate-law expression: Rate = k [A] × [B].

Spectator ions – ions in a solution that do not participate in a chemical reaction.

Spectral line – any of several lines corresponding to definite wavelengths of an atomic emission or absorption spectrum; marks the energy difference between two energy levels.

Spectrochemical series – arrangement of ligands in order of increasing ligand field strength.

Spectroscopy – the study of radiation and matter, such as X-ray absorption and emission spectroscopy.

Spectrum – display of component wavelengths (colors) of electromagnetic radiation.

Speed of light – the speed at which radiation travels through a vacuum (299,792,458 m/sec).

Square planar – describes molecules and polyatomic ions with one atom in the center and four atoms at the corners of a square.

Square planar complex – relationship with metal in the center of a square plane, with ligand donor atoms at each of the four corners.

Standard conditions for temperature and pressure (STP) – used to compare experimental results (25 °C and 100.000 kPa).

Standard electrodes – half-cells in which the oxidized and reduced forms of a species are present at the unit activity (1.0 M solutions of dissolved ions, 1.0 atm partial pressure of gases, pure solids, and liquids).

Standard electrode potential – by convention, the potential (Eo) of a half-reaction as a reduction relative to the standard hydrogen electrode when species are present at unit activity.

Standard entropy – the absolute entropy of a substance in its standard state at 298 K.

Standard molar enthalpy of formation – the amount of heat absorbed in forming one mole of a substance in a specified state from its elements in their standard states.

Standard molar volume – the space occupied by one mole of an ideal gas under standard conditions; 22.4 liters.

Standard reaction – a process where the numbers of moles of reactants in the balanced equation, in their standard states, are entirely converted to the numbers of moles of products in the balanced equation, at their standard state.

State of matter – a homogeneous, macroscopic phase (e.g., gas, plasma, liquid, solid) in increasing concentration.

Stoichiometry – the quantitative relationships among elements and compounds undergoing chemical changes.

Strong electrolyte – a substance that conducts electricity well in a dilute aqueous solution.

Strong field ligand – ligand that exerts a strong crystal or ligand electrical field and generally forms low spin complexes with metal ions when possible.

Structural isomers – compounds that contain the same number and kinds of atoms with different geometry.

Subatomic particles – comprise an atom (e.g., protons, neutrons, electrons).

Sublimation – the direct vaporization of a solid by heating without passing through the liquid state; a phase transition from solid to limewater fuel or gas.

Substance – any matter, specimens with the same chemical composition and physical properties.

Substitution reaction – a reaction in which another atom or group of atoms replaces an atom or a group of atoms.

Supercooled liquids – liquids that, when cooled, apparently solidify but continue to flow very slowly under the influence of gravity.

Supercritical fluid – a substance at a temperature above its critical temperature.

Supersaturated solution – contains a higher than saturation concentration of solute; slight disturbance or seeding causes crystallization of excess solute.

Suspension – a heterogeneous mixture in which solute-like particles settle out of the solvent-like phase sometime after their introduction. A mixture of a liquid and a finely divided insoluble solid.

T

Talc – a mineral representing the Mohs Scale composed of hydrated magnesium silicate with the chemical formula $H_2Mg_3(SiO_3)_4$ or $Mg_3Si_4O_{10}(OH)_2$.

Temperature – a measure of heat intensity (i.e., the hotness or coldness of a sample); a measure of the kinetic energy of an object. Unit is degrees, and scales are Celsius, Fahrenheit, and Kelvin.

Temporary hardness – hardness in water that can be removed by boiling; caused by calcium hydrogen carbonate.

Ternary acid – a ternary compound containing H, O and another element, often a nonmetal.

Ternary compound – a compound consisting of three elements; may be ionic or covalent.

Tetrahedral – a term used to describe molecules and polyatomic ions with one atom in the center and four atoms at the corners of a tetrahedron.

Theoretical yield – the maximum amount of a specified product that could be obtained from specified amounts of reactants, assuming complete consumption of the limiting reactant according to only one reaction and complete recovery of the product; see *actual yield*.

Theory – a model describing the nature of a phenomenon.

Thermal conductivity – a property of a material to conduct heat (often noted as k).

Thermal cracking – decomposition by heating a substance in the presence of a catalyst and the absence of air.

Thermochemistry – the study of absorption/release of heat within a chemical reaction; studies heat energy associated with chemical reactions and physical transformations.

Thermodynamics – studying the effects of changing temperature, volume, or pressure (or work, heat, and energy) on a macroscopic scale.

Thermodynamic stability – when a system is in its lowest energy state with its environment (equilibrium).

Thermometer – a device that measures the average energy of a system.

Thermonuclear energy – energy from nuclear fusion reactions.

Third Law of Thermodynamics – entropy of a pure crystalline substance at absolute zero temperature is zero.

Titration – a procedure in which one solution is added to another solution until the chemical reaction between the two solutes is complete; the concentration of one solution is known, and that of the other is unknown. The process of adding one solution to a measured amount of another to find out exactly how much of each is required to react.

Torr – a unit to measure pressure; 1 Torr is equivalent to 133.322 Pa or 1.3158×10^{-3} atm.

Total ionic equation – the expression for a chemical reaction written to show the predominant form of species in aqueous solution or contact with water.

Transition elements (metals) – B Group elements except IIB in the periodic table; sometimes called transition elements, elements with incomplete d sub-shells; the d-block elements.

Transition state theory –reactants pass through high-energy transition states before forming products.

Transuranic element – an atomic number greater than 92; none of the transuranic elements are stable.

Triple bond – the sharing of three pairs of electrons within a covalent bond (e.g., N_2).

Triple point – where the temperature and pressure of three phases are the same; water has a unique phase diagram.

Tyndall effect – results from light scattering by colloidal particles (a mixture where one substance is dispersed evenly throughout another) or by suspended particles.

U

Uncertainty –any measurement that involves estimating any amount that cannot be precisely reproducible.

Uncertainty principle – knowing the location of a particle makes the momentum uncertain, while knowing the momentum of a particle makes the location uncertain.

Unit cell – the smallest repeating unit of a lattice.

Unit factor – statements used in converting between units.

Universal (or ideal) gas constant – proportionality constant in the ideal gas law (0.08206 L·atm/(K·mol)).

UN number – a four-digit code used to note hazardous and flammable substances.

Unsaturated hydrocarbons – hydrocarbons that contain double or triple carbon-carbon bonds.

V

Valence bond theory – proposes that covalent bonds are formed when atomic orbitals on different atoms overlap and the electrons are shared.

Valence electrons – outermost electrons of atoms; usually those involved in bonding.

Valence shell electron pair repulsion theory (VSEPR) – assumes electron pairs are arranged around the central element of a molecule or polyatomic ion with maximum separation (and minimum repulsion) among regions of high electron density.

Valency – the number of electrons an atom wants to gain, lose, or share to have a full outer shell.

Van der Waals' equation – a quantitative relationship of a state extending the ideal gas law to real gases by including two empirically determined parameters, which are specific for different gases.

Van der Waals force – one of the forces (attraction/repulsion) between molecules.

Van't Hoff factor – the ratio of moles of particles in solution to moles of solute dissolved.

Vapor – when a substance is below the critical temperature in the gas phase.

Vaporization – the phase change from liquid to gas.

Vapor pressure – the particle pressure of vapor at the surface of its parent liquid.

Viscosity – the resistance of a liquid to flow (e.g., oil has a higher viscosity than water).

Volt – one joule of work per coulomb; the unit of electrical potential transferred.

Voltage – the potential difference between two electrodes; a measure of the chemical potential for a redox reaction.

Voltaic cells – electrochemical cells in which spontaneous chemical reactions produce electricity; known as *galvanic cells*.

Voltmeter – an instrument that measures the cell potential.

Volumetric analysis – measuring the volume of a solution (of known concentration) to determine the substance's concentration within the solution; see *titration*.

W

Water equivalent – the amount of water absorbing the same heat as the calorimeter per degree of temperature increases.

Weak electrolyte – a substance that conducts electricity poorly in a dilute aqueous solution.

Weak field ligand – a ligand exerting a weak crystal or ligand field and generally forms high spin complexes with metals.

X

X-ray – electromagnetic radiation between gamma and UV rays.

X-ray diffraction – a method for establishing structures of crystalline solids using single wavelength X-rays and studying the diffraction pattern.

X-ray photoelectron spectroscopy – a spectroscopic technique used to measure the composition of a material.

Y

Yield – the amount of product produced during a chemical reaction.

Z

Zone melting – remove impurities from an element by melting and slowly traveling it down an ingot (cast).

Zone refining – a method of purifying a metal bar by passing it through an induction heater; this causes impurities to move along a melted portion.

Zwitterion (formerly called a dipolar ion) – a neutral molecule with a positive and negative electrical charge; multiple positive and negative charges can be present, distinct from dipoles at different locations within that molecule; known as *inner salts*.

If you benefited from this book, please leave a review on Amazon so others can learn from your input. Reviews help us understand our customers' needs and experiences while keeping our commitment to quality.

Customer Satisfaction Guarantee

Your feedback is important because we strive to provide the highest quality educational materials. Email us comments or suggestions.

info@sterling-prep.com

We reply to emails – check your spam folder

Everything You Always Wanted to Know About…

Chemistry

Physics

Cell & Molecular Biology

Organismal Biology

Human Anatomy & Physiology

American History

American Law

American Government & Politics

Comparative Government & Politics

World History

European History

Psychology

Sociology

Environmental Science

Human Geography

Visit our Amazon store

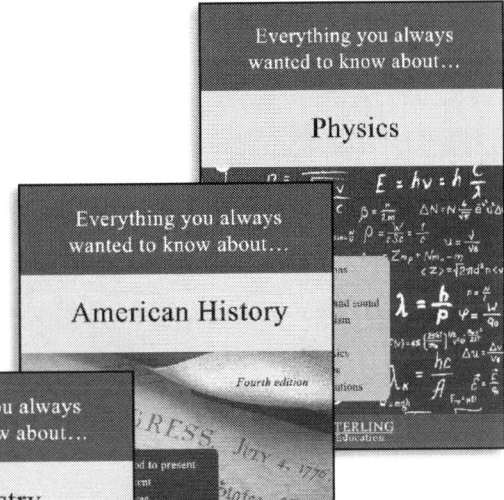

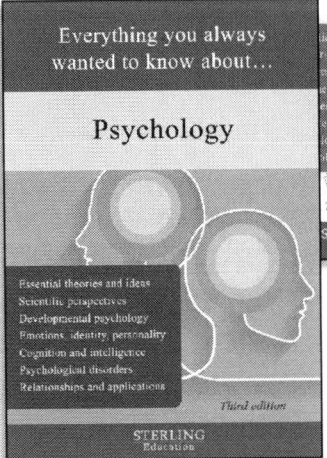

Made in the USA
Columbia, SC
20 May 2025

58219103R00354